ELEVENTH EDITION

Laboratory Fundamentals of MICROBIOLOGY

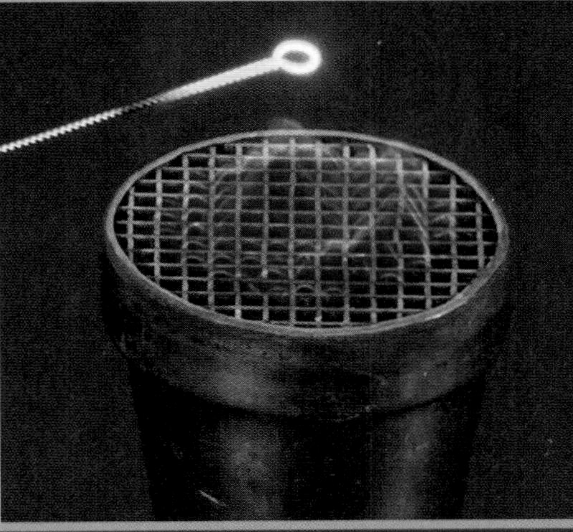
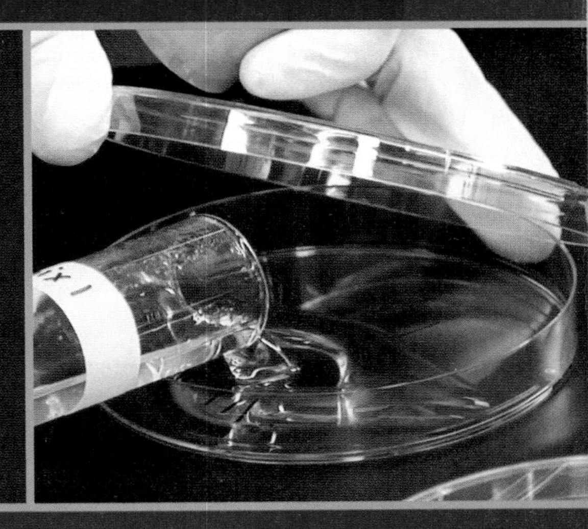

Jeffrey C. Pommerville
Glendale Community College

JONES & BARTLETT
LEARNING

World Headquarters
Jones & Bartlett Learning
5 Wall Street
Burlington, MA 01803
978-443-5000
info@jblearning.com
www.jblearning.com

Jones & Bartlett Learning books and products are available through most bookstores and online booksellers. To contact Jones & Bartlett Learning directly, call 800-832-0034, fax 978-443-8000, or visit our website, www.jblearning.com.

Substantial discounts on bulk quantities of Jones & Bartlett Learning publications are available to corporations, professional associations, and other qualified organizations. For details and specific discount information, contact the special sales department at Jones & Bartlett Learning via the above contact information or send an email to specialsales@jblearning.com.

To order this product, use ISBN: 978-1-284-10097-6

Production Credits
VP, Executive Publisher: David D. Cella
Executive Editor: Matthew Kane
Associate Editor: Audrey Schwinn
Senior Development Editor: Nancy Hoffmann
Production Manager: Dan Stone
Director of Marketing: Andrea DeFronzo
Production Services Manager: Colleen Lamy
Manufacturing and Inventory Control Supervisor: Amy Bacus
Composition: SourceHOV LLC
Cover Design: Kristin E. Parker
Director of Rights & Media: Joanna Gallant
Rights & Media Specialist: Wes DeShano
Media Development Editor: Troy Liston
Cover Image: (Bottom Photo) © M.I.T. Case No. 17683C, "C Food", by Roman Stocker, Steven Paul Smriga, and Vicente Ignacio Fernandez
Printing and Binding: LSC Communications
Cover Printing: LSC Communications

Library of Congress Cataloguing-in-Publication Data Not Available at Time of Printing

ISBN13: 978-1-284-12924-3

6048

Printed in the United States of America
21 20 19 18 17 10 9 8 7 6 5 4 3 2 1

Contents

Laboratory Videos XI
Preface XII
Acknowledgments XIII
Video Credits XIII
Reviewers XIV

PART I Laboratory Safety 1

Exercise 1: Safety Overview: An Important Message About Safety 3
A Word About Chemical Hazard Labels 3
Exercise 1: Results 7

Exercise 2: Best Practices in Safety 9

Exercise 3: Student Safety Contract 11
Laboratory Rules and Regulations 11

PART II Laboratory Techniques and Skills 13

Basic Growth Media 14
Culture Equipment 15
Transfer Instruments 15
Growth and Incubation Appliances 16

Exercise 4: Culture Transfer Techniques 17
Subculture Transfer Techniques 17
Exercise 4: Results 21

Exercise 5: Pure Culture Techniques: Streak Plate Method 25
Streak Plate Method 25
Exercise 5: Results 31

Exercise 6: Pure Culture Techniques: Pour Plate Method 33
Pour Plate Method 33
Exercise 6: Results 37

Exercise 7: Solution Transfer 41
Solution Transfer 41
Exercise 7: Results 43

Exercise 8: Serial Dilutions 45
Serial Dilutions 45
Exercise 8: Results 47

PART III Microscopy 49

Exercise 9: The Compound Microscope 51
Parts of the Microscope and Their Use 51
Exercise 9: Results 55

Exercise 10: Observations of Prepared Slides 57
Microscopic Examination of Prepared Slides 57
Exercise 10: Results 61

Exercise 11: Observations of Motile Bacteria 65
Bacterial Motility 65
Exercise 11: Results 69

PART IV Bacterial Staining Techniques 71

Exercise 12: Preparation of a Bacterial Smear 73
Preparation of a Bacterial Smear 73
Exercise 12: Results 77

Exercise 13: Simple Stain Technique 79
The Simple Stain Technique 79
Exercise 13: Results 83

Exercise 14: Negative Stain Technique 85
Negative Stain Technique 85
Exercise 14: Results 89

Exercise 15: Gram Stain Technique 91
Gram Stain Technique 91
Exercise 15: Results 97

Exercise 16: Bacterial Structures: Spore Stain Technique 99
Spore Stain Technique 99
Exercise 16: Results 103

Exercise 17: Bacterial Structures: Capsule Stain Technique 105

Capsule Stain Technique 105

Exercise 17: Results 109

PART V Viruses and Eukaryotic Microorganisms 111

Exercise 18: Viruses: The Effect of Bacteriophages on Bacteria . . .113

The Effect of Bacteriophages on Bacteria 113

Exercise 18: Results 115

Exercise 19: Viruses: Plaque Formation . . . 117

Plaque Formation 117

Exercise 19: Results 119

Exercise 20: Identification of Bacteriophages from Sewage 121

Identification of Bacteriophages from Sewage 121

Exercise 20: Results 123

Exercise 21: Fungi: Molds. 125

Fungi. 125

Exercise 21: Results 129

Exercise 22: Fungi: Yeasts 131

Fungi: Yeasts 131

Exercise 22: Results 135

Exercise 23: Protists and Multicellular Parasites 137

Protists and Multicellular Parasites 137

Exercise 23: Results 139

PART VI Control of Microorganisms 141

Exercise 24: The Effect of Physical Agents on Bacteria: Heat Tolerance 143

Heat Tolerance 143

Exercise 24: Results 147

Exercise 25: The Effect of Physical Agents on Bacteria: Ultraviolet Light. 149

Ultraviolet Light 149

Exercise 25: Results 153

Exercise 26: The Effect of Physical Agents on Bacteria: Drying 155

Drying 155

Exercise 26: Results 157

Exercise 27: The Effect of Physical Agents on Bacteria: Other Physical Agents 159

Other Physical Agents 159

Exercise 27: Results 161

Exercise 28: The Effect of Chemical Agents on Bacteria: Antiseptics and Disinfectants. 163

Antiseptics and Disinfectants 163

Exercise 28: Results 167

Exercise 29: The Effect of Chemical Agents on Bacteria: Metals 169

Metals. 169

Exercise 29: Results 171

Exercise 30: The Effect of Chemical Agents on Bacteria: Other Chemical Substances. 173

Other Chemical Substances 173

Exercise 30: Results 175

Exercise 31: The Effect of Chemical Agents on Bacteria: Lysozyme 177

Lysozome. 177

Exercise 31: Results 179

Exercise 32: Evaluation of Disinfectants and Antiseptics: Effectiveness of Disinfectants on Inanimate Objects 181

Effectiveness of Disinfectants on Inanimate Objects 181

Exercise 32: Results 183

Exercise 33: Evaluation of Disinfectants and Antiseptics: Effectiveness of Disinfectants and Hand Washing on the Skin Surface 185

Effectiveness of Disinfectants and Hand Washing on the Skin Surface 185

Exercise 33: Results 187

Exercise 34: Evaluation of Disinfectants and Antiseptics: Mouthwashes and Oral Bacteria 189
Mouthwashes and Oral Bacteria 189
Exercise 34: Results 193

Exercise 35: Evaluation of Disinfectants and Antiseptics: The Effect of Antibiotics on Bacteria 195
The Effect of Antibiotics on Bacteria 195
Exercise 35: Results 199

PART VII Estimating Virus and Bacterial Cell Numbers and Measuring Population Growth 201

Exercise 36: Estimating the Number of Viruses and Bacterial Cells in a Population 203
Estimating the Number of Bacteriophages 203
Using a Petroff-Hausser Chamber 206
Exercise 36: Results 209

Exercise 37: Microbial Growth: Analysis of a Bacterial Growth Curve. . .211
Bacterial Growth Dynamics 212
Plotting a Growth Curve and Determining the Generation Time 213
Exercise 37: Results 215

PART VIII Medical Microbiology 219

Exercise 38: The Genus *Mycobacterium*: Colony and Cellular Morphology 223
Colony and Cellular Morphology 223
Exercise 38: Results 225

Exercise 39: The Genus *Mycobacterium*: Acid-Fast Stain Technique. . . .227
Acid-Fast Stain Technique 227
Exercise 39: Results 231

Exercise 40: The Genus *Mycobacterium*: Cold Acid-Fast Stain Technique 233
Cold Acid-Fast Stain Technique 233
Exercise 40: Results 235

Exercise 41: The Genus *Streptococcus*: Streptococci from the Upper Respiratory Tract 237
Streptococci from the Upper Respiratory Tract 237
Exercise 41: Results 241

Exercise 42: The Genus *Streptococcus*: Streptococci from the Oral Cavity 243
Streptococci from the Oral Cavity 243
Exercise 42: Results 245

Exercise 43: The Genus *Neisseria* 247
The Genus *Neisseria* 247
Exercise 43: Results 251

Exercise 44: The Genus *Staphylococcus*: Isolation of Staphylococci. . . .255
Isolation of Staphylococci 255
Exercise 44: Results 259

Exercise 45: The Genus *Staphylococcus*: Differentiation between Staphylococcal Species. 261
Differentiation between Staphylococcal Species . . . 261
Exercise 45: Results 265

Exercise 46: The Enteric Bacteria: Isolation of Enteric Bacteria. . .267
Isolation of Enteric Bacteria 267
Exercise 46: Results 269

Exercise 47: The Enteric Bacteria: Differentiation of Enteric Bacteria on TSI Agar 271
Differentiation of Enteric Bacteria on TSI Agar 271
Exercise 47: Results 273

Exercise 48: The Enteric Bacteria: The IMViC Series 275
The IMViC Series 276
Exercise 48: Results 279

Exercise 49: The Enteric Bacteria: Rapid Identification of Enteric Bacteria 281
Rapid Identification of Enteric Bacteria 281
Exercise 49: Results 283

Exercise 50: The Genus *Bacillus* 285
The Genus *Bacillus* 285
Exercise 50: Results 287

Exercise 51: The Genus *Clostridium* 291
The Genus *Clostridium* 291
Exercise 51: Results 295

Exercise 52: The Genus *Lactobacillus*: Isolation of Lactobacilli 299
Isolation of Lactobacilli 299
Exercise 52: Results 301

Exercise 53: The Genus *Lactobacillus*: Caries Susceptibility Test 303
Caries Susceptibility Test. 303
Exercise 53: Results 305

PART IX Identification of a Bacterial Unknown 307

Part IX: Results 311

Exercise 54: Bacterial Structural Characteristics 315
Bacterial Structural Characteristics 315
Exercise 54: Results 317

Exercise 55: Bacterial Culture Characteristics 321
Bacterial Culture Characteristics 321
Exercise 55: Results 325

Exercise 56: Biochemical Characteristics of Bacteria: Carbohydrate Fermentation 329
Carbohydrate Fermentation. 329
Exercise 56: Results 333

Exercise 57: Biochemical Characteristics of Bacteria: Starch Hydrolysis. 335
Starch Hydrolysis 335
Exercise 57: Results 339

Exercise 58: Biochemical Characteristics of Bacteria: Catalase Production 341
Catalase Production 341
Exercise 58: Results 345

Exercise 59: Biochemical Characteristics of Bacteria: DNA Hydrolysis. 347
DNA Hydrolysis 347
Exercise 59: Results 349

Exercise 60: Biochemical Characteristics of Bacteria: Hydrogen Sulfide Production 351
Hydrogen Sulfide Production. 351
Exercise 60: Results 353

Exercise 61: Biochemical Characteristics of Bacteria: IMViC Series. 355
IMViC Series 355
Exercise 61: Results 357

Exercise 62: Biochemical Characteristics of Bacteria: Urea Hydrolysis 359
Urea Hydrolysis 359
Exercise 62: Results 361

Exercise 63: Biochemical Characteristics of Bacteria: Fat (Triglyceride) Hydrolysis 363
Lipid Digestion. 363
Exercise 63: Results 365

Exercise 64: Biochemical Characteristics of Bacteria: Casein Hydrolysis. 367
Casein Hydrolysis 367
Exercise 64: Results 369

PART X Bacterial Genetics 371

Exercise 65: Selecting for Antibiotic Resistant Bacteria: Gradient Plate Technique 373
Gradient Plate Technique. 373
Exercise 65: Results 377

Exercise 66: Induced Mutations in Bacteria: Mutagenic Effect of Ultraviolet Light 379
Mutagenic Effect of Ultraviolet Light 379
Exercise 66: Results 383

Exercise 67: Mutations and DNA Repair 385
DNA Repair of UV Light Damage 385
Exercise 67: Results 387

Exercise 68: Carcinogens and the Ames Test 389
The Ames Test 389
Exercise 68: Results 393

Exercise 69: Bacterial Conjugation 395
Bacterial Conjugation 395
Exercise 69: Results 399

Exercise 70: Bacterial Transformation 401
Bacterial Transformation 402
Exercise 70: Results 407

PART XI **Immunology** 411

Exercise 71: Serology: Slide Agglutination 413
Slide Agglutination 413
Exercise 71: Results 415

Exercise 72: Serology: Tube Agglutination 417
Tube Agglutination 417
Exercise 72: Results 419

Exercise 73: Serology: Determination of Blood Type 421
Hemagglutination: Determination of Blood Type 421
Exercise 73: Results 425

Exercise 74: Serology: Blood Cell Identification-Blood Smear 427
Blood Smear 427
Exercise 74: Results 431

Exercise 75: Enzyme-Linked Immunosorbent Assay (ELISA) 433
Direct ELISA 434
Exercise 75: Results 435

PART XII **Public Health and Environmental Microbiology** 437

Exercise 76: A Simulated Epidemic: Microbial Transmission via Fomites 441
Microbial Transmission via Fomites 441
Exercise 76: Results 445

Exercise 77: A Microbial Hunt 447
A Microbial Hunt 447
Exercise 77: Results 451

Exercise 78: Transmission of Microorganisms through Toilet Paper 453
Transmission of Microorganisms through Toilet Paper 453
Exercise 78: Results 455

Exercise 79: Disinfection of Drinking Water 457
Disinfection of Drinking Water 457
Exercise 79: Results 459

Exercise 80: Microbiology of Foods: Food Preservation with Salt and Garlic 461
Food Preservation with Salt and Garlic 461
Exercise 80: Results 463

Exercise 81: Microbiology of Foods: Standard Plate Count 467
Standard Plate Count 467
Exercise 81: Results 471

Exercise 82: Microbiology of Foods: Fermentation of Wine and Beer 475
Fermentation of Wine and Beer 475
Exercise 82: Results 477

Exercise 83: Microbiology of Milk and Dairy Products: Standard Plate Count of Milk 479
Standard Plate Count of Milk 479
Exercise 83: Results 483

Exercise 84: Microbiology of Milk and Dairy Products: Coliform Plate Count of Milk. .485
Coliform Plate Count of Milk 485
Exercise 84: Results 487

Exercise 85: Microbiology of Milk and Dairy Products: Methylene Blue Reduction Test 489
Methylene Blue Reduction Test. 489
Exercise 85: Results 491

Exercise 86: Microbiology of Milk and Dairy Products: Preparation of Cheese 493
Preparation of Cheese 493
Exercise 86: Results 495

Exercise 87: Microbiology of Milk and Dairy Products: Preparation of Yogurt. 497
Preparation of Yogurt 497
Exercise 87: Results 499

Exercise 88: Microbiology of Milk and Dairy Products: Natural Bacterial Content of Milk. 501
The Natural Bacterial Content of Milk 501
Exercise 88: Results 503

Exercise 89: Microbiology of Water: Presumptive Test by the MPN Method 507
Presumptive Test by the MPN Method. 507
Exercise 89: Results 509

Exercise 90: Microbiology of Water: Confirmed Test. 511
Confirmed Test 511
Exercise 90: Results 513

Exercise 91: Microbiology of Water: Completed Test 515
Completed Test 515
Exercise 91: Results 517

Exercise 92: Microbiology of Water: Membrane Filter Technique. . .519
Membrane Filter Technique 519
Exercise 92: Results 521

Exercise 93: Microbiology of Water: Preparation of a Biofilm. 523
Preparation of a Biofilm 523
Exercise 93: Results 525

Exercise 94: Microbiology of Soil: Isolation of *Rhizobium* from Legume Roots 527
Isolation of *Rhizobium* from Legume Roots 527
Exercise 94: Results 529

Exercise 95: Microbiology of Soil: Ammonification by Soil Microorganisms 531
Ammonification by Soil Microorganisms 531
Exercise 95: Results 533

Exercise 96: Microbiology of Soil: Isolation of *Streptomyces* from Soil. 535
Isolation of *Streptomyces* from Soil 535
Exercise 96: Results 537

Exercise 97: Microbiology of Soil: Antibiotic Production by *Streptomyces* 539
Antibiotic Production by *Streptomyces*. 539
Exercise 97: Results 541

Exercise 98: Microbiology of Soil: Plate Count of Soil Bacteria 543
Plate Count of Soil Bacteria. 543
Exercise 98: Results 545

Exercise 99: Microbiology of Soil: Microbial Ecology in the Soil—The Winogradsky Column 547
Microbial Ecology in the Soil—The Winogradsky Column 547
Exercise 99: Results 551

Appendix A Preparation of Stains and Diagnostic Reagents A-1

Appendix B Using Enterotube® II. B-1

Appendix C C-1

Glossary G-1

Index I-1

Laboratory Videos

List of Laboratory Videos Referenced in Text

Laboratory Safety 9

Cleaning up a Culture Spill 10

How to Use a Gas Burner/Bacti-cinerator 16

Broth to Broth Transfer 19

Broth to Agar Transfer 19

Agar to Agar Transfer 19

How to Make a Four-Way Streak Plate 26

Preparing the Pour Plate 33

Transferring a Solution with a Pipette 41

How to Make a Serial Dilution 45

How to Use the Light Microscope: I. Identifying the Parts 52

How to Use the Light Microscope: II. Focusing with the Microscope 53

Measuring Cell Size 58

How to Make a Hanging Drop Preparation 65

How to Make a Bacterial Smear 73

How to Make a Simple Stain Preparation 79

How to Make a Negative Stain Preparation 85

Preparing a Gram Stain 91

Preparing an Endospore Stain 102

How to Make a Capsule Stain Preparation 105

How to Prepare a Disk Diffusion Assay 163

How to Prepare the Kirby-Bauer Test 195

How to Prepare a Standard Plate Count 203

How to Use a Petroff-Hausser Chamber 206

How to Use a Spectrophotometer 212

Preparing the Acid-fast Stain 227

Oxidase Test 249

IMViC: Indole Test 275

IMViC: Methyl Red Test 275

IMViC: Voges-Proskauer Test 275

IMViC: Citrate Test 275

Biochemical Tests: Carbohydrate Fermentation Test 330

Biochemical Tests: Starch Hydrolysis Test 335

Biochemical Tests: Catalase Test 341

Biochemical Tests: Hydrogen Sulfide Test 351

IMViC: Indole Test 355

IMViC: Methyl Red Test 355

IMViC: Voges-Proskauer Test 355

IMViC: Citrate Test 355

Urease Test 359

Preface

For years, *Laboratory Fundamentals of Microbiology* has been the trusted resource for providing undergraduate students with a solid foundation of microbiology laboratory skills. Now, the completely modernized *Eleventh Edition* represents a lab manual revolution built for today's learners, focusing on the student's experience in the lab.

In preparation for the revision, the editorial team at Jones & Bartlett Learning asked instructors what would most help their students prepare for and succeed in a microbiology laboratory course. The overwhelming answer was video that *demonstrated* proper techniques and exercises. We listened and delivered the ultimate microbiology laboratory experience! We are excited to offer 110 minutes of 34 instructor-chosen, high-quality videos of actual students performing the most common lab skills, procedures, and techniques. From lab safety to microscopy to bacterial staining techniques, nowhere else will you find a more comprehensive and customized microbiology video program. Best-selling author and expert Jeffrey C. Pommerville provides narration, context, and rationale for the on-screen action, and *Skills Checklists* are available online to record progress and for assignability.

The laboratory videos are fully integrated into *Laboratory Fundamentals of Microbiology, Eleventh Edition* to provide a seamless experience for the user. Within the manual, Sections and Exercises open with a list of relevant videos, and icons identify where students should refer to them to best prepare for each exercise. This encourages students to read, see, do, and connect with the material.

Preparing the Pour Plate

Our belief that these videos will improve students' laboratory skills and safety, coupled with our commitment to fully supporting instructors and students on their path to educational excellence, inspired us to provide access to the video content free with every new print copy of *Laboratory Fundamentals of Microbiology, Eleventh Edition*.

In addition to the integration of videos, we made several other significant updates to this edition. Perhaps most noticeable is the new, full-color interior design, with images from the videos found throughout the manual. The microbial world and the laboratory where we study and learn are vibrant and colorful, and it is important that instructional tools reflect real life. We retained the convenient lay-flat spiral binding—a simple but practical choice that will make students' lab experience easier as they read and write on the pages. Perforated pages will simplify assigning and grading as students can tear out their work for submission to their instructor. Furthermore, the area for responses and sketching has been better defined for ease of use.

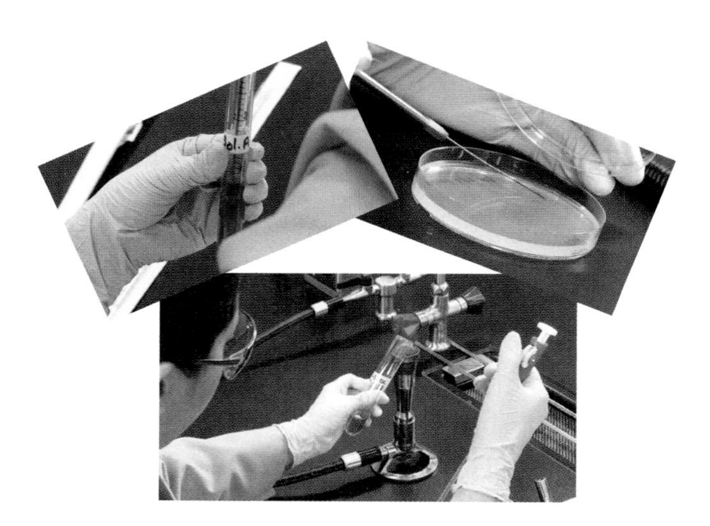

Finally, all labs have been expanded and reorganized into new sections, such as "Laboratory Safety," "Measuring Population Growth," or "Immunology," and are presented from easiest to most challenging. Previously, multiple lab skills were grouped together in 34 exercises. For this edition we separated each skill into 99 distinct exercises for instructors to more easily identify and assign topics for which their lab is equipped. In addition, a new section on lab safety emphasizes an important "culture of safety" approach to the microbiology lab.

All of this makes *Laboratory Fundamentals of Microbiology, Eleventh Edition* the perfect companion to any modern microbiology course.

Acknowledgments

Laboratory Fundamentals of Microbiology, Eleventh Edition was revised, quite simply, to address microbiology education needs as communicated by microbiology instructors. Extensive market feedback solicited by Jones & Bartlett Learning identified gaps in student learning and garnered suggestions about how a publisher could support instructor efforts to close those gaps with a product solution. In that sense, *Laboratory Fundamentals of Microbiology* and the associated *Fundamentals of Microbiology Laboratory Videos* reflect the collective choices instructors communicated to us about the optimal delivery medium, the content and labs chosen to be executed through that medium, and the manner in which the videos should be created.

The instructor feedback we collected also echoed many of the American Association for the Advancement of Science "Vision and Change" recommendations for undergraduate education, as identified by science instructors and stakeholders. Among the goals of the laboratory manual with videos was to create a dynamic and engaging student solution that connected action with rationale, promoted student competence, and ignited interest in the sciences. Jones & Bartlett Learning has been responsive to the call for change in education, and is committed to strategically build solutions that emphasize mastery of concepts, the practical over the theoretical, and the experiential over the abstract.

Video Credits

Creating the videos was no small undertaking. Jones & Bartlett Learning, would like to thank the many people who made it possible for us to bring this educational resource to life:

- The administration of Glendale Community College in Glendale, Arizona, who graciously permitted us to utilize their facilities. We could not have asked for a better place to film.
- The student volunteers, who donated their time and talents to perform each of the laboratory procedures you see on the videos. We were so impressed with your poise and professionalism, and wish you the best in your careers:
    - Maryann Babeti
    - Daisy Castellano
    - Alec Goding
    - Yamin Ko
    - Christina Mitchell
- Rhonda Lee, who allowed us to occupy her lab for a week, prepared everything we needed for each video, and patiently answered our questions as they came up.
- Rhonda Fabian, Jerry Baber, Bridget Yachetti, and Dane Smith at Fabian-Baber, who worked tirelessly to "get it right", from the preparation to the lab shoot in Arizona to post-production in Pennsylvania. Every student and instructor who uses this series will benefit from your skill and creativity, and we benefitted from your humor and camaraderie for a week in the desert.
- Jeffrey Pommerville, the author, subject matter expert, and instructor/narrator on the videos. You always go the extra mile and that has made all the difference. Our hope is that the final product makes you proud, because you've certainly earned it.

Reviewers

There are many people to thank for this edition of *Laboratory Fundamentals of Microbiology*. The feedback from the following instructors had a direct influence on what exercises to bring to life through video, and how:

Khaled Abou-Aisha, German University in Cairo, Cairo, Egypt

Rich Adler, University of Michigan-Dearborn, Dearborn, MI

Patricia Alfing, Davidson County Community College, Lexington, NC

Mary Allen, Hartwick College, Oneonta, NY

Carrie Arnold, Mount Wachusett Community College, Gardner, MA

Brianne Barker, Drew University, Madison, NJ

Barbara Baumeister, Rose State College, Midwest City, OK

Sue Bergs, Madison College, Madison, WI

Jennifer Bess, Hillsborough Community College, Tampa, FL

Kari Brossard Stoos, Ithaca College, Ithaca, NY

Kristen Butela, Seton Hill University, Greensburg, PA

Christie Canady, Middle Georgia State University, Cochran, GA

Donna Catapano, Briarcliffe College, Patchogue, NY

Moria Chambers, Muhlenberg College, Allentown, PA

Carlos Cuervo, Florida National University, Miami, FL

Marc Daniels, William Carey University, Biloxi, MS

Shelley des Estages, Mitchell College, New London, CT

Linda DeVeaux, South Dakota School of Mines and Technology, Rapid City, SD

Sondra Dubowsky, McLennan Community College, Waco, TX

Veronica Ernst, Delaware State University, Dover, DE

Yiwen Fang, Loyola Marymount University, Los Angeles, CA

Susan Galindo, Idaho State University, Pocatello, ID

Julie Gibbs, College of DuPage, Glen Ellyn, IL

David Gilmore, Arkansas State University, Jonesboro, AR

Maureen Gutzweiler, Harrisburg Area Community College, Lancaster, PA

Julie Harless, Lone Star College—Montgomery, Conroe, TX

Arlene Hoogewerf, Calvin College, Grand Rapids, MI

Patricia Jaacks, Jefferson Community College, Watertown, NY

Rita King, The College of New Jersey, Ewing, NJ

Laurieann Klockow, Marquette University, Milwaukee, WI

Dubear Kroening, University of Wisconsin-Fox Valley, Menasha, WI

Cayle S. Lisenbee, Arizona State University, Phoenix, AZ

Kelly R. Lynn Johnson, Somerset Community College, Somerset, KY

Kristin Malm, Washington State University, Pullman, WA

Elizabeth Mansberger, Juniata College, Huntingdon, PA

Tina Marshall, Marshall County School District, Calvert, KY

Thomas McGuire, Penn State—Abington, Abington, PA

Pamela McNutt, National American University, Wichita, KS

Karin Melkonian, LIU-Post, Brookville, NY

Jennifer Metzler, Ball State University, Muncie, IN

Janie Milner, Santa Fe Community College, Santa Fe, NM

Pamela Monaco, Molloy College, Rockville Centre, NY

Jon Morris, Manchester Community College, Manchester, CT

Susan Morrison, College of Charleston, Charleston, SC

Valerie Narey, Santa Monica College, Santa Monica, CA

Steve Openshaw, Iowa Central Community College, Fort Dodge, IA

Robert Osgood, Rochester Institute of Technology, Rochester, NY

Joan Petersen, Queensborough Community College, Bayside, NY

Mark Pilgrim, Lander University, Greenwood, SC

Sarah Richart, Azusa Pacific University, Azusa, CA

Meredith Rogers, Wright State University, Dayton, OH

Brenda Royals, Park University, Parkville, MO

Lorraine Sanassi, SUNY Downstate Medical Center, Flushing, NY

Rajesh Sani, South Dakota School of Mines and Technology, Rapid City, SD

Beatriz Santos, University of Salamanca, Salamanca, Spain

Harold Schreier, University of Maryland—Baltimore County, Baltimore, MD

Alyssa Schuck, Yeshiva University, New York, NY

David Singleton, York College of Pennsylvania, York, PA

Robert Sizemore, Alcorn State University, Alcorn State, MS

Julie Stacey, University of Cincinnati, Cincinnati, OH

Emily Stowe, Bucknell University, Lewisburg, PA

Don Takeda, College of the Canyons, Santa Clarita, CA

Elizabeth Tatersall, Western Nevada College, Minden, NV

David Turnbull, Lake Land College, Mattoon, IL

David J. Wartell, Harrisburg Area Community College, Harrisburg, PA

Cindy Williams, Northpoint Christian School (formerly SBEC), Southaven, MS

George Wistreich, East Los Angeles College (ELAC), Los Angeles, CA

Joe Wolf, ECTC, Elizabethtown, KY

Yunxuan Yang, Allegany College of Maryland, Cumberland, MD

Brenda Zink, Northeastern Junior College, Sterling, CO

LABORATORY FUNDAMENTALS OF MICROBIOLOGY, ELEVENTH EDITION

Read. See. Do. Connect.

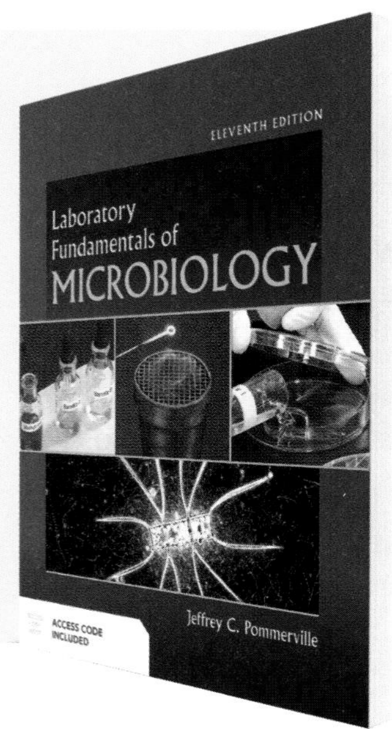

Packaged With Access To NEW Videos That Teach Common Lab Skills

WHAT'S NEW?

- Every new lab manual is packaged with access to over 110 minutes of NEW videos that teach common lab skills and are tied to the labs in the manual. Students clearly see how to work safely in the lab setting, how to use operating equipment, how to swab cultures, and many more valuable lab skills.
- NEW - all labs have been expanded and reorganized to fit neatly into new sections
- Features NEW introductions for each section
- Contains NEW full-color photos and micrograph examples
- Updated with NEW exercises and assessments

BUNDLE AND SAVE

BUNDLE *Fundamentals of Microbiology, Eleventh Edition* WITH *Laboratory Fundamentals of Microbiology, Eleventh Edition* and SAVE

This all-inclusive bundle contains the textbook and the laboratory manual with access to *Fundamentals of Mircobiology Laboratory Videos*, all for only $25 more than the printed textbook alone.

Bundle ISBN: 978-1-284-14396-6

LABORATORY EXERCISES

Visit **go.jblearning.com/PommervilleLabManual** to view a sample video, sample lab exercises, and to see the complete list of 99 laboratory exercises.

FUNDAMENTALS OF MICROBIOLOGY LABORATORY VIDEOS

The Ultimate Microbiology Laboratory Experience

Welcome to the ultimate microbiology laboratory experience, with 110 minutes of instructor chosen, high-quality videos of actual students performing the most common lab skills, procedures, and techniques. From lab safety to microscopy to bacterial staining techniques, nowhere else will you find a more comprehensive and customized microbiology video program. Jeffrey Pommerville provides narration, context, and rationale for the on-screen action, and *Skills Checklists* are available online to record progress and for assignability. ***Fundamentals of Microbiology Laboratory Videos*** is the perfect companion to any modern microbiology laboratory course.

- 34 videos and over 110 minutes of quality production-value video!
- Visually demonstrates common skills typically covered in an introductory undergraduate microbiology lab (safety protocols, operating equipment, swabbing cultures, etc.)
- Jeffrey Pommerville provides the educational context for each lab skill, and actual students perform the techniques in a typical undergraduate lab setting
- ***Skills Checklists*** on PDF allows students to record progress and instructors to assign lab activities

BUNDLE AND SAVE

BUNDLE *Fundamentals of Microbiology, Eleventh Edition* WITH *Fundamentals of Microbiology Laboratory Videos* and SAVE

This bundle includes the textbook and standalone access to the lab skills videos for only $25 more than the printed textbook alone.

Bundle ISBN: 978-1-284-14435-2

STANDALONE ACCESS

Get the complete *Fundamentals of Microbiology Laboratory Videos* series for $99.95/user for one year's access.

LABORATORY VIDEOS

Section: Lab Safety
Laboratory Safety
Cleaning Up a Culture Spill

Section: Laboratory Techniques and Skills
How To Use a Gas Burner / Bacti-Cinerator to Sterilize an Inoculating Loop / Needle
Broth-To-Broth
Broth-To-Agar
Four-Way Streak Plate
Preparing the Pour Plate
Transferring a Solution with a Pipette
How To Make a Serial Dilution

Section: Microscopy
How To Use the Light Microscope: Identifying the Parts
Measuring Cell Size

Section: Bacterial Staining Techniques
How To Make a Bacterial Smear
How To Make a Simple Stain Technique
How To Make a Negative Stain Preparation
Preparing a Gram Stain
Preparing an Endospore Stain
How To Make a Capsule Stain Preparation
How To Make a Hanging Drop Preparation

Section: Control of Microorganisms
How To Prepare a Disk Diffusion Assay
How To Prepare the Kirby-Bauer Test

Section: Measuring Population Growth
How To Use a Petroff-Hausser Chamber
How To Prepare a Standard Plate Count
How To Use a Spectrophotometer

Section: Medical Microbiology
Preparing the Acid-Fast Stain
Oxidase Test
IMViC: Indole Test
IMViC: Methyl Red Test
IMViC: Voges-Proskauer Test
IMViC: Citrate Test

Section: Identification of a Bacterial Unknown
Biochemical Tests: Carbohydrate Fermentation
Starch Hydrolysis Test
Catalase Test
Hydrogen Sulfate
Urease Test

Visit **go.jblearning.com/PommervilleLabVideos** to view a sample video 

Laboratory Safety

PART I

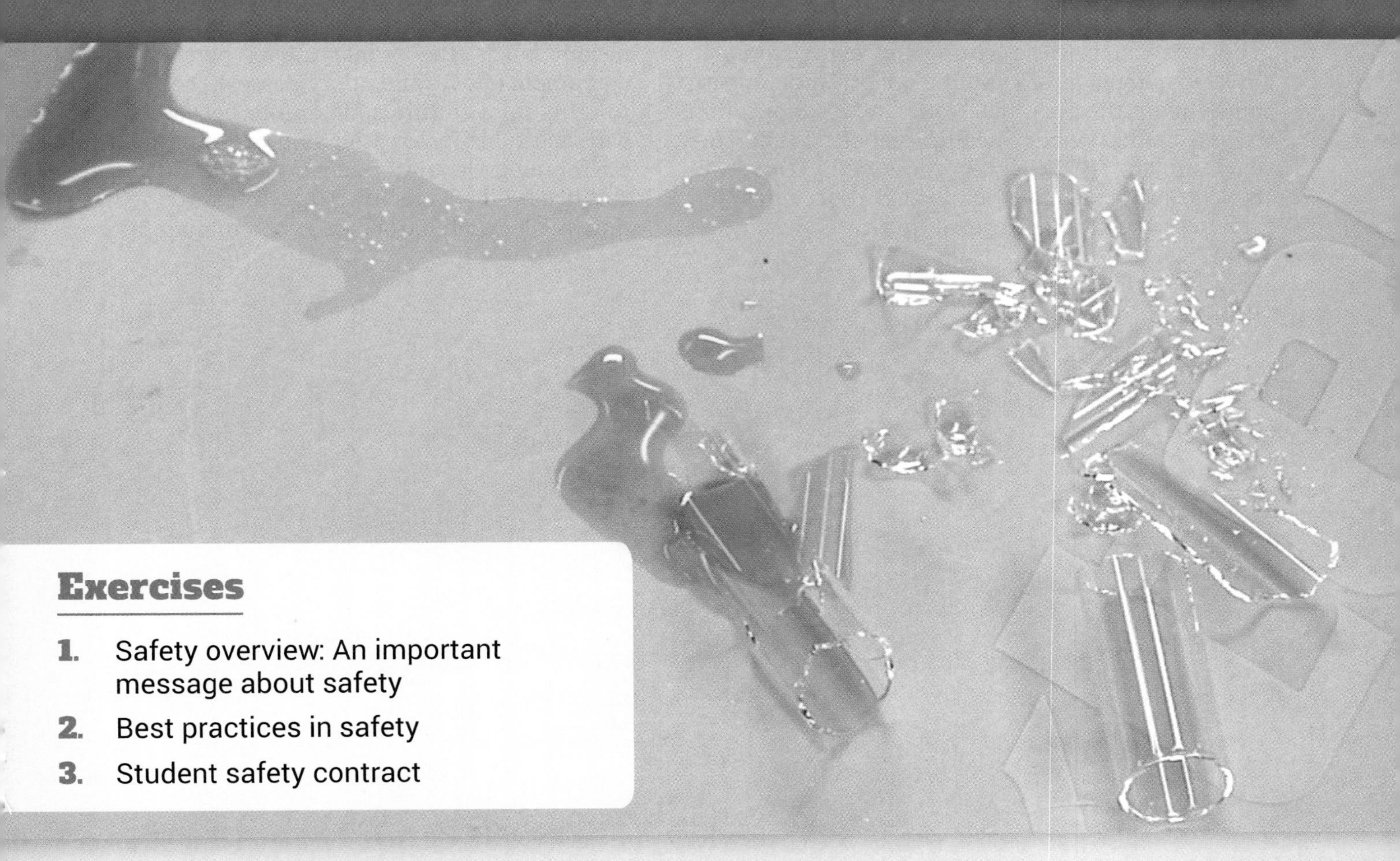

Exercises

1. Safety overview: An important message about safety
2. Best practices in safety
3. Student safety contract

Learning Objectives

When you have completed the exercises in Part I, you will be able to:

- Interpret the information on a chemical hazard label.
- Explain biosafety levels.
- Describe how to clean up a culture spill.
- Summarize laboratory safety rules and regulations.

Watch These Videos

- Laboratory Safety
- Cleaning up a Culture Spill

There cannot be too much emphasis on safety in a microbiology lab. When performing lab experiments, you may be interacting with bacteria, viruses, fungi, protista, and multicellular parasites. In addition, you may be exposed to more normal hazards, including chemical agents and the open flame of the Bunsen burner, and may be handling breakable glassware. The lab exercises, activities, and experiments in this lab manual afford you the opportunity to learn much about the microbial world and its application to medicine, public health, genetics, and the environment. However, to successfully carry out the lab exercises, microbiological procedures and practices must be performed in a careful and safe manner to limit the transmission of microorganisms. Disregarding safety rules creates a dangerous lab environment that may lead to personal injury. At the very least, it is likely to impact your, or another student's, experiments.

In this section, you are going to learn about, or review, the information contained on chemical hazard labels, the proper use of personal safety equipment (gloves and safety glasses), the procedure to clean up a culture spill, and to become familiar with the rules for creating and maintaining a safe microbiology laboratory environment. Good safety practices will ensure any risks while working in the microbiology laboratory are kept a minimum.

Safety Overview: An Important Message About Safety

EXERCISE 1

Exercises and experiments in the microbiology laboratory course involve exposure to living organisms, which, although very remote, conceivably may pose a threat to health. In past decades, microorganisms were classified as pathogenic or nonpathogenic, but in more recent years, scientists have demonstrated that even the most harmless organisms can be infectious under certain circumstances.

A WORD ABOUT CHEMICAL HAZARD LABELS

The Occupational Safety and Health Administration (OSHA), a federal organization that is part of the U.S. Department of Labor, requires that all hazardous areas in the laboratory should be marked with a biohazard symbol (FIGURE 1.1). Also, many teaching labs have instituted a policy whereby all chemical bottles are marked with a color-coded labeling system that allows you to be aware of any known hazards when working with various chemicals. Therefore, you must be able to read and interpret **hazard labels** on chemical bottles used in the microbiology laboratory.

All chemicals are rated by a standard system using a 0–4 scale (TABLE 1.1), where 0 is of minimal risk and 4 is of extreme risk. These numbers are obtained from the Material Safety Data Sheet (MSDS) provided with the chemical. The MSDS should be available in the laboratory area. The risks fall into three categories of concern regarding chemical agents.

FIGURE 1.1 Biohazard symbol.

First, what is the relative **health hazard** of the chemical? That is, is the chemical dangerous if inhaled, ingested, or absorbed onto or into the skin?

TABLE 1.1 Summary of Information Contained on Chemical Hazard Labels

Category (Number Rating)*	Color Code	Hazard	Example
Health (0–4)	Blue	Inhale, ingest, or absorb through skin	Microbiological stains Diagnostic reagents Mutagens and carcinogens
Flammability (0–4)	Red	Ability to catch fire	Alcohols
Reactivity (0–4)	Yellow	Explodes or reacts violently with water, air, or another chemical	Corrosives (acids)

*Rating index: 0 = minimal; 1 = slight; 2 = moderate; 3 = serious; 4 = extreme.

This health category is perhaps most important in the microbiology lab because there is a chance that a stain of a microorganism or diagnostic reagent might be spilled on the skin.

The second category is **flammability**. What is the likelihood that a chemical will catch fire? In the microbiology lab, an example is ethyl alcohol—if spilled and lit with a match or Bunsen burner flame, it will catch on fire!

The third category is **reactivity**. This refers to the possibility of a chemical reacting in some way with another chemical resulting in an explosion or violent reaction. In the microbiology laboratory, there are few if any known chemicals that would react in this fashion. Adding water to a concentrated acid would be one example that results in a "spattering-type" reaction.

The hazard labels found on chemical bottles are one of two shapes (FIGURE 1.2). There are small, diamond-shaped labels and larger, rectangular-shaped labels. Both are color coded by category (blue, red, and yellow) (FIGURE 1.3). Within the label category is the rating (0–4). For example, Figure 1.2 shows a hazard label for Gram's iodine (a) and another label for 95% ethyl alcohol (b). Recognizing and interpreting these labels can prepare you for a lab accident, should one occur.

Importantly, the assignment of a number rating for each category is based on an industrial-strength concentration of the chemical. However, the concentrations used in the microbiology laboratory are substantially more dilute, so their potential adverse affect is less of a risk than the biohazard label might indicate.

Criteria for Laboratory Biosafety

All experiments and exercises in this laboratory manual can be done with microorganisms assigned

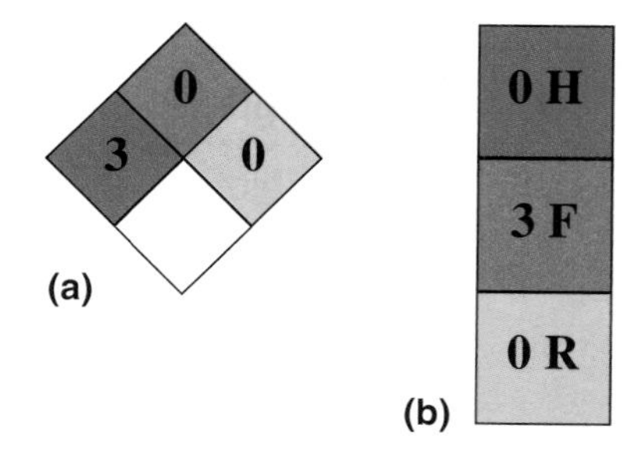

FIGURE 1.2 Chemical hazard labels for **(a)** Gram's iodine and **(b)** 95 percent ethyl alcohol. The small diamond-shaped labels are color-coded but, unlike the rectangular labels, are without the category letter.

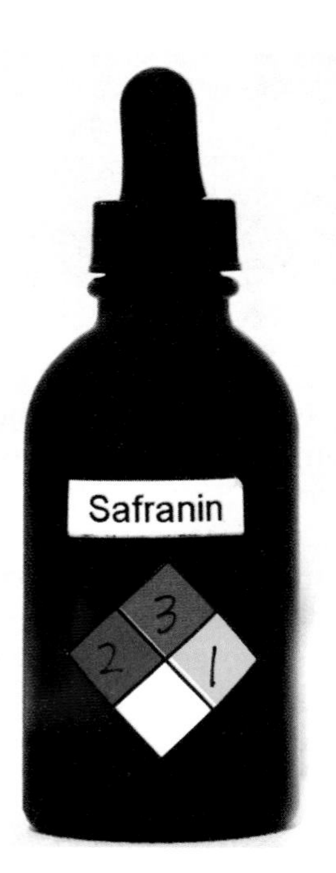

FIGURE 1.3 Chemical hazard label for the staining reagent safranin.

to biosafety level 1 (BSL-1). However, some instructors may have a need to use organisms assigned to BSL-2. Most commercial supply houses, including the American Type Culture Collection and MicroBioLogics, identify the BSL level of the organisms they provide. The biosafety levels are organized, in part, by the degree of protection needed by personnel using those organisms and the facilities required. A summary of BSL-1 and BSL-2 are provided in the following sections and in TABLE 1.2. A complete description can be found in *Biosafety in Microbiological and Biomedical Laboratories (BMBL), 5th Edition*, available at the Centers for Disease Control and Prevention (CDC) Web site (https://www.cdc.gov/biosafety/publications/bmbl5/).

Biosafety Level 1 (BSL-1)

Biosafety Level 1 organisms are well-characterized agents not known to consistently cause disease in healthy adult humans. In addition, these nonpathogenic organisms are of minimal hazard to laboratory personnel and the environment. In terms of the physical laboratory setting, the laboratory does not have to be separated from the general traffic patterns in the building. Students can carry out their experiments and exercises on the open bench tops by using the standard microbiological

TABLE 1.2 Summary of Recommended Biosafety Levels for Infectious Agents

BSL	Agent	Practice	Primary Barriers	Secondary Barriers
1	Not known to consistently cause disease in healthy adults	Standard microbiological practices	Safety glasses/goggles; gloves recommended	Open bench top; sink required
2	Associated with human disease; hazard: percutaneous injury, ingestion, mucous membrane exposure	Standard microbiological practices; limited access; biohazard warning signs; "sharps" precautions; a biosafety manual defining any needed waste decontamination or medical surveillance policies	Physical containment devices used for all manipulations of agents that cause splashes or aerosols of infectious materials; laboratory coats; gloves; face protection as needed	Open bench top; sink and autoclave required

Modified from Centers for Disease Control and Prevention. (2009). *Biosafety in microbiological and biomedical laboratories* (5th ed). Bethesda, MD: NIH.

techniques described in this lab manual. These standard practices are outlined in Exercise 2 and in the Student Safety Contract in Exercise 3.

Biosafety Level 2 (BSL-2)

Biosafety Level 2 is similar to Biosafety Level 1, but is also suitable for using microorganisms assigned to BSL-2. Such organisms are potentially a moderate health hazard to personnel and the environment. If using such organisms: (1) the laboratory and teaching assistants should have specific training in handling such pathogenic agents; (2) access to the laboratory should be limited when working with these agents; and (3) precautions should be taken to prevent the spread of infectious aerosols or splashes by conducting the critical steps of the experiment or procedure in a biological safety cabinet or other physical containment equipment.

A biohazard sign must be posted on the entrance to the laboratory where BSL-2 agents are used. All cultures and wastes should be decontaminated outside of the laboratory, and they should be transported in strong, leak-proof, closed containers.

Name: ______________________ Date: __________ Section: __________

Safety Overview: An Important Message About Safety

EXERCISE RESULTS

1

Questions

1. Using the chemical labeling system shown in Figure 1.2, fill in the number rating for (a) a chemical that has minimal flammability, slight reactivity, and is a moderate health hazard and (b) a chemical that has a moderate health risk, minimal reactivity, and slight flammability

a.

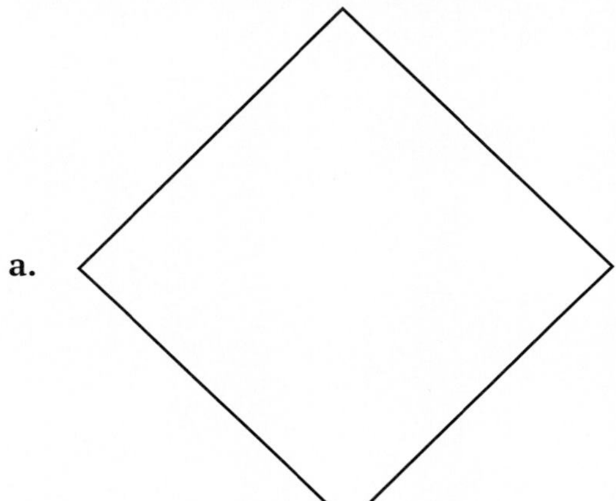

b.

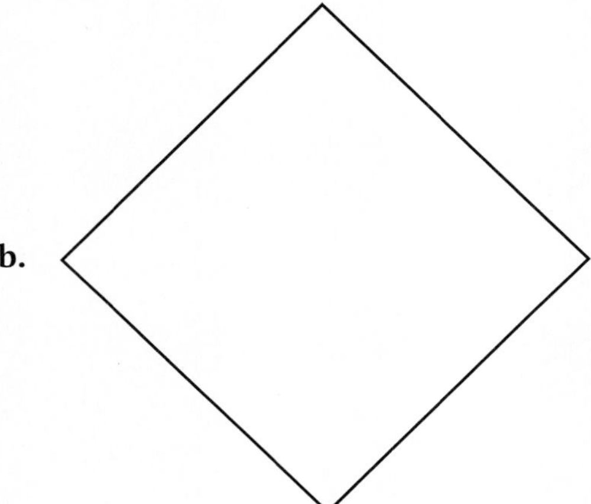

2. Based on Figure 1.3, determine the potential hazards when working with safranin.

____

____

3. What is the laboratory biosafety level (BSL-1 or BSL-2) for each of the following scenarios?

   a. A student in a microbiology course is doing an experiment using an open bench top, wearing a lab coat, gloves, and eye protection to reduce accidental infection, working in a biological safety cabinet, and using a moderately pathogenic strain of *Staphylococcus aureus*.

   b. A student in a microbiology course is doing an experiment using an open bench top, wearing eye protection, and using a nonpathogenic strain of *Escherichia coli*.

Best Practices in Safety

EXERCISE 2

Materials and Equipment

- Sink
- Antibacterial soap
- Paper towels
- Disinfectant
- Safety glasses
- Gloves
- Biohazard receptacle

Watch These Videos

- Laboratory Safety
- Cleaning up a Culture Spill

The exercises in this manual provide an opportunity to learn the techniques of microbiology, to study the activities of microorganisms discussed in class, and to apply laboratory principles to the solution of practical problems. Because many experiments are intended to yield a range of possible outcomes, it is good to remember that the results will be only as reliable as the methods used to obtain them. A close adherence to established procedures and careful attention to detail, together with a keen and discriminating eye, will ensure valuable results and a successful laboratory experience. You may well go on to a career in a government, academic, or health services laboratory, so it is never too soon to adopt a "culture of safety" attitude.

Review these laboratory best practices:

- Handwashing should occur before every lab period, as well as when you leave or enter the lab, and at the conclusion of a lab period. There is a proper technique for washing your hands. Be sure to use soap and warm or cold water, and to wash between your fingers. Scrub your hands for at least 20 seconds by humming the "Happy Birthday" song from beginning to end twice. Then, rinse your hands with clean, running water. Dry your hands with a clean paper towel. For tips on how to wash your hands properly, watch **Microbiology Video: Laboratory Safety**.

Laboratory Safety

- Disinfect your lab bench to minimize the chance of contaminating a culture and to eliminate the possibility of transmitting microorganisms to lab partners and the lab environment. Spray the bench with disinfectant and wait one minute for it to take effect. Wipe the area clean with paper towels.
- Wear safety glasses/goggles in the lab, except when looking at organisms through the microscope (FIGURE 2.1).
- Wear gloves as required by the instructor. Change your gloves if they become contaminated.
- Follow your instructor's or institution's policy for disposal of sharps (FIGURE 2.2).
- Clean up spills as soon as they occur and alert your instructor. To learn the proper technique, watch **Microbiology Video: Cleaning up a Culture Spill**.

FIGURE 2.1 Wear safety glasses/goggles to protect your eyes.

FIGURE 2.3 No eating in the laboratory!

FIGURE 2.2 Sharps containers.

Cleaning up a Culture Spill

- Do not eat or drink in the lab (FIGURE 2.3).
- Know the location of the exits, emergency shower/eye washing station (FIGURE 2.4), and first aid kits, as well as what to do in an emergency.
- Never smoke in the lab!!

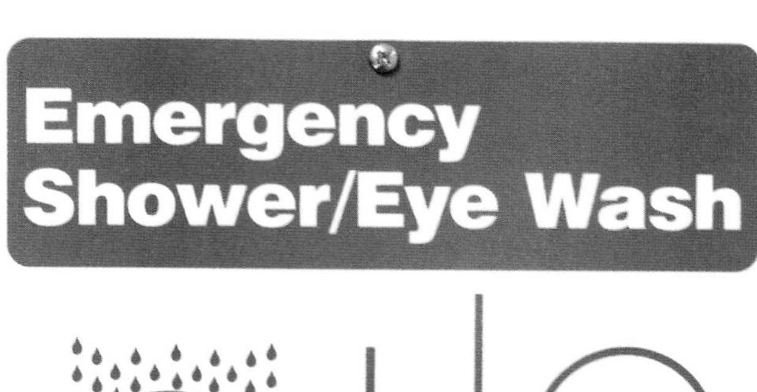

FIGURE 2.4 Eye washing station.

Student Safety Contract

EXERCISE

3

LABORATORY RULES AND REGULATIONS

Personal recognition of safety and the acceptance of certain precautions are necessary prerequisites to successful work in microbiology. It is important for your safety and the safety of those around you to follow standard safety rules. These rules are designed to create a safe learning environment.

1. Review exercises before the laboratory period and plan your work. Watch relevant microbiology lab videos to understand the processes you'll encounter.
2. Wear safety glasses/goggles as required by your instructor.
3. If necessary, wear a laboratory coat to protect against stains, spills, or splattering.
4. Tie back long hair. Bandage any existing cuts.
5. Wear gloves as required by the instructor, especially if handling body fluids.
6. Clean the laboratory desk or bench with disinfectant before working to limit the spread of dust-borne organisms. Disinfect it again at the end of the lab period to destroy any organisms that may have contaminated the work surface.
7. Eating, drinking, or smoking in the laboratory is prohibited, and objects such as pencils should be kept away from the mouth. Writing instruments (pens, pencils) may be provided to you for use in the lab.
8. If gummed labels are used, moisten them with tap water rather than saliva (pressure-sensitive labels are preferred). Pipetting should be done with mechanical pipetters – never pipette by mouth!
9. Keep laboratory doors closed during sessions.
10. Report any damage and all culture spills immediately to the instructor. Treat spills by surrounding and flooding the area with disinfectant, working from the outside in. Place paper toweling on top of the spill, and allow a period of 15 minutes to elapse before wiping the area clean. Place contaminated towels in a biohazard or other designated disposal container. Do not place contaminated materials in normal trash cans.
11. Promptly treat cuts and abrasions suffered in the laboratory with an antiseptic.
12. Thoroughly wash hands with soap and cold or warm water before leaving the laboratory during the period or at its end.
13. The fire extinguisher and eyewash stations should be clearly marked, and everyone should be able to locate them without difficulty. Be sure to check the whereabouts of the exits, fire extinguisher, first aid kit, and biohazard container.
14. If you or a classmate smells gas, notify the instructor without delay.
15. Hang up coats, excess clothing, and book bags away from the work area to avoid contamination.
16. To avoid spilling materials or injuring yourself with a flame, always gather your materials and place them in front of you on the bench top before beginning your work.
17. An area for placing cultures and used materials should be readily apparent. At the end of the lab period, place all cultures and used materials in the designated disposal area.
18. Let your instructor know if you are taking immunosuppressive drugs, are pregnant, or have any medical condition that might require special attention.

19. Use a complete and carefully thought-out label on all materials for incubation or storage to ensure that you and others know their contents.

20. Do not remove any materials, however innocuous, from the laboratory. This applies especially to microbial cultures.

I have read these rules on laboratory safety and agree to abide by them.

Name: ____________________________________

Date: ____________________________________

Laboratory Techniques and Skills

PART II

Exercises

4. Culture transfer techniques
5. Pure culture techniques: Streak plate method
6. Pure culture techniques: Pour plate method
7. Solution transfer
8. Serial dilutions

Learning Objectives

When you have completed the exercises in Part II, you will be able to:

- Complete the aseptic transfer of microorganisms.
- Carry out pure culturing procedures to separate bacteria in a mixed population.

Watch These Videos

- How to Use a Gas Burner / Bacti-cinerator to Sterilize an Inoculating Loop / Needle
- Broth to Broth Transfer
- Broth to Agar Transfer
- Agar to Agar Transfer
- Four-way Streak Plate
- Preparing the Pour Plate
- Transferring a Solution with a Pipette
- How to Make a Serial Dilution

In this section, you are going to practice standard laboratory techniques and skills that you will build on as you progress through the exercises in this lab manual. These techniques and skills are used in laboratories around the world. Specifically, you will learn how to isolate, grow, and study microorganisms (specifically bacteria, fungi, and protists). To do so in the lab, each species must be grown in culture. A **culture** consists of one or more species of microorganisms growing in a nutrient growth medium usually under controlled temperatures. A culture containing only one species of microorganism is called a **pure culture**. A culture that contains more than one species is a **mixed culture**. To isolate and grow microorganisms in pure culture, you will need a variety of materials and equipment and the skills to carry out a few standard procedures (microbiological techniques). You will also need to learn the skills to separate bacteria in a mixed culture.

BASIC GROWTH MEDIA

Like all living organisms, microbes need nutrients, such as a carbon and nitrogen source as well as vitamins and other growth factors, to survive, grow, and reproduce. The source for these nutrients usually comes from the enzymatic degradation of other complex nutrients derived from plant and animal sources. The composition of the nutrients represents a **culture medium** (FIGURE II.1).

Such culture media may be in the form of a liquid or solid (FIGURE II.2). A liquid solution in which microbes, especially bacteria and protists, will grow is called a **nutrient broth**. A broth medium is one way to grow microorganisms to a high cell density. However, if the broth contains more than one species of microorganism, each individual species cannot be distinguished with the naked eye.

A broth supplemented with a solidifying agent, called **agar**, produces a relatively solid medium on which many bacteria and fungi easily grow. Agar is a complex polysaccharide extracted from a seaweed (red algae); agar is physiologically inert and has no nutritional value in an agar culture. In solution, powdered agar liquefies at 100°C and then will solidify at 40°C as the solution cools. This means microorganisms isolated from humans can be cultivated at human body temperature (37°C) without fear of the medium liquefying.

Different types of agar media can be prepared. While in the liquefied state, the culture medium is **sterilized**; that is, all living organisms and viruses are killed and no spores exist. If such a sterile medium is placed in sterile test tubes and allowed to cool and harden in a slanted position, **agar slants** are produced. These are useful for maintaining pure cultures. Similar tubes allowed to harden in the upright position are called **agar deeps**. If sterile liquid agar medium is poured into sterile culture (Petri) dishes and allowed to harden, **agar plates** are formed. Agar plates provide a large surface area for the isolation and study of microorganisms.

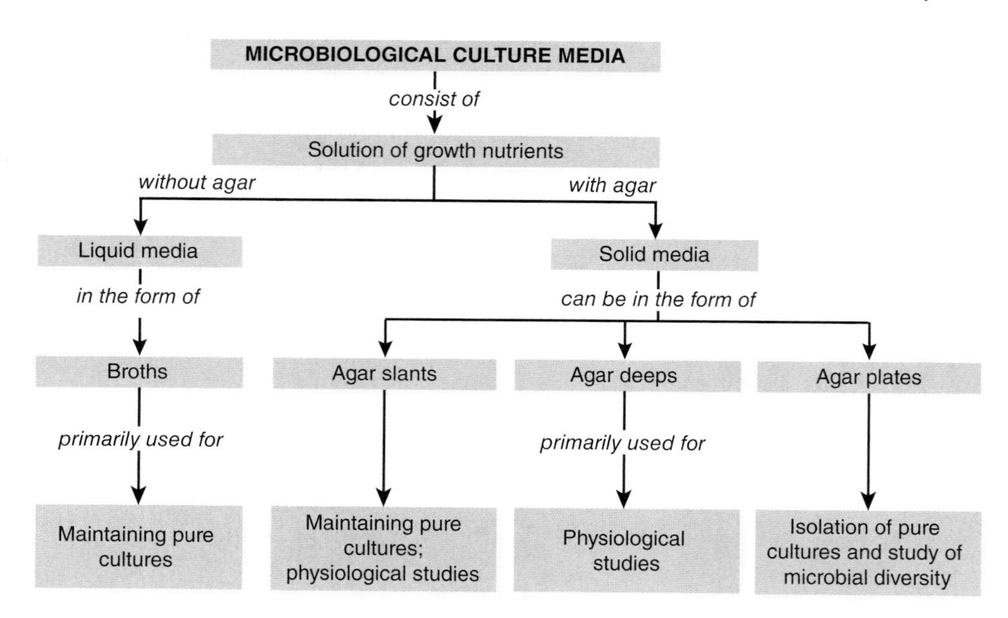

FIGURE II.1 A Concept Map for the different types of microbiological media and their uses are shown.

(a)

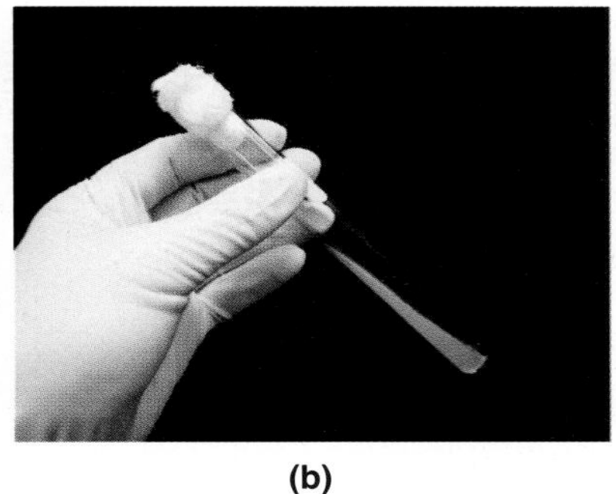

(b)

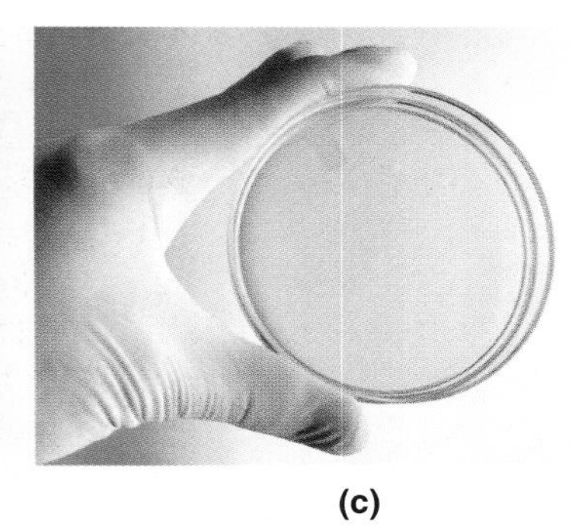

(c)

FIGURE II.2 **Different Types of Media: Solid and Broth.** Media can be in the form of **(a)** a broth, or solid media can be prepared using agar to prepare **(b)** agar slants, agar deeps, and **(c)** agar plates.

(a) © Khamkhlai Thanet/Shutterstock; (b) © Trinset/Shutterstock; (c) © angellodeco/Shutterstock.

The agar medium has many advantages:

- It represents a hardened surface on which bacteria or fungi can grow as individual **colonies**; that is, clusters of cells derived through multiplication from a single cell or spore. For example, a soil sample placed in a sterile broth medium would produce a dense population of cells. The same sample on an agar medium will show the diversity of bacteria and fungi in the sample.
- Different species often produce unique colony characteristics that can aid in identification. Therefore, most bacterial species can be visually separated from fungi, and even different species of bacteria can be identified by the way the colonies appear on the agar surface (**Exercise 5**).
- Since the colony represents a single species, a sample from a well-isolated colony can be transferred to a sterile broth or agar medium to form a pure culture.

CULTURE EQUIPMENT

During your work in the microbiology laboratory, cultures will be studied or maintained in different types of culture equipment.

- **Culture tubes.** Broth cultures typically use glass tubes with a suitable sleeve-like cap to keep out contaminating microbes. Built-in ridges on tube closures allow entry of air. Some broth media may use screw-cap tubes. Agar slants and agar deeps also use culture tubes either with a sleeve-like cap or a screw cap.
- **Culture plates.** Plastic **culture (Petri) dishes** are used to cultivate microorganisms. The solidified medium is in the bottom half of the plate, while a slightly larger, loose-fitting lid covers the plate and again keeps out contaminants. Once inoculated with a microorganism, the dishes usually are incubated in an inverted position (lid down) to prevent condensation forming on the cover. Left right-side up, the condensation could drop onto the agar surface and make the identification of unique colonies difficult.

TRANSFER INSTRUMENTS

Microorganisms often have to be transferred from one culture medium to another or from a stock culture to various media for maintenance, growth, and study (**Exercise 4**). Such a transfer is called **subculturing** and the practice must be carried out under **aseptic conditions**; that is, transfer between culture media without contamination from other microorganisms. The most commonly used instruments are described below.

- **Wire inoculating loops** and **needles** are used to perform transfers (FIGURE II.3A, B). They are easily sterilized by incineration in the blue (hottest) portion of the **Bunsen burner** flame or in a Bacti-cinerator (Watch **Microbiology Video: How to Use a Gas Burner / Bacti-cinerator**). After slight cooling, a sample of organism is aseptically placed on the loop and streaked or suspended in an appropriate agar or broth medium; needles with an organism sample usually are stabbed into an agar deep. Disposable loops or needles also may be used.

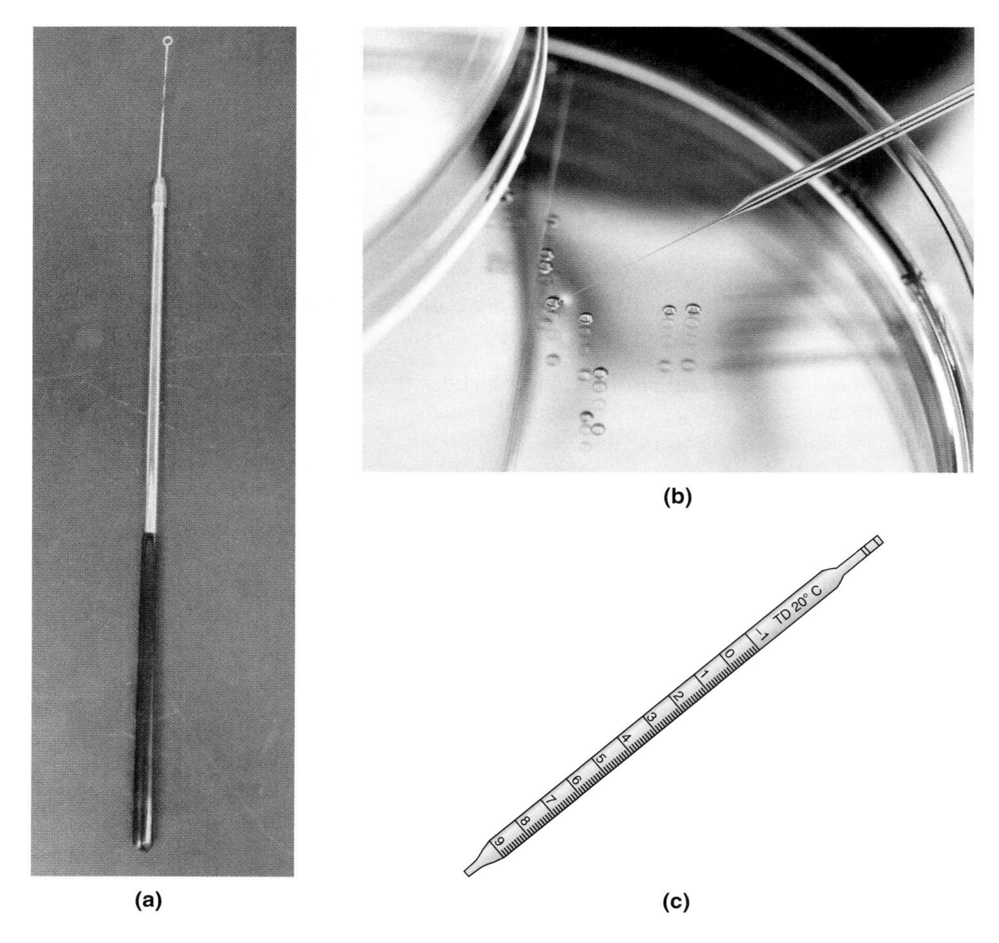

FIGURE II.3 Microbiological Transfer Instruments. Inoculating loops **(a)** and needles **(b)**, and pipettes **(c)**, are among the more common items used.

(b) © Science Photo Library/Shutterstock.

How to Use a Gas Burner/Bacti-cinerator

- **Pipettes** can be used for sterile transfers (FIGURE II.3C, D). Sterile plastic or glass pipettes draw up liquids into the graduated (calibrated) column (Exercise 7). The graduations mark the different volumes that can be delivered.

Never pipette chemical solutions or broths by mouth! All pipetting should be done with the aid of mechanical pipette devices.

GROWTH AND INCUBATION APPLIANCES

Many microorganisms have specific temperature requirements for growth. In the entry-level course, most organisms can be grown at either room temperature (20–22°C) or human body temperature (37°C). Maintaining the latter temperature for long periods of time, such as between lab sessions, requires putting the cultures in an **incubator** that maintains the optimum temperature necessary for growth. Since most incubators use dry heat, which can dehydrate the agar or broth medium, a beaker of water can be placed in the incubator during the growth period to maintain a moist environment.

For short periods of time, such as during a lab exercise or experiment, cultures may be incubated in a **waterbath**, where the appropriate temperature is maintained by a heating device. A waterbath can be used only for cultivation of organisms in a broth medium, as agar will liquefy in the waterbath.

Since your lab session might meet only once a week, cultures incubated at 37°C for seven days might overgrow the tube or plate. Often, after 24 to 48 hours at 37°C, the cultures will be moved to a refrigerator, which will stop or greatly slow down microbial growth.

Culture Transfer Techniques

EXERCISE 4

Materials and Equipment

- Inoculating loop
- Nutrient broth and nutrient agar slant cultures
- One 24-hour nutrient broth of *Serratia marcescens* (or another bacterial species that is colorfully pigmented) per student
- One 24-hour nutrient agar slant of *Serratia marcescens* (or another bacterial species) per student

Watch These Videos

- Broth to Broth Transfer
- Broth to Agar Transfer
- Agar to Agar Transfer

In several microbiology exercises and experiments, you will need to transfer microorganisms from one culture medium to another by **subculturing**. This aseptic technique also is commonly used when preparing and maintaining stock cultures, and when carrying out a number of microbiological test procedures.

Microorganisms are everywhere, including in the air of your course lab, as well as on the floors, bench tops, and equipment. If appropriate precautions are not taken, these microbes could end up in one of your subcultures; that is, microbes could **contaminate** your materials. To prevent contamination by unwanted microbes, the microbes you want must be transferred using proper aseptic techniques.

Aseptic transfer techniques are not difficult to learn and perform, although some eye-hand coordination and practice may be needed. To simplify the process:

- Make sure all needed materials, cultures, and media are ready, and the wire of the transferring device (inoculating loop or needle) is straight.
- Label appropriately all media into which microbes will be transferred so you can identify your cultures during the lab period or, if incubated, when the next lab period meets.

SUBCULTURE TRANSFER TECHNIQUES

PURPOSE: to transfer aseptically a pure culture of bacteria from culture tube to another sterile culture tube.

The three steps of the subculture process are described below and shown in FIGURE 4.1. The description and figure show the method for transferring a pure-culture sample of bacteria between culture media in tubes. **Read through the process completely before attempting the procedures described below.**

Step 1. The loop sterilization process

- Light the Bunsen burner.
- If you are right-handed, hold the pure-culture tube (the broth or slant from which you will make the transfer) and the sterile agar slant or broth tube to be inoculated in the palm of your left hand (Figure 4.1a). (If you are left-handed, the tubes should be in your right hand).

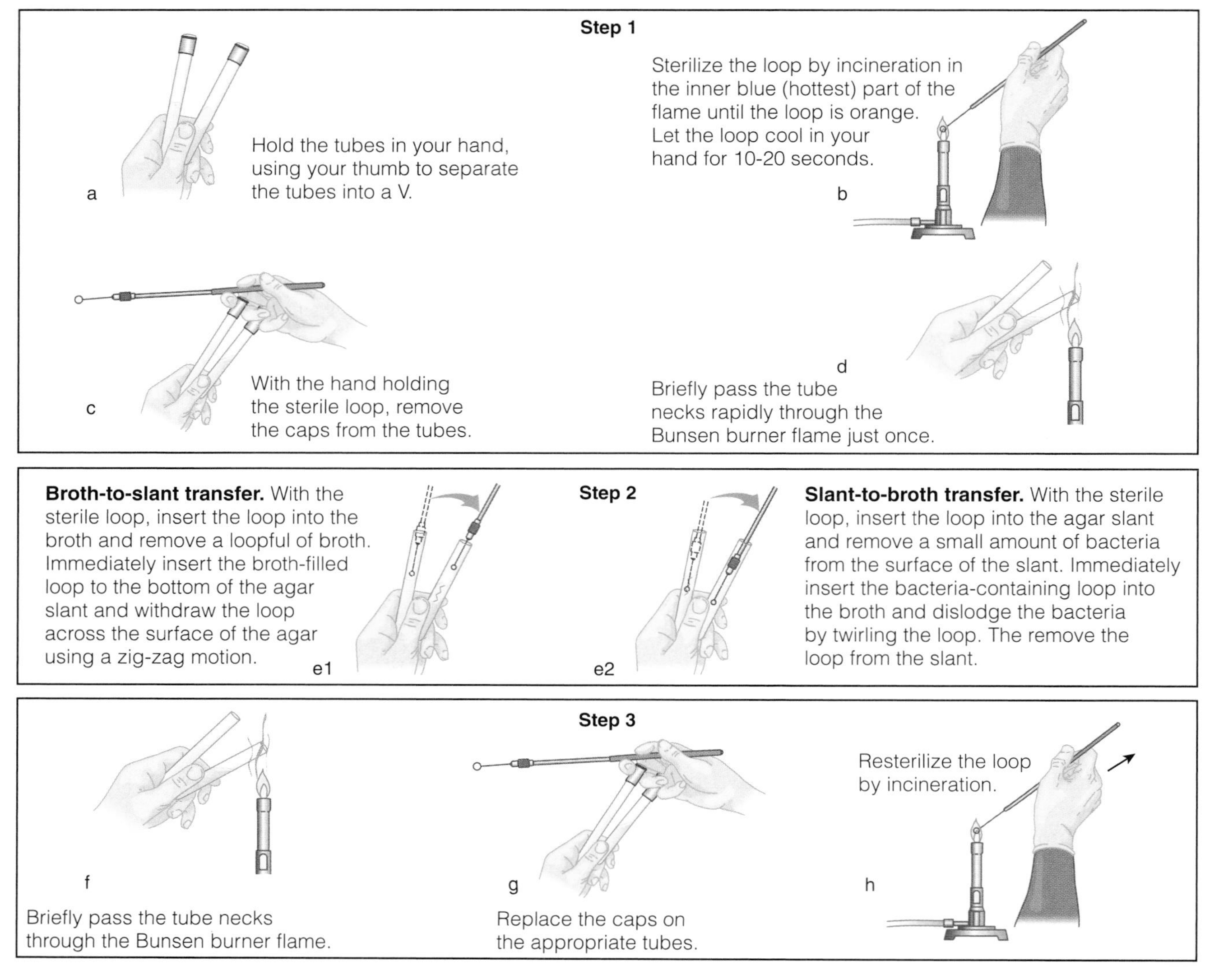

FIGURE 4.1 Subculturing techniques are used to transfer aseptically a pure-culture sample from tube to tube.

- The inoculating loop is not sterile. To sterilize it, hold the handle of the instrument in your free hand like you would a pen or pencil.
- Place the loop in the hottest portion of the Bunsen burner flame, which is the top of the blue flame (Figure 4.1b). (If you are using incinerators, your instructor will describe the sterilization process.) In a few seconds, the entire loop will turn orange. When it does, move the rest of the wire rapidly through the flame.
- Once sterilized, do not set the loop down; rather hold the loop in your hand and allow the loop to cool for 10 to 20 seconds.

Step 2. The transfer process

You are now ready to carry out the tube-to-tube transfer.

- Remove the caps or closures of the two tubes by grasping the cap of the left tube between your little finger and ring finger; and the cap of the right tube between the ring finger and middle finger (Figure 4.1c).
- Now, remove the caps and briefly heat the necks of the tubes in the Bunsen burner flame (Figure 4.1d). Watch **Microbiology Video: Broth-to-Broth Transfer** to see how this is done.

Broth to Broth Transfer

- Insert the loop into the broth tube with organisms and withdraw a loopful of liquid (Figure 4.1e1 and FIGURE 4.2). You should see a watery film on the loop.
- Immediately insert the broth-containing loop into the receiving broth tube. Shake the loop gently to dislodge the bacteria and suspend them in the broth. Remove the loop.

Broth to Agar Transfer

Broth to Agar Slant

- Insert the loop into the broth tube and withdraw a loopful of liquid (Figure 4.1e1). You should see a watery film on the loop (Figure 4.2).
- Immediately insert the broth-containing loop into the bottom agar slant tube. Place the loop flat on the agar surface and move the loop rapidly in a zigzag motion gently across the agar surface from bottom to top.
- Remove the loop from the agar slant. Watch **Microbiology Video: Broth to Agar Transfer** for a demonstration.

Agar slant to broth

- If you are transferring a bacterial sample from an agar slant to a broth, repeat step 1.
- Now, insert the sterile loop into the agar slant and touch the loop to the agar surface where the bacteria are growing. Remove just a minute sample of bacteria (Figure 4.1e2).
- Immediately insert the loop into the sterile broth tube. Shake the loop gently to dislodge the bacteria and suspend them in the broth.
- Withdraw the loop from the broth tube.

Agar slant to agar slant

- Insert the loop into the agar slant and touch the loop to the agar surface where the bacteria are growing. Remove just a minute sample of bacteria.

Do not fill the entire loop with bacteria when transferring bacteria from an agar culture medium.

- Immediately insert the loop into the sterile agar slant tube. Place the loop flat on the agar surface and move the loop rapidly in a zigzag motion gently across the agar surface from bottom to top.
- Remove the loop from the agar slant. Watch **Microbiology Video: Agar-to-Agar Transfer** to see this in action.

Agar to Agar Transfer

Agar slant to agar slant

- Insert the loop into the agar slant and touch the loop to the agar surface where the bacteria are growing. Remove just a minute sample of bacteria.

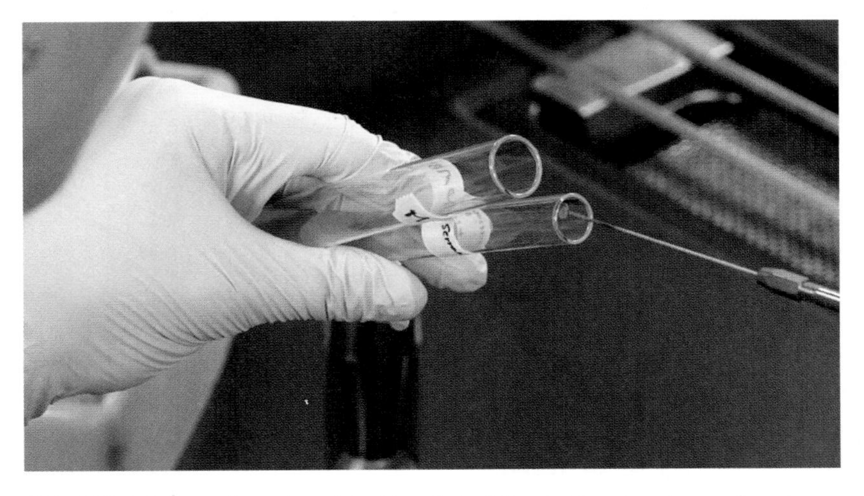

FIGURE 4.2 Transfer a loopful of broth, holding both tubes in one hand.

- Immediately insert the loop into the sterile agar slant tube. Place the loop flat on the agar surface and move the loop rapidly in a zigzag motion gently across the agar surface from bottom to top.
- Remove the loop from the agar slant. Watch **Microbiology Video: Agar-to-Agar Transfer** to see this in action.

Step 3. Capping tubes and re-sterilizing the loop

- Following the transfer, briefly reflame the necks of the tubes and place the caps back on their respective tubes (Figure 4.1f, g).
- The loop is again flamed to incinerate any remaining bacteria still on the loop (Figure 4.1h).
- Place the tubes in the test tube rack or appropriate tube holder.

PROCEDURE

1. Label all tubes of sterile media.
2. Following the procedure outlined and illustrated, perform the following aseptic transfers:
   a. Transfer *S. marcescens* from a broth culture to a nutrient agar slant.
   b. Transfer *S. marcescens* from an agar slant culture to a nutrient broth.
3. Incubate all cultures at 25°C for 24 to 48 hours. (If lab does not meet until the following week, arrangements should be made to move the cultures to a refrigerator after the incubation period.)
4. In the next lab period:
   - Examine the agar slant culture for the appearance of growth. Growth on agar can be seen by noting the appearance of a thick, red film of bacteria on the surface of the slant.
   - Examine the broth culture for the appearance of growth. Growth is indicated by the broth appearing fairly **turbid**; that is, the liquid should be dense and cloudy.
   - Record your observations in TABLE 4-1 in the Results section.

Quick Procedure

Subculture Transfer Technique

1. Sterilize inoculating loop.
2. Transfer broth film or small bacterial sample on loop from stock tube (broth or agar slant) or agar plate to appropriate sterile tube (mix in broth; streak on agar slant or plate).
3. Resterilize loop.

Name: ______________________ Date: __________ Section: __________

Culture Transfer Techniques

EXERCISE RESULTS 4

RESULTS

SUBCULTURE TRANSFER TECHNIQUES

TABLE 4.1 Subculture Transfer Results

	Medium Subcultured	
	Nutrient Agar Slant	**Nutrient Broth**
Growth (+) or (−)		
Draw the distribution of growth.		
Describe the appearance of bacterial growth		

Questions

1. Why must you use aseptic techniques when carrying out subculturing?

2. Explain why it is necessary to:
   a. Flame the inoculating loop before and after each inoculation.
   b. Cool the inoculating loop prior to obtaining the bacterial sample.

3. Why was *S. marcescens* used in this exercise?

4. From your answer to question 3, can you say with a high degree of certainty that
   a. The broth culture is a pure culture of *S. marcescens?*
   b. The agar slant is a pure culture of *S. marcescens?*

5. Which of the subculture transfer processes was more difficult to perform? Explain.

____

____

____

____

Pure Culture Techniques: Streak Plate Method

EXERCISE

5

Materials and Equipment

- A 24-hour nutrient broth mixed culture (3 parts *Micrococcus luteus*, 1 part *Serratia marcescens*, and 2 parts *Escherichia coli*
- 1 sterile nutrient agar plate
- Inoculating loop and/or needle
- Ruler
- Wax pencil

Watch This Video

- How to Make a Four-Way Streak Plate

Bacteria usually exist in **mixed populations** in soil, water, and some parts of the human body. It is not feasible to identify or study the characteristics of a particular species when it is mixed with other species. Therefore, a pure culture must be obtained in preparation for further work. As the name implies, a **pure culture** contains a large group of bacterial cells of a single species of organism.

In this exercise, the **streak plate** method will be used to separate and isolate bacterial cells in a mixed population. The fundamental principle underlying this technique is that a mixture of bacteria is gradually thinned out in a growth medium, giving individual cells room to form separate masses of bacteria called **colonies**. Presumably, each colony arises from the growth and reproduction of a single bacterium, and therefore all the cells in the colony are genetically identical. A sample of the cells may then be selected and subcultured to form a pure culture.

Bacterial colony characteristics also will be studied in this exercise. Once you have a pure culture, you can observe certain bacterial characteristics that reflect the genetic composition of a bacterium and serve as markers for different bacterial species. The laboratory microbiologist often uses colony characteristics as one criterion in the identification of an unknown bacterium.

STREAK PLATE METHOD

PURPOSE: to perform the streak plate method for isolating a single bacterial species in pure culture.

The **quadrant (four-way) streak plate method** is a relatively inexpensive and straight forward method for separating bacterial cells in a mixed population of high cell density. It requires only a single plate of growth medium and, with practice, it yields excellent distribution of bacterial colonies. For these reasons, the method is standard practice in many research, industrial, and clinical laboratories.

The following section describes the process to streak and dilute bacterial cells over the surface of a sterile agar plate. The flaming of the loop that is described is done to achieve the dilution of the cells in the mixed culture so that fewer cells are streaked into each area, resulting in the desired cell separation. Watch **Microbiology Video: How to Make a Four-Way Streak Plate** to familiarize yourself with the technique. The method is illustrated in FIGURE 5.1.

How to Make a Four-Way Streak Plate

- Using a flame-sterilized loop, place a loopful of the mixed culture on the agar surface in Area 1 (FIGURE 5.1a). Flame the loop, and cool it by touching an unused part of the agar surface close to the periphery of the plate, and then drag the loop rapidly several times across the surface of quadrant 1 (FIGURE 5.2).
- Resterilize and cool the loop. Turn the culture dish 90°. Then, touch the loop to a corner of the culture in quadrant 1 and drag the loop several times across the agar in Area 2 (FIGURE 5.1b). The loop should never enter quadrant 1 again.
- Once again, sterilize and cool the loop and again turn the dish 90°. Now, streak quadrant 3 in the same manner as how quadrant 2 was streaked (FIGURE 5.1c).
- Now, without sterilizing the loop, again turn the dish 90° and then drag the loop from a corner of quadrant 3 across quadrant 4, using a wider streak (FIGURE 5.1d). Don't let the loop touch any of the previously streaked quadrants.
- Resterilize the loop.

PROCEDURE

1. Select a nutrient agar plate or, at the direction of the instructor, pour a plate of the medium using a Petri dish and melted nutrient agar. With a wax pencil or felt marker, label the bottom

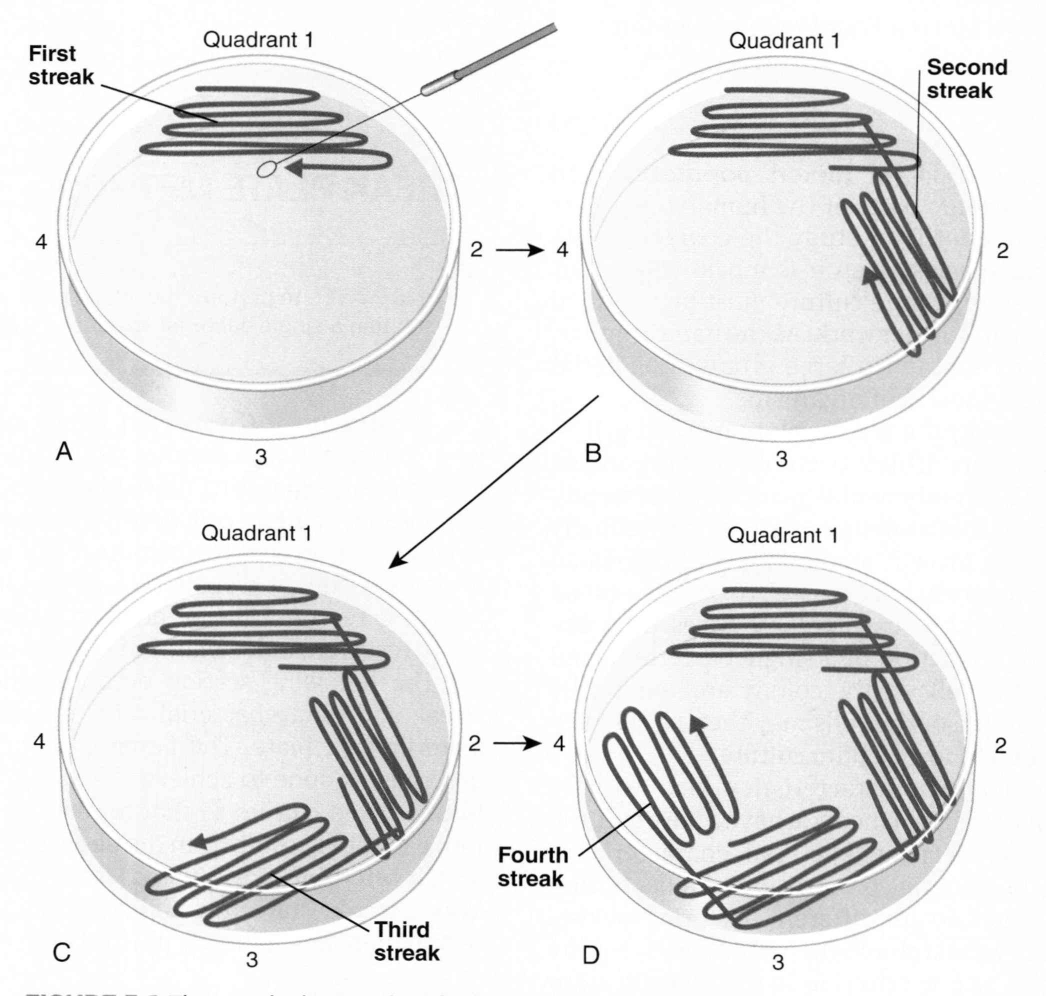

FIGURE 5.1 The streak plate method for isolating discrete colonies

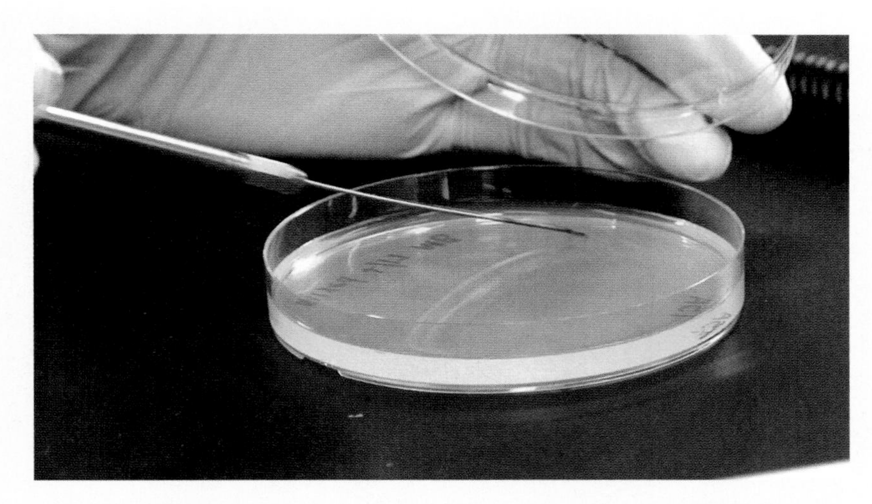

FIGURE 5.2 Lift the lid of the Petri dish only enough to permit entry of the loop.

side of the plate with your name, the date, and the designation "streak plate." Obtain a broth culture containing the mixed population of bacteria.

2. Aseptically obtain a loopful of the mixed population and lightly streak it several times along one area (quadrant) of the plate. Try not to cut into the agar surface and avoid airborne contamination by lifting the lid of the Petri dish only enough to permit entry of the loop.
3. Replace the lid. Sterilize the loop and then cool it by touching the loop to the center of the plate or between the agar and the edge of the plate.

Never touch the bacterial colonies on a plate with your fingers. Each colony contains millions of live organisms.

4. Pass the loop one time across the previous streaks to pick up some bacteria, and continue streaking into a second area of the plate. Replace the lid.
5. Sterilize the loop as before, then be certain that it is cool.
6. Pass the sterile loop one time through the second area, and continue streaking into a third area. Replace the lid.
7. Finally, *without* sterilizing the loop, streak some bacteria from the third area of the plate into the fourth area. Be sure to use up the remaining space on the plate. Replace the lid. Sterilize the loop.
8. Invert the plate and incubate it for 24–48 hours at the appropriate temperature.
   - After incubation, refrigerate the plates in the inverted position until the next laboratory session to preserve the bacterial growth and prevent drying of the medium.
9. Examine the plates for well-isolated and separated colonies, as shown in FIGURE 5.3, and enter a representation of a good streak plate in the Results section.
10. Add appropriate labels and brief explanations to produce a "talking picture."
    - Describe several isolated colonies by numbering the colonies and noting their size, color, and characteristics, with reference to the standard terminology provided in FIGURE 5.4 and TABLE 5.1. Size may be determined by measuring the colony diameter in millimeters if rulers are available.
    - Record your observations in TABLE 5.2 in the Results section.

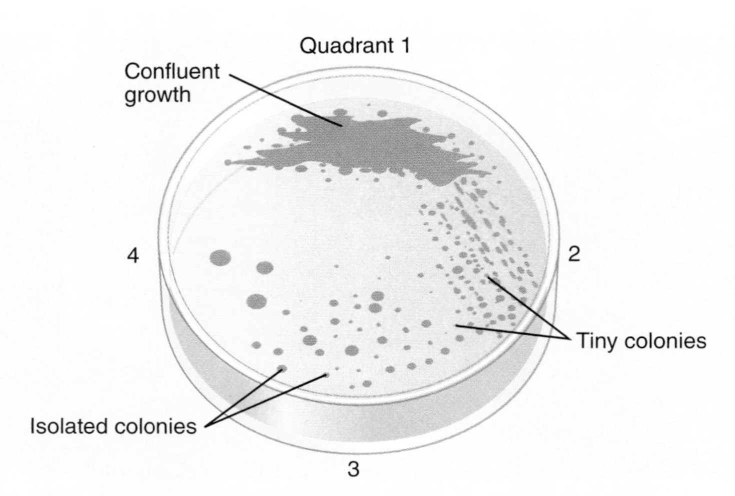

FIGURE 5.3 Isolated colonies of bacteria on a well-executed streak plate.

11. At the direction of the instructor, use a bacteriological needle or loop to select samples from various colonies, and inoculate nutrient agar slants to obtain pure cultures. Stained smears may also be made from the colonies to determine the morphological characteristics of the organisms. If your instructor wishes you to do this, record your observations in the Results section, following Table 5.2.

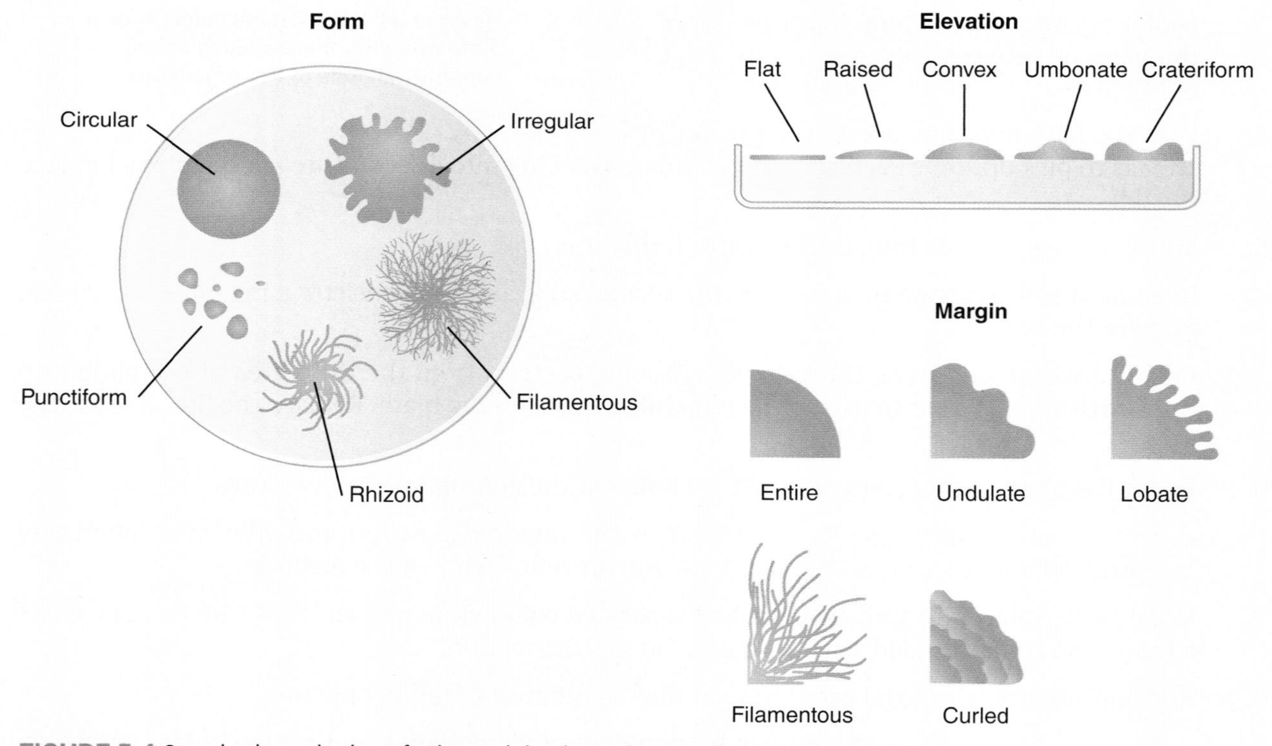

FIGURE 5.4 Standard terminology for bacterial colony characteristics.

TABLE 5.1 Descriptors for Bacterial Colony Characteristics

Characteristic	Descriptors
Size	Pinpoint, small, medium, large
Form	Circular, irregular, filamentous, rhizoid, punctiform
Elevation	Flat, raised, convex, umbonate (raised center), crateriform (or umbilicate; depressed center)
Margin	Smooth/entire, undulate, lobate, filamentous, curled, rough
Surface	Dull, glistening, moist
Consistency	Mucoid, creamy, viscous, brittle, sticky, dry, waxy
Density	Translucent, transparent, opaque
Color	White, golden, tan, red, green, gray, ecru

Modified from Delost, M. (2015). *Introduction to diagnostic microbiology for the laboratory sciences*. Burlington, MA: Jones & Bartlett Learning.

Quick Procedure

Streak Plate Method

1. Streak a loopful of bacteria in one area of an agar plate.
2. Flame the loop and streak some bacteria from the first area into a second area.
3. Flame the loop and streak some bacteria from the second area into a third area.
4. Do not flame the loop before streaking some bacteria from the third area into a fourth area. Incubate.

Name: ______________________ Date: __________ Section: __________

Pure Culture Techniques: Streak Plate Method

EXERCISE RESULTS 5

RESULTS

STREAK PLATE METHOD

Streak Plate 1

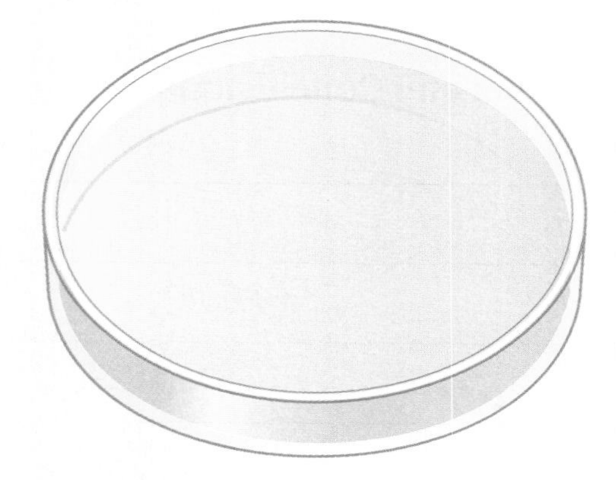

Streak Plate 2

Source: ______________________ ______________________

Table 5.2 Summary of Colony Characteristics

Colony Number and Source	Colony Characteristics				
	Size	Color	Form	Elevation	Margin

If your instructor has required you to stain a sample of bacterial cells, your light microscope observations can be recorded in the areas below.

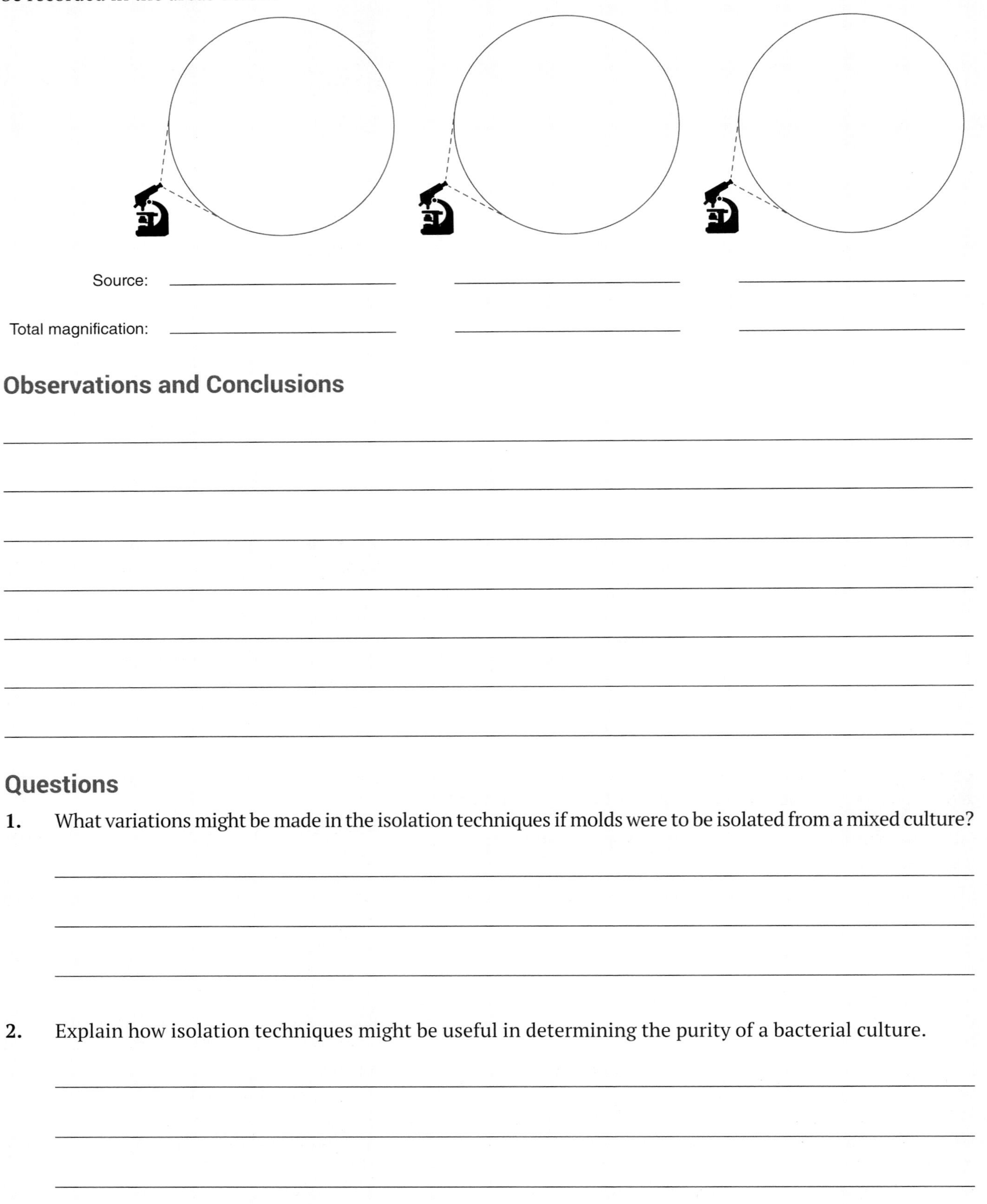

Observations and Conclusions

Questions

1. What variations might be made in the isolation techniques if molds were to be isolated from a mixed culture?

2. Explain how isolation techniques might be useful in determining the purity of a bacterial culture.

Pure Culture Techniques: Pour Plate Method

EXERCISE

6

Materials and Equipment

- A 24-hour nutrient broth mixed culture (3 parts *Micrococcus luteus*, 1 part *Serratia marcescens*, and 2 parts *Escherichia coli*
- Deep tubes of liquid nutrient agar in a 45°C water bath
- Sterile culture (Petri) plates
- Inoculating loop and/or needle

Watch This Video

- Preparing the Pour Plate

In this exercise, another pure culture technique, called the pour plate method, will be used to separate and isolate the bacteria in a mixed population. As with Exercise 5, the principle underlying this method is that a mixture of bacteria is gradually thinned out in a growth medium, giving individual organisms room to form colonies. Presumably, each colony arises from the growth and reproduction of a single bacterium, and therefore all the cells in the colony are genetically identical. A sample of the cells may then be selected and subcultured to form a pure culture. Bacterial colony characteristics also will be studied in this exercise.

POUR PLATE METHOD

PURPOSE: to perform the pour plate method for isolating a single bacterial species in pure culture.

In the **pour plate method**, a mixed population of bacteria is progressively diluted in three "deep" tubes of liquid nutrient agar. The medium is then poured into sterile culture plates for incubation. Colonies of bacteria subsequently appear on the surface as well as within the medium. Once it is started, the method must be performed without delay because the liquid agar solidifies quickly in the tubes.

The following section describes the process to dilute the cells in molten agar poured into a sterile culture dish plate. Watch **Microbiology Video: Preparing the Pour Plate** to familiarize yourself with the technique. The method is illustrated in FIGURE 6.1.

Preparing the Pour Plate

- Using a flame-sterilized loop, aseptically transfer a loopful of the mixed culture into a tube of molten agar deep labeled #1 (FIGURE 6.1A). Gently swirl the tube to distribute evenly the bacterial cells. Replace the lid on the tube and put the tube back in the 45°C water bath. Flame the loop.
- Resterilize and cool the loop for 10-15 seconds. Now aseptically transfer a loopful of material from tube #1 to a molten agar deep tube labeled #2 (FIGURE 6.1B). Gently swirl the tube to distribute evenly the bacterial cells. Replace the lid on the tube and put the tube back in the 45°C water bath. Flame the loop.

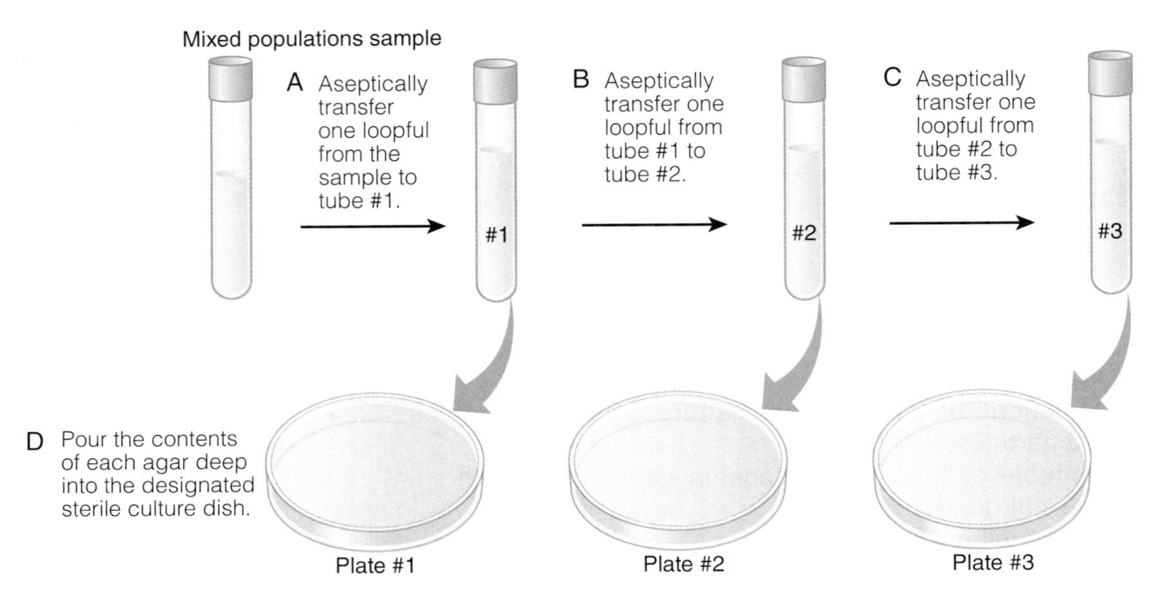

FIGURE 6.1 The pour plate isolation method.

- Resterilize and cool the loop for 10-15 seconds. Now aseptically transfer a loopful of material from tube #2 to a molten agar deep tube labeled #3 (FIGURE 6.1C). Gently swirl the tube to distribute evenly the bacterial cells. Replace the lid on the tube and put the tube back in the 45°C water bath. Flame the loop.
- Now, pour the molten agar from each tube into its designated sterile culture dish (plate #1, plate #2, and plate #3; FIGURE 6.1D). Raise the dish lid just high enough so the molten agar can be poured easily. Gently swirl each dish to ensure the molten agar has covered the dish.
- Place the empty tubes in the disposal area.

PROCEDURE

1. The pour plate technique must be performed rapidly to prevent premature hardening of the agar medium in the test tubes. Therefore, review steps 2 through 6 before beginning the procedure.
2. Obtain three sterile culture (Petri) dishes and label the bottom sides with your name, the date, and the designations “Pour Plate #1,” “Pour Plate #2,” and “Pour Plate #3.”
3. The instructor will explain the preparation of the nutrient agar deep tubes and how they are melted and maintained at approximately 45°–50° C in a water bath to prevent solidification. Obtain three tubes for the technique. A mixed population of bacteria should be available.
4. Aseptically obtain a loopful of the mixed population. Inoculate it into the first melted agar deep tube by placing the loop below the surface and shaking the loop lightly. Resterilize the loop and replace it in its holder.
5. Mix the contents of the tube by rolling it between the palms of your hands or by vigorously tapping its bottom with your index finger.
6. Sterilize the loop and be certain that it is cool. Obtain one loopful of liquid agar from the first deep tube and inoculate it into the second deep tube. Sterilize the loop, then mix the tube contents well by rolling or tapping the tube. Place tube #1 back in the water bath.

7. Sterilize the loop, ensure that it is cool, and then obtain one loopful of liquid agar from the second deep tube and inoculate it into the third deep tube. Mix the contents well, again, by rolling or tapping the tube. Resterilize the loop and replace it in its holder. Place both agar deeps back in the water bath.
8. Aseptically pour the contents of each tube into the designated Petri dish (FIGURE 6.2), being sure to flame the neck of the tube after removing the cap. Gently rotate the dishes in a wide arc on the laboratory desk to distribute the medium evenly over the bottom of the plates.
9. Allow the agar to solidify for several minutes, then incubate the plates in the inverted position.
10. Examine the plates for well-isolated colonies. Surface as well as subsurface colonies should be visible.
11. Enter a well-labeled and explained representation of the plate series in the appropriate space in the Results section, and describe the colonies according to size, color, form, elevation, and margin in TABLE 6.1. If you have already completed Exercise 5, compare the number and variety of colonies appearing in this technique with those appearing in the streak plate technique as part of your notes. At the instructor's direction, transfer samples from the plate to nutrient agar slants for pure culture cultivation, and prepare smears for staining. Subsurface colonies may be reached by applying a warm inoculating needle to melt the agar down to the level of the colony.

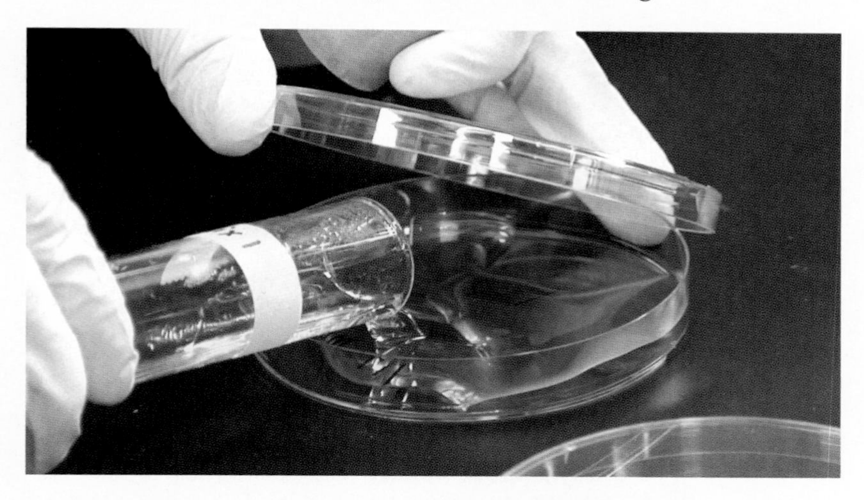

FIGURE 6.2 Pour the contents of the tube into the designated culture dish.

Quick Procedure

Pour Plate Method

1. Inoculate a loopful of bacteria to a melted agar deep tube.
2. Transfer a loopful from the first deep tube to a second deep tube.
3. Transfer a loopful from the second deep tube to a third deep tube.
4. Pour the three deep tubes of agar into Petri dishes. Incubate.

Name: ______________________ Date: __________ Section: __________

Pure Culture Techniques: Pour Plate Method

EXERCISE RESULTS 6

RESULTS

POUR PLATE METHOD

Pour Plate #1

Pour Plate #2

Source: ______________________ ______________________

Pour Plate #3

Source: ______________________

TABLE 6.1 Summary of Colony Characteristics

Colony Number and Source	Colony Characteristics				
	Size	Color	Form	Elevation	Margin

If your instructor has required you to stain a sample of bacterial cells, your light microscope observations can be recorded in the areas below.

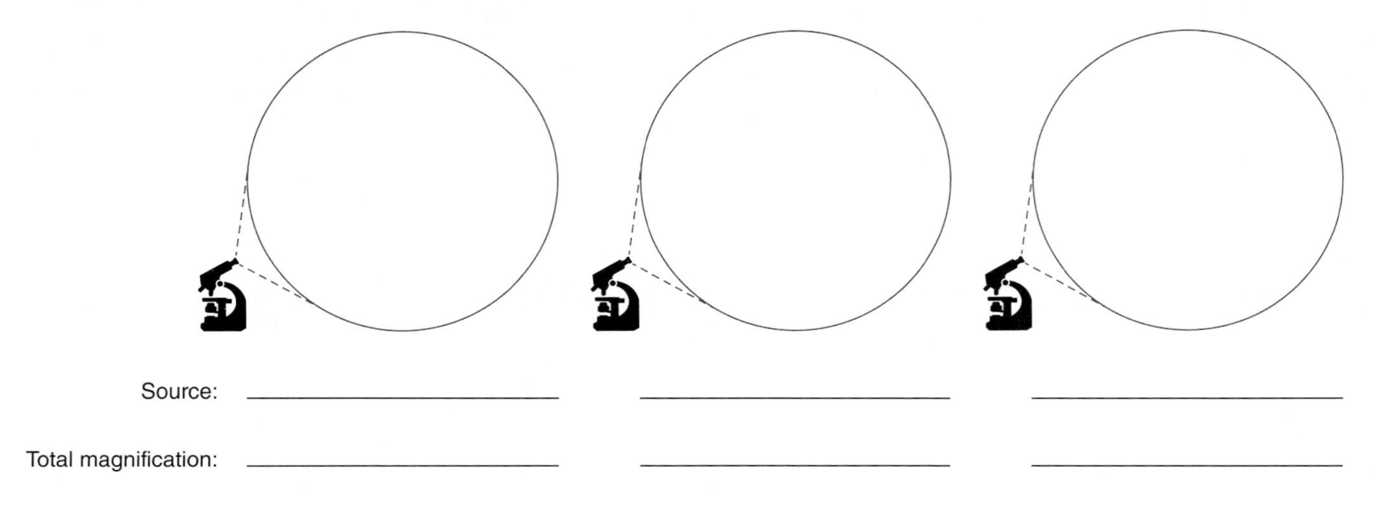

Observations and Conclusions

Questions

1. Provide several advantages of the streak plate method over the pour plate method, and several reasons why the pour plate method might be preferred to the streak plate method.

2. Predict the outcome of incubating inoculated streak and pour plates in the upright rather than the inverted position.

3. While performing the pour plate method, a student inadvertently dipped a hot loop into molten agar tube #1 to obtain bacteria for molten agar tube #2. What will result?

Solution Transfer

EXERCISE 7

Materials and Equipment

- Solution in test tube
- Serological (volumetric) pipettes
- Gloves
- Empty receiving test tube
- Pipette pump

Watch This Video

- Transferring a Solution with a Pipette

In several microbiology exercises and experiments, you will need to transfer microorganisms using proper aseptic techniques. In some experiments, it will be necessary to transfer solutions from one tube to another. When specific amounts are called for, the best transfer method is with an instrument called a pipette. Pipettes can be sterile and plastic, which are disposable, or sterilizable glass pipettes that are reusable.

The liquid volume dispensed by a pipette is indicated in milliliters. For example, 10 ml in 1/10 means that the pipette has been calibrated to measure a maximum of 10 ml in 0.1-ml increments (FIGURE 7.1). Calibration marks are indicated on the side of the pipette. In this exercise, you will practice transferring a solution using a pipette.

SOLUTION TRANSFER

PURPOSE: to perform aseptic transfers of solutions or culture broths organisms between test tubes, using a pipette.

The steps of this exercise are shown in **Microbiology Video: Transferring a Solution with a Pipette.**

Transferring a Solution with a Pipette

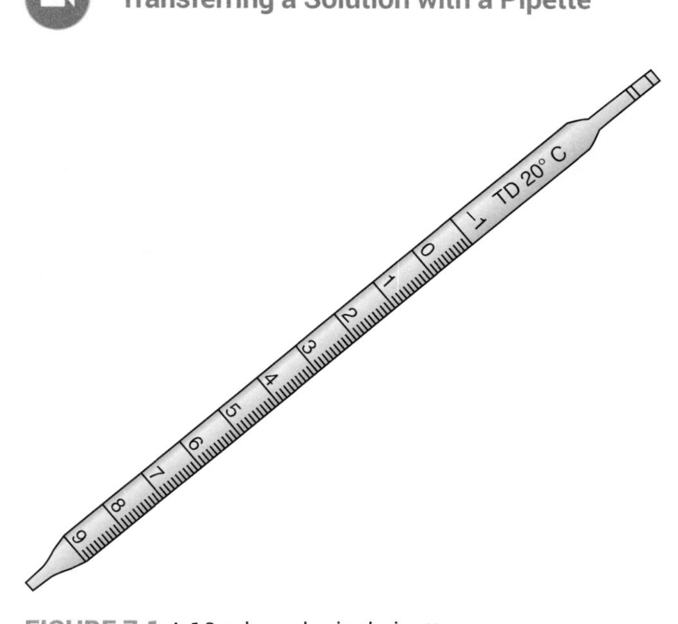

FIGURE 7.1 A 10-ml serological pipette.

PROCEDURE

1. Select a pipette with the appropriate volume scale, as indicated by your instructor.
2. If using disposable pipettes, open the pipette wrapper, remove the sterile pipette, and insert the top wide end into a mechanical pipette pump.
3. Hold the pipette pump with one hand, placing your thumb on the wheel. Carefully place the pipette tip into the solution tube to transfer. Use your thumb to rotate the wheel, causing the liquid to rise into the pipette (FIGURE 7.2).

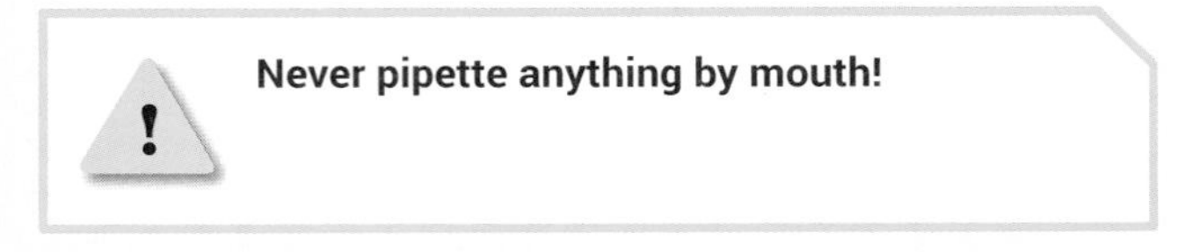

4. Fill the pipette above the desired amount of solution and then slowly lower the meniscus (the curve in the upper surface of the liquid contained in the pipette) to the desired volume. Make certain the pipette tip stays below the surface of the liquid (FIGURE 7.3).
5. Remove the pipette from the tube. Allow the outside of the pipette tip to gently touch the inner lip of the tube to remove any adherent liquid. Place the solution tube in the holding tray.
6. Pick up the receiving tube in one hand, keeping the pump in your other hand.
7. Use your thumb to rotate the wheel upward, dispensing the liquid from the pipette into the receiving tube. Lower the white top of the pump all the way to the pump shaft to expel the contents of the pipette.
8. Remove the pipette aseptically and discard into an appropriate discard receptacle.

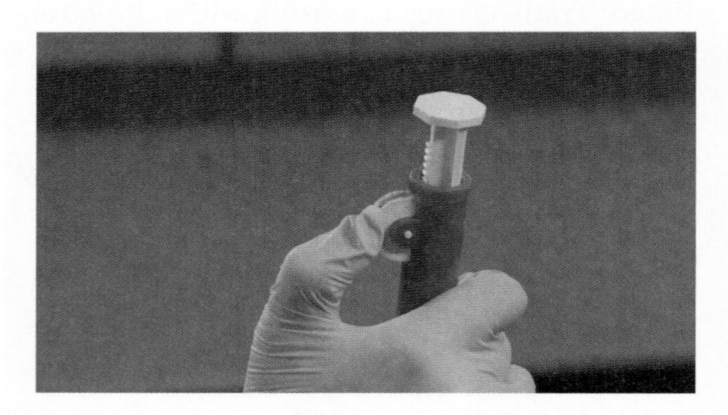

FIGURE 7.2 Use your thumb to rotate the wheel.

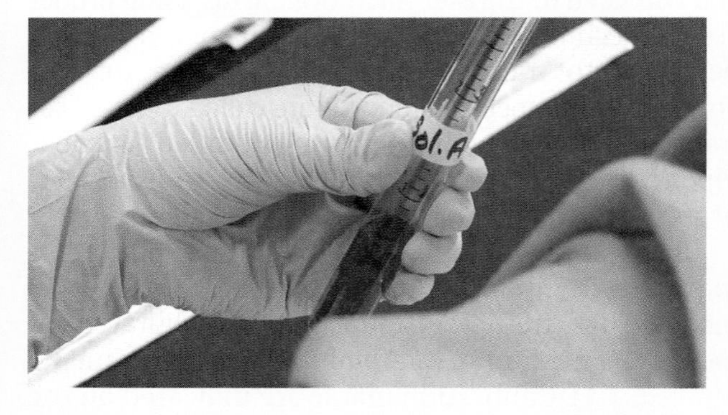

FIGURE 7.3 Keep the pipette tip below the surface of the liquid.

Name: ______________________ Date: __________ Section: __________

Solution Transfer

EXERCISE RESULTS **7**

RESULTS

SOLUTION TRANSFER

Observations and Conclusions

____

____

____

____

____

Questions

1. How can you ensure aseptic transfer with a pipette?

____

____

____

2. Are there solutions that could not be transferred via pipette? Can you give examples?

____

____

Serial Dilutions

EXERCISE

8

Materials and Equipment

- 5–1-ml sterile serological pipettes
- Pipette pump
- 5 test tubes with 9 ml of sterile nutrient broth
- 1 24-hour nutrient broth culture of *Escherichia coli*

- How to Make a Serial Dilution

It is often important to know how many bacterial cells are present in a broth culture sample or some other liquid sample. There are several methods to estimate cell number. One method for doing this is to make **serial dilutions**; that is, a process that, through dilution, sequentially reduces the concentration of a substance or the concentration of microorganisms in a sample.

In this exercise, you will be transferring bacteria from one broth tube to another, further the diluting the number of cells with each step. You can familiarize yourself with this procedure by viewing **Microbiology Video: How to Make a Serial Dilution**.

How to Make a Serial Dilution

SERIAL DILUTIONS

PURPOSE: to perform transfers of bacteria to dilute the cell concentration.

PROCEDURE

1. Label all of the test tubes before starting the procedure. Each tube will be a tenfold dilution, starting from the undiluted nutrient broth culture of *E. coli*.
2. Label the first tube of the five sterile broth tubes -1 (10^{-1} dilution). Label the second nutrient broth tube -2 (10^{-2} dilution). Label the third tube -3 (10^{-3} dilution). Repeat with the two remaining tubes, labeling them -4 (10^{-4} dilution) and -5 (10^{-5} dilution) (FIGURE 8.1).

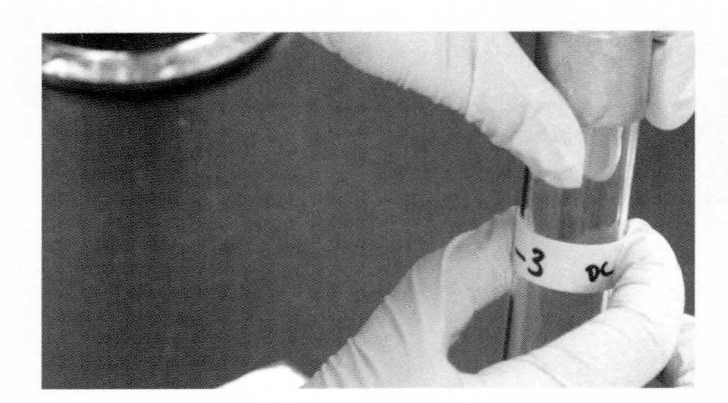

FIGURE 8.1 Label each tube according to the level of dilution.

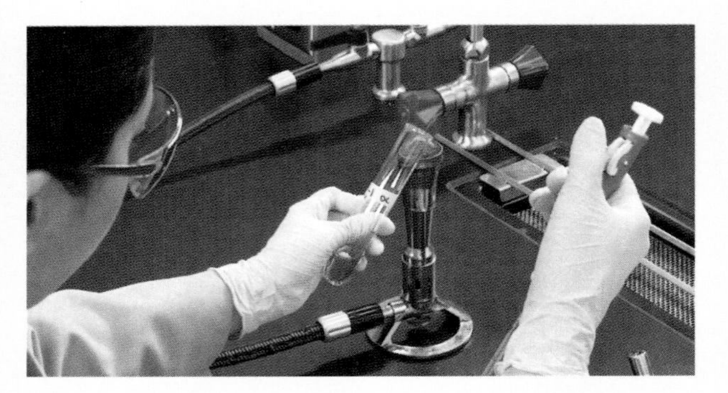

FIGURE 8.2 Be careful to perform these transfers in an aseptic manner.

3. Gently swirl the *E. coli* broth to completely suspend the bacterial cells and then, using a 1-ml sterile serological pipette (see Exercise 7), aseptically transfer 1 ml of the *E. coli* broth culture to the 10^{-1} tube (FIGURE 8.2). Discard the pipette into an appropriate container. Mix the solution by gently shaking the tube.
4. Using a new pipette, aseptically transfer 1 ml of the 10^{-1} solution into the 10^{-2} tube. Discard the pipette. Mix the solution by gently shaking the tube.
5. Using a third pipette, aseptically transfer 1 ml of the 10^{-2} solution into the 10^{-3} tube. Discard the pipette. Mix the solution by gently shaking the tube.
6. Repeat this 1-ml transfer process for the two remaining tubes (10^{-4} and 10^{-5} dilutions).
7. Examine the three tubes and enter representations in the Results section.

Name: ______________________ Date: __________ Section: __________

Serial Dilutions

EXERCISE RESULTS **8**

RESULTS

SERIAL DILUTIONS

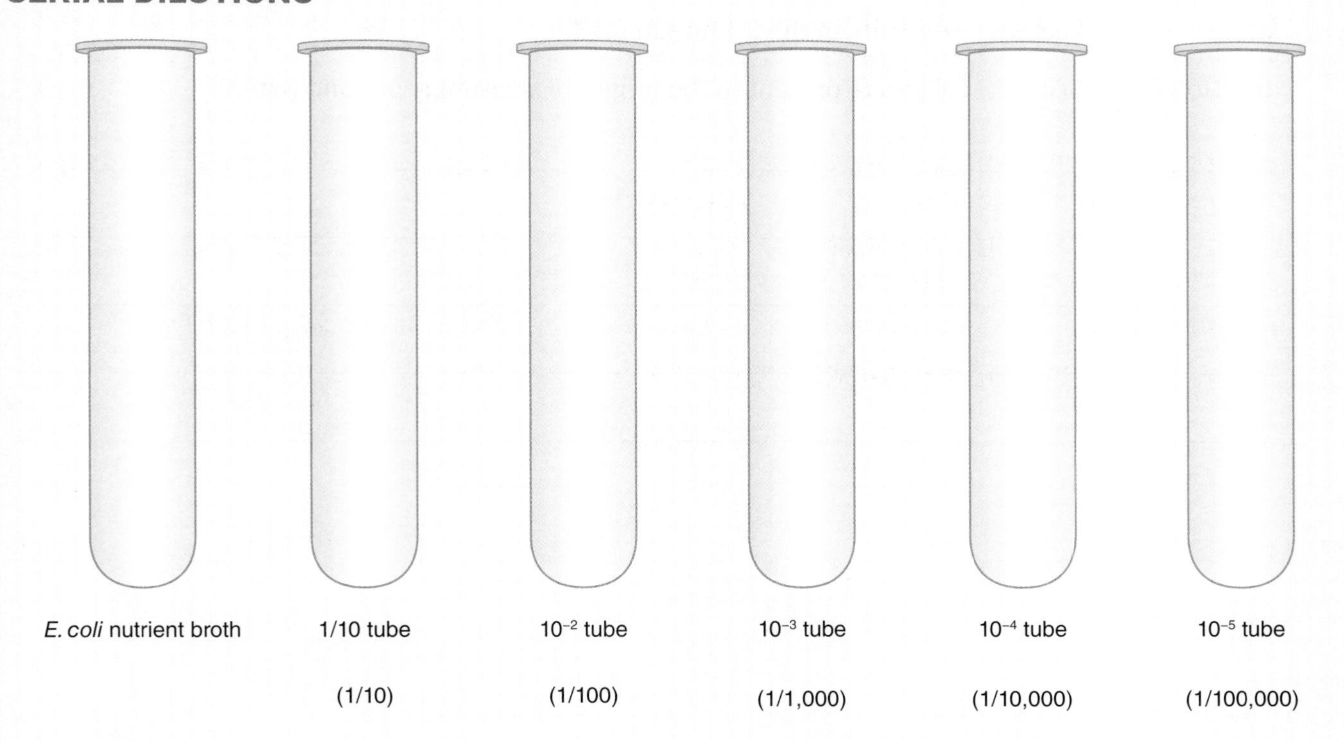

Observations and Conclusions

____

____

____

____

____

Questions

1. What is the dilution factor used in this exercise?

2. A 1 ml sample of pond water is diluted 10-times with sterile water. Then, 0.1 ml of the diluted sample is spread on a nutrient agar plate and incubated fro 48 hours at room temperature. When the plate is observed after the 48-hour incubation, a total of 92 bacterial colonies are counted on the plate.

   a. How many bacterial cells were in the 0.1 ml sample?

   b. How many bacterial cells were present in the original 1 ml sample of pond water?

Microscopy

PART III

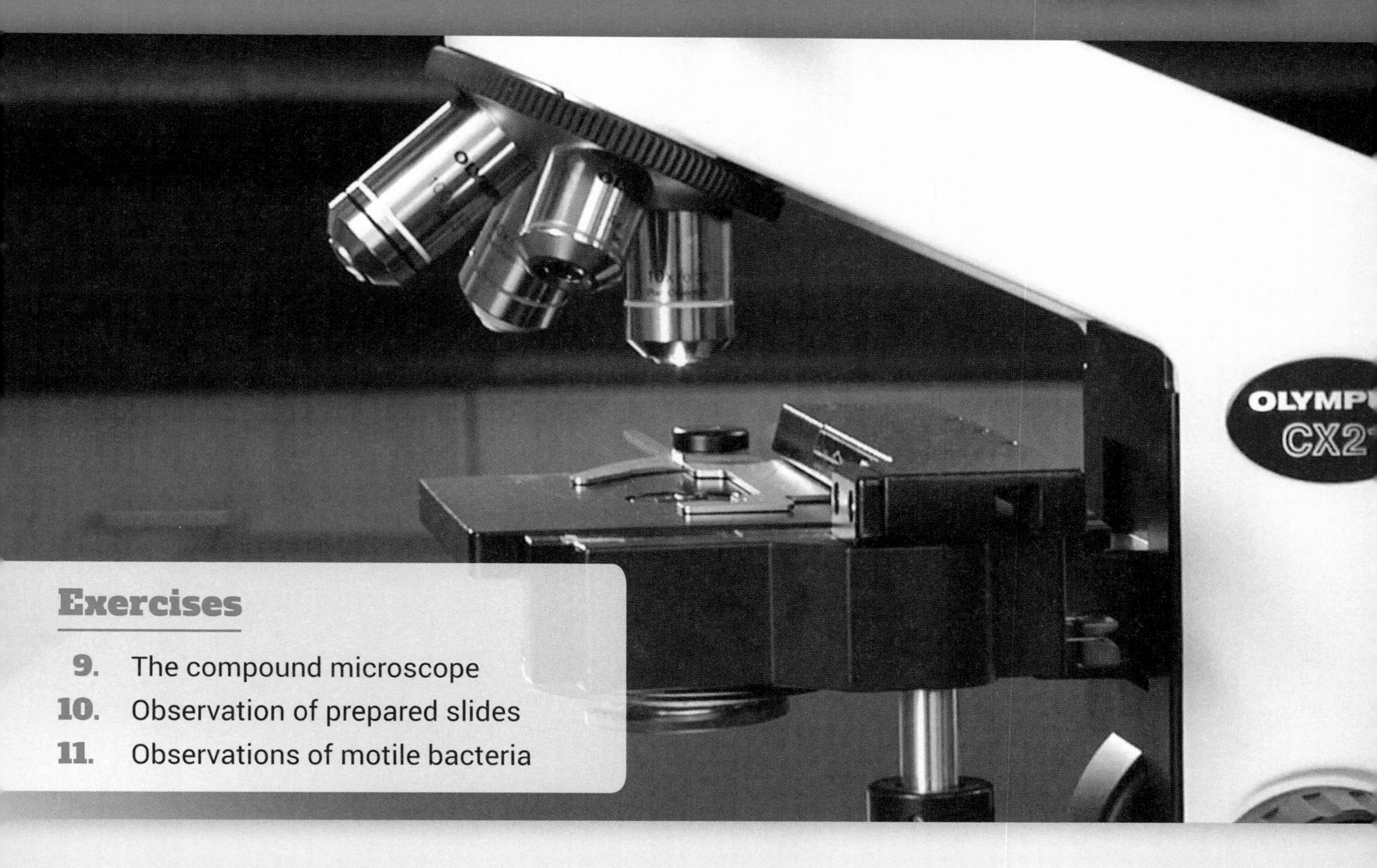

Exercises

9. The compound microscope
10. Observation of prepared slides
11. Observations of motile bacteria

Learning Objectives

When you have completed the exercises in Part III, you will be able to:

- Operate the light microscope correctly to measure and observe living and dead microorganisms.

Watch These Videos

- How to Use the Light Microscope: Identifying the Parts
- Measuring Cell Size

Microscopy had its beginnings with the ability of Robert Hooke and Anton van Leeuwenhoek to see small objects that often were invisible to the naked eye. In fact, Leeuwenhoek was the first to see bacteria with his "homemade" microscopes (hand lenses). Since then, great strides have been made in microscope design, magnification, and resolution.

Today's **light (compound) microscope**, which uses visible light to magnify and resolve objects, is a core instrument of many microbiology research labs. **Brightfield microscopy**, introduced in Exercise 9, is what you will use with your light microscope in many of the laboratory exercises in this manual.

The Compound Microscope

EXERCISE **9**

Materials and Equipment

- Compound microscope
- Glass slides and coverslips
- Various samples for viewing

Watch This Video

- How to Use the Light Microscope: I. Identifying the Parts
- How to Use the Light Microscope: II. Focusing with the Microscope

The **compound microscope** is a basic tool of the microbiology laboratory. This precision instrument contains a series of lenses allowing a specimen to be magnified up to a thousand times (1,000×). Mechanical adjustments and optical features of the microscope afford a broad range of possibilities for viewing various types of microorganisms.

You will be using a brightfield compound microscope that combines the principles of an optical system and an illumination system to achieve **magnification**, which refers to increasing the apparent size of the specimen being observed. Light, projected toward an object, passes through the object and is collected by the objective lens (near the object) to form a magnified image. This magnified image becomes an object for a second lens, the ocular lens (near the eye), which magnifies the image further for viewing by the observer.

An understanding of certain aspects of microscopy is essential for optimal use of the microscope. For example, the **resolving power** determines the size of the smallest object that can be seen clearly under specified conditions. Another aspect, the **working distance**, refers to the distance between the slide and the bottom of the objective lens. The **refractive index** pertains to the light-bending ability of glass, oil, and air—materials through which light must pass during image formation.

PARTS OF THE MICROSCOPE AND THEIR USE

PURPOSE: to identify the parts of the light microscope and explain their functions by focusing on a sample specimen.

In this exercise, you will explore the basic features of the microscope and the function of the parts will be explained. You will gain experience in using the instrument through the observation of various objects. You may also watch **Microbiology Video: How to Use the Light Microscope: Identifying the Parts** before you begin.

How to Use the Light Microscope: I. Identifying the Parts

PROCEDURE

1. Each student should secure a microscope for use and place it on the bench top. When carrying the microscope to and from the lab bench, hold it upright with two hands, one hand holding the arm of the microscope with the other hand supporting its base.

   At the bench top, after the microscope has been placed in front of you, **lens paper** should be used to clean the lenses and stage area before work is begun. Coarser types of tissue are not useful because they leave lint and may scratch the lens.

   DO NOT use paper towels or tissues to clean the lenses. Use lens paper only.

2. If you are using a binocular microscope, the major parts of the microscope and their functions include: two **ocular lens** (the eyepieces, 10×), a set of **objective lenses** (low power = 10×, high power = 40× or 45×, and oil immersion = 100×) **(FIGURE 9.1)**, the **stage** and **mechanical stage** (slide holder), the **condenser** and **iris diaphragm,** the **coarse** and **fine adjustments**, and the light source. Note the position of each of these parts in FIGURE 9.2, and enter the functions in TABLE 9.1 in the Results section.
3. Binocular microscopes have a mechanism for adjusting the oculars to match the distance between your eyes.
4. The **total magnification** of the microscope's lens system is calculated by multiplying the ocular magnification (10×) by the objective magnification (10×, 40× or 45×, or 100×). The total magnification possible with each objective should be determined and recorded in the Results section in TABLE 9.2.

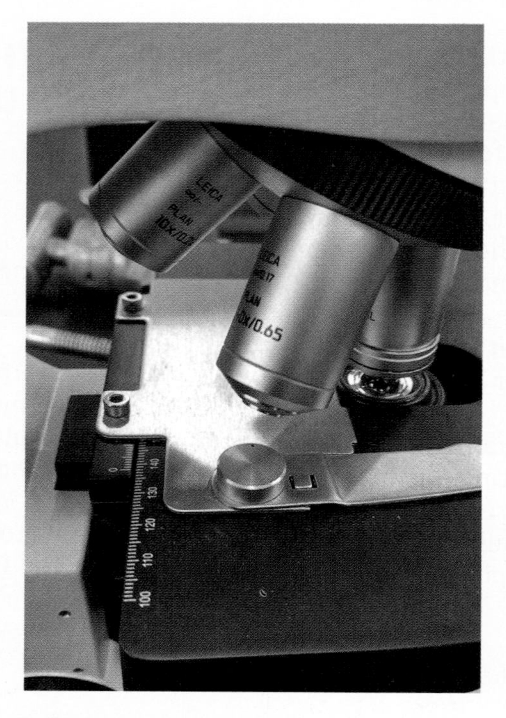

FIGURE 9.1 Objective lenses.

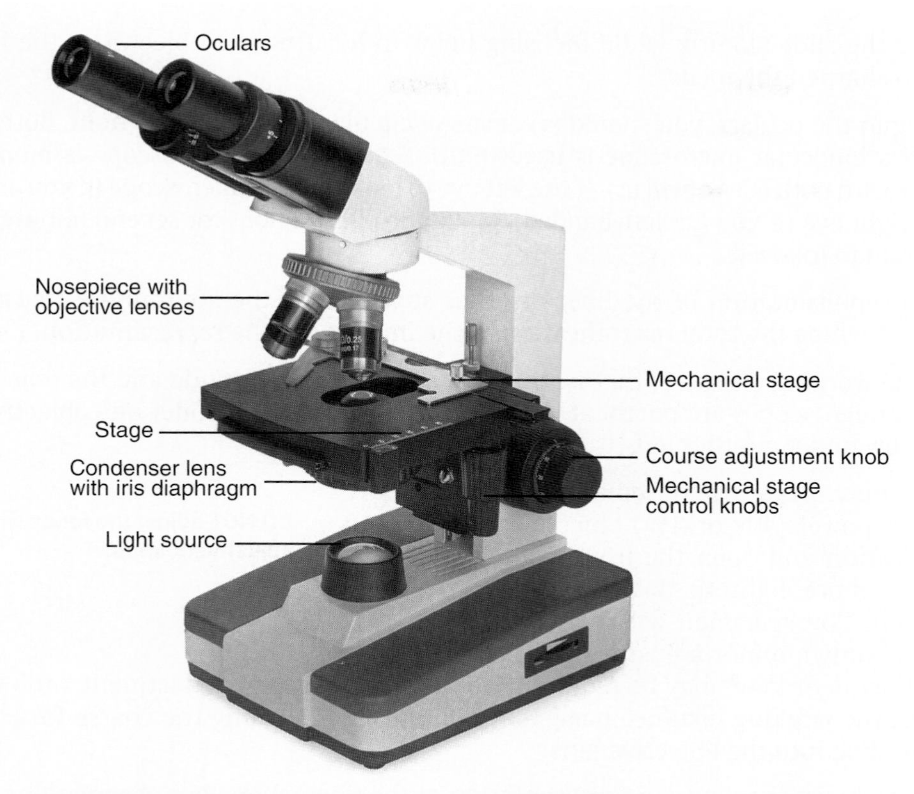

FIGURE 9.2 The parts of the compound microscope.

How to Use the Light Microscope: II. Focusing with the Microscope

PROCEDURE

1. After you have placed the microscope on the lab bench, make sure you have cleaned off the lenses with lens paper and adjusted the oculars for your eyes.
2. Plug in the microscope power cord and turn on the microscope light so that it is as bright as possible. [The light switch will be in different places depending on the brand of microscope you have.] Make sure the iris diaphragm allows in only a the minimal amount of light.
3. To begin your microscope work, prepare a **wet mount** of a hair fiber [Your instructor may substitute another specimen for viewing.]
   - Place a small drop of water on a clean glass slide, add a hair fiber, and cover the preparation with a clean coverslip.
   - Remove any excess water by touching a piece of paper toweling at the edge of the coverslip.
4. With the stage all the way down and the low power (10×) objective lens in the viewing position, place the slide right-side-up (coverslip on upper surface of slide) in the slide holder or mechanical stage of the microscope. Now move the slide to place the specimen directly over the hole in the stage.
5. With the coarse adjustment knob, raise the stage all the way up and then, very slowly lower the stage to locate the hair fiber.

If you break a slide while working, be sure to place it in the designated sharps container in the lab or in the glass receptacle indicated by the instructor.

Moving the slide slightly while focusing helps in locating the object. Use the fine adjustment knob to sharpen the focus.

6. Looking in the oculars, you should see a single circular area called the **field**. Both eyes should be open if a binocular microscope is used (FIGURE 9.3). If your microscope is monocular, practice viewing with both eyes open, using the left eye to look into the microscope (if you are right-handed) or the right eye (if you are left-handed). Move the object about for several minutes while training your brain to follow it.
7. Draw a representation of the fiber or other specimen in the appropriate space in the Results section. Include the total magnification of the image with the representation.
8. Note the working distance; that is, the distance between the slide and the objective lens. Most modern microscopes are **parfocal**, meaning an object in view under one objective lens will still be in view under another lens.
9. Now, simply rotate the nosepiece to swing the high power (40× or 45×) objective lens into position and open the iris diaphragm to allow more light to pass through the specimen. The specimen will be almost in focus, so only a minor adjustment with the fine adjustment knob may be required. Do not use the coarse adjustment knob because, as you can see, the working distance has greatly diminished. Turning the coarse focusing knob could jam the slide into the objective lens.

DO NOT adjust the focus with the coarse adjustment knob.

10. Scan the object, then enter a representation of the fiber when your observations are complete.
11. Do not use the oil immersion (100×) objective lens with this specimen.
12. To remove the specimen, rotate the nosepiece back to the low power (10×) objective, lower the stage with the coarse adjustment knob and then remove the slide.
13. Clean or recycle the slide and coverslip as directed by your instructor.
14. Repeat steps 4–12 for any other specimens supplied by your instructor.
15. When your observations are completed, make sure the low power (10×) lens is in the viewing position, the stage is lowered, and the light has been turned off.

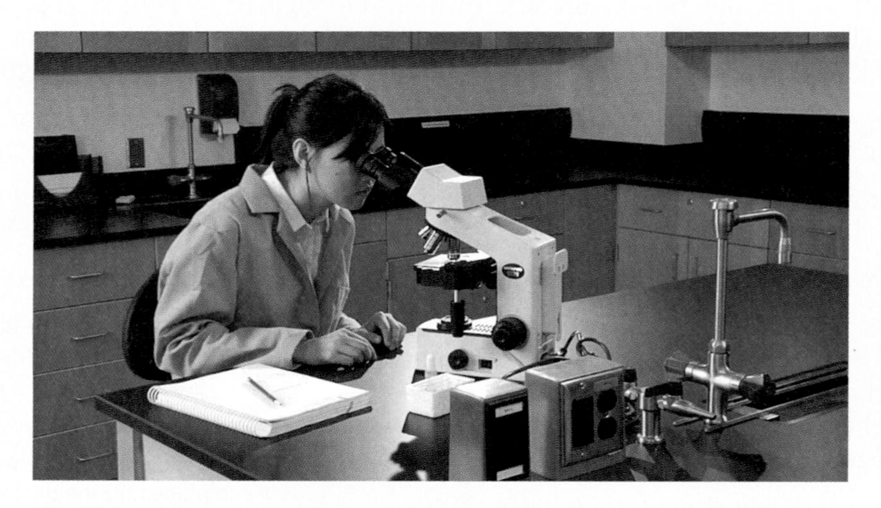

FIGURE 9.3 Keep both eyes open if a binocular microscope is used.

Name: ______________________ Date: __________ Section: __________

The Compound Microscope

EXERCISE RESULTS 9

RESULTS

PARTS OF THE MICROSCOPE AND THEIR USE

TABLE 9.1 Function of the Microscope Parts

Microscope Part	Function
Ocular s (eyepieces)	
Objective lenses	
Stage	
Lens	
Iris diaphragm	
Coarse adjustment knob	
Fine adjustment knob	

TABLE 9.2 Calculating Total Magnification

Objective Lens	Objective Lens Magnification	Ocular Lens Magnification	Total Magnification
Low power			
High power			
Oil immersion			

Observation of Wet Mount

Make a representative drawing of the specimen at each magnification.

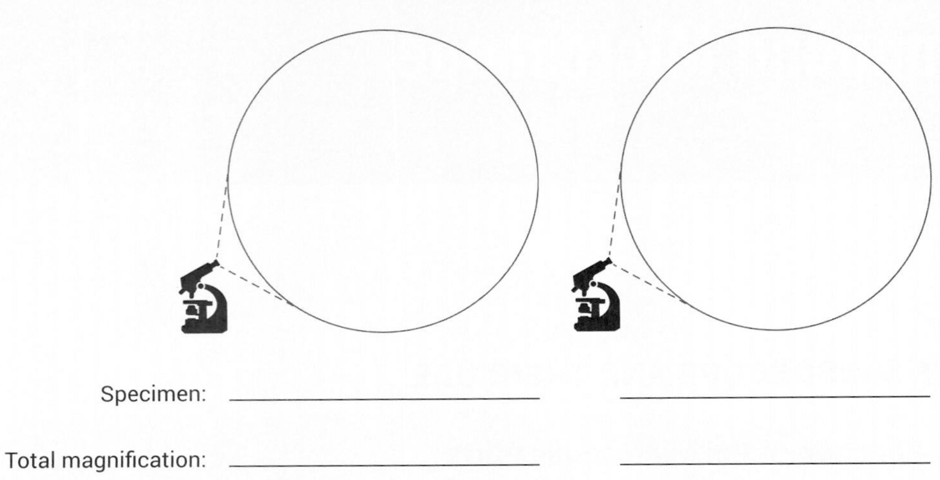

Specimen: ______________________ ______________________

Total magnification: ______________________ ______________________

Questions

1. Describe the important steps that should be taken to care for the microscope during and after use in the laboratory.

____

____

____

2. What does the adjective *parfocal* mean when applied to a microscope? How is it a valuable asset in the use of the microscope?

____

____

____

3. What controls the amount of light reaching the specimen and ocular lens?

____

____

____

4. What happens to the working distance as you increase magnification with the objective lenses?

____

____

____

Observations of Prepared Slides

EXERCISE 10

Materials and Equipment

- Brightfield compound microscope
- Prepared slides of selected microorganisms
- Immersion oil
- Lens paper

Watch This Video

- Measuring Cell Size

By now you should have gained some experience with the microscope and observing specimens. However, much of your microscope observations will entail use of the oil immersion objective lens (100×) because many microorganisms, especially bacteria, are so small.

In this exercise, you will be using **prepared slides**. These are microscope slides on which dead, and often stained, specimens have been placed previously and sealed in place with an adhesive and a coverslip. Your instructor will provide you with several prepared slides for microscope examination.

Because most bacterial cells are so small, you need to get the highest magnification and resolution possible from the compound microscope. That means using the oil **immersion lens** (100× objective lens). When using this lens, the air between the slide and lens must be replaced by a special type of synthetic oil called **immersion oil**, which has the same refractive index (or light-bending ability) as glass. Therefore, the oil keeps light in a straight line as it passes from the glass slide to the oil and then to the glass of the objective lens. Immersion oil improves the resolving power of the microscope and provides enough light to let you see your specimen clearly and distinctly.

MICROSCOPIC EXAMINATION OF PREPARED SLIDES

PURPOSE: to observe specimens with the oil immersion objective lens.

PROCEDURE

1. Obtain a prepared slide of stained microorganisms. Clean the slide with soap and water, and dry it with paper towels. Remove any remaining "towel lint" with a piece of lens paper.
2. Place the slide on the slide holder or mechanical stage and center the specimen over the hole in the stage. Adjust the diaphragm lever and focus on the specimen with the low-power (10×) objective lens.

3. At these magnifications, do not expect to see individual cells. Most bacteria are far too small to be resolved. Rather, focus on the areas containing stain used to stain the cells. Watch **Microbiology Video: Measuring Cell Size** to see how this exercise is done.

Measuring Cell Size

4. Once you have located the stained specimen and have focused on it, center the specimen in the field. Rotate the high power (40×) objective lens into position. Use the fine focusing knob only to sharpen the focus.

5. With the high power (40×) lens, you may be able to see individual bacteria. Still, most are too small to be observed or for their size, shape, and arrangement to be determined. The oil immersion lens (100×) must be used. Center the specimen in the field before proceeding.

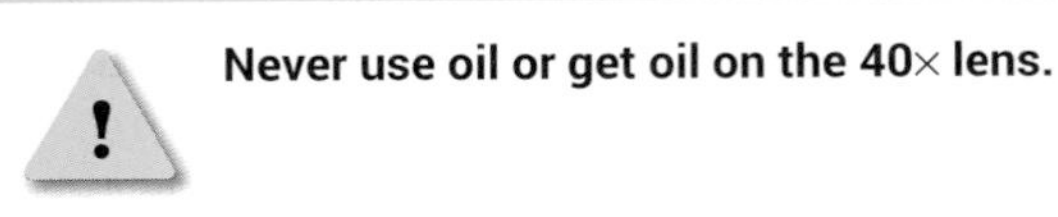

6. To use the oil immersion lens, swing away the high-power objective (40×), and apply a drop of immersion oil on the slide area being viewed.

7. Open the iris diaphragm fully and swing the oil immersion objective into position. To bring the object into view, a minor adjustment with the fine adjustment knob may be needed.

8. Once the microorganisms have been located, spend a few minutes scanning the slide. Scanning allows you to determine the general pattern of the microorganisms as well as their shape, size, and arrangement. It also helps your mind eliminate the debris that may be mixed in with the specimen; and it helps you locate a thin area most suitable for your drawing.

9. Drawings and documentation of the specimens viewed are almost always made at 1000× total magnification. Using a sharp pencil, enter representations of the microorganisms in the Results section along with the magnifications used. Consult with your instructor on any special preferences he or she may have.

10. A rough estimate of the size of an object can be made in one of two ways. Your microscope may have an **ocular micrometer** mounted in one ocular. At 1,000× total magnification, each ocular division is equal to one micrometer (μm) (FIGURE 10.1). For example, if the length of a bacterial

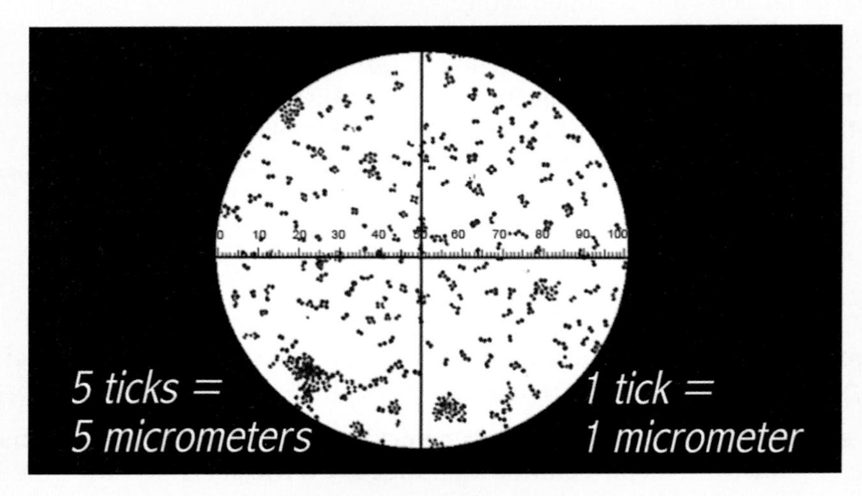

FIGURE 10.1 An oil-immersion lens (total magnification = 1000×) is required to see distinctly the bacterial cells.

cell spans three divisions, it is 3 μm long; if a spherical yeast cell is 6 divisions, it is 6 μm in diameter.

11. If an ocular micrometer is not provided, consider that the distance across the field of view with the low-power lens is approximately 1,600 micrometers. With the high-power lens, it is approximately 400 micrometers. And with the oil immersion lens, it is approximately 100 micrometers. By determining how wide the object is relative to the diameter of the field of view, you can estimate the object's size.

12. Additional practice with the microscope may be obtained with other specimens supplied by the instructor.

13. *See it–Draw it.* This exercise will give you an appreciation of how difficult it is to accurately describe microscopic organisms to someone else without the benefit of a detailed drawing. It requires the participation of two persons and one microscope. One person is the viewer, the other person the recorder. The first person locates a group of bacteria under oil immersion. Now he or she describes the view to the second person, who makes a representation of the bacteria on a sheet of drawing paper (preferably hidden from the view of the first person). When completed, compare the drawing with the microscope image. It will soon become apparent that drawing an accurate picture is far superior to trying to explain what bacteria look like.

14. *Microscope troubleshooting.* Beginning students as well as experienced ones often encounter difficulties with the microscope that can be resolved by relatively simple adjustments or procedures. TABLE 10-1 shows a list of ten possible problems and the solutions you might find useful.

TABLE 10.1 Troubleshooting Microscope Problems

	Problem	Solution
1.	The object is observed clearly under low power, but you lose it when moving to high power.	Be sure to place the object in the center of the field before moving to the next higher objective. The microscope field becomes smaller as the magnification increases, so you should always center your object before switching objectives.
2.	You see a black half-moon or quarter-moon off to the side of the microscope field.	The objective must be "clicked" into position before using it. Check to see that it is in the viewing position.
3.	You can see the objects at 400×, but they cannot be found when you move to oil immersion (1,000×).	After fine focusing at 1,000× without finding the objects, rotate the nosepiece forward to the low power (10×) objective. **Do not go backward to the 40× lens, as the lens will touch the oil: never get oil on the 40× lens.** Re-find the objects at 10× and then rotate the nosepiece backward to the 100× oil immersion lens. Use the fine focus to find the objects. (See also Problem 4).
4.	You cannot focus down to the level where the specimen is.	Check the slide to ensure that it is not upside down on the stage.
5.	You constantly encounter a fuzzy image under high power	Someone using the microscope before you may have left oil on the lens. Moisten a piece of lens tissue with lens cleaner and vigorously clean the objective, then dry it with a clean piece of lens tissue. If you are still having problems, get help from the instructor.

(*Continued*)

TABLE 10.1 Troubleshooting Microscope Problems (Continued)

	Problem	Solution
6.	The image is too dark to see clearly.	Try opening the diaphragm to let in more light. If the results are not satisfactory, adjust the condenser up or down until the quality and quantity of light improve.
7.	You have cleaned the ocular and objective lenses thoroughly, and still the image is not clear.	Try cleaning the lens of the condenser where it meets the opening of the stage. There may be a layer of oil on the lens.
8.	You are not sure whether the lint and debris you see are on the slide or in the lens system.	As you view through the microscope, slowly move the mechanical stage knobs to move position of the specimen. If the specimen does not move, the material is in the lens system. Ask for assistance from your instructor.
9.	You wonder whether you should wear your glasses while using the microscope.	Most microscopes are adjusted to compensate for the viewer's vision and glasses. Indeed, the most comfortable viewing point without glasses is farther back than it is with glasses. Also, wearing glasses while you view allows you to move from the microscope to your notebook and back without putting on or removing your glasses.
10.	You have a binocular microscope, and you have trouble focusing with both eyes open.	You should check your oculars to see which one has an ocular adjustment (not a focus adjustment); if it is the right ocular, then close the right eye, and while looking with the left eye, focus the ocular adjustment on the right ocular (do the reverse if the adjustment is on the left ocular); now use the knob between the oculars to adjust the distance between your pupils.

16. When your work with the microscope is completed for the lab period, rotate the nosepiece to bring the low power (10×) lens into position. Remove the slide and clean the slide with soap and water. Dry the slide. Also, remove any excess oil from the oil immersion lens by wiping it with lens paper.

17. The other lenses and parts of the microscope should also be cleaned well. For microscope storage, move the stage to the safety stop (i.e., all the way down), turn off the microscope light, and wind the power cord neatly around the base or arm of the instrument.

Quick Procedure

Microscopic Examination of Prepared Slides

1. Place prepared or stained slide on microscope stage and center specimen. Make sure the iris diaphragm is allowing in only a minimal amount of light.
2. Focus at low power (10×) on the specimen or areas containing stain.
3. Center specimen in the field and rotate the high-power (40×) lens into position. Open the iris diaphragm to allow in more light.
4. Use the fine focus to resolve the specimen and center it in the field.
5. Rotate the 40× lens away from the specimen and add a drop of immersion oil onto the slide surface.
6. Rotate the 100× lens into position, open the iris diaphragm completely, and use the fine focus to resolve the specimen.
7. Make a drawing or record results from your observations.

Name: ______________________ Date: __________ Section: __________

Observations of Prepared Slides

EXERCISE RESULTS **10**

RESULTS

OBSERVATIONS OF PREPARED SLIDES

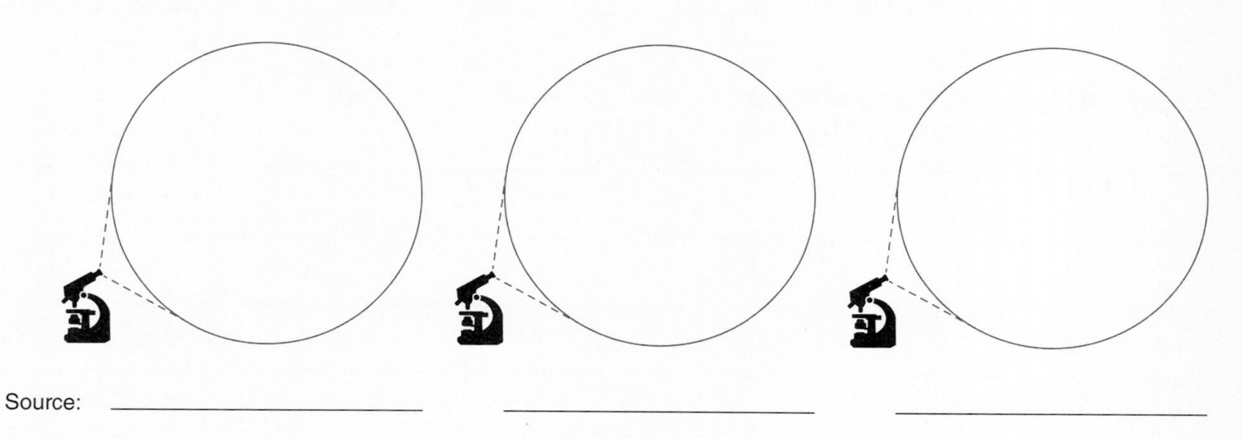

Source: ______________ ______________ ______________

Total magnification: ______________ ______________ ______________

Approximate cell size (μm) ______________ ______________ ______________

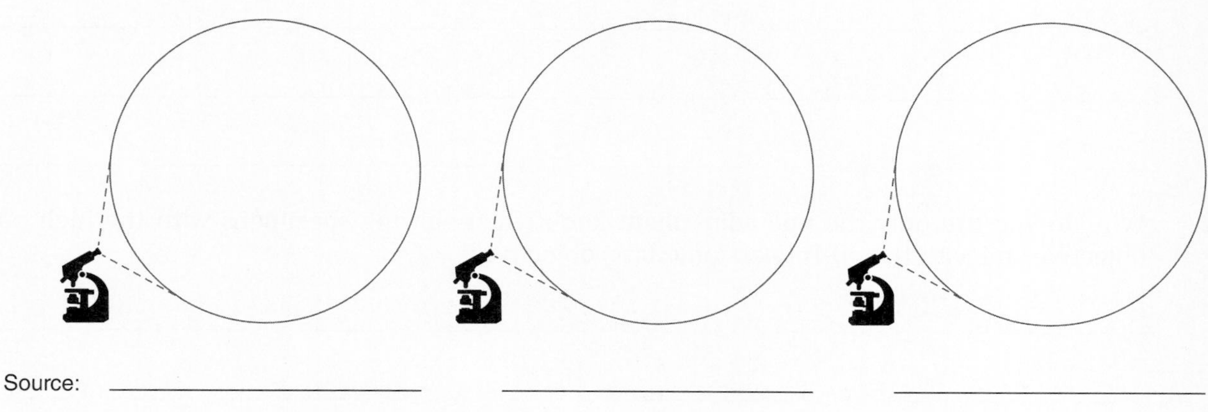

Source: ______________ ______________ ______________

Total magnification: ______________ ______________ ______________

Approximate cell size (μm) ______________ ______________ ______________

Source: ______________________ ______________________

Total magnification: ______________________ ______________________

Approximate cell size (μm) ______________________ ______________________

Observations and Conclusions

____

____

____

____

____

Questions

1. Explain the problem encountered when magnifying an object with the oil immersion lens and indicate how immersion oil helps solve the problem.

____

____

____

2. Why do you use only the fine adjustment knob when viewing specimens with the high power (40×) objective and with the oil-immersion (100×) objective?

____

____

____

3. In Figure 10.1, estimate the cell size (diameter) of one of the stained bacterial cells.

____

____

____

4. Determine the resolving power of your microscope, using the oil immersion objective. Can a spherical bacterium measuring 3 µm in diameter be seen with the oil immersion lens? Explain.

$$\text{The resolving power (RP)} = \frac{\lambda \text{ (wavelength of light in µm)}}{2 \text{ NA (numerical aperture)}}$$

Assume $\lambda = 500$ nm and NA = 1.25

____

____

____

EXERCISE 11

Observations of Motile Bacteria

Materials and Equipment

- 24-hour broth cultures of *Pseudomonas fluorescens*
- 24-hour broth cultures of *Bacillus cereus* and *Staphylococcus epidermidis*
- Concave depression slides
- Coverslips
- Petroleum jelly
- Applicator sticks
- Tubes of motility test agar
- Inoculating needles
- Brightfield compound microscope

Watch This Video

- How to Make a Hanging Drop Preparation

Bacteria are complex organisms with intricate structural details. Flagella are among the important bacterial structures.

Flagella are protein appendages that facilitate motion (motility) of bacteria. Many species of bacterial rods and spirilla, and a few species of cocci, possess flagella. Although flagella often are many times the length of the cell, these external appendages normally cannot be seen with the light microscope because they are extremely thin.

In this exercise, the presence of flagella will be inferred by observing evidence of bacterial motion.

BACTERIAL MOTILITY

PURPOSE: to determine if a bacterial cell is motile (has one or more flagella).

Evidence for the presence of bacterial flagella is obtained by observing motility. Two methods are available. In the first method, live, unstained bacteria are seen moving about in the hanging drop technique. Watch **Microbiology Video: How to Make a Hanging Drop Preparation** to familiarize yourself with the technique.

How to Make a Hanging Drop Preparation

In the second method, bacteria are inoculated into a semisolid medium. During the incubation, they migrate from the inoculation site and form a characteristic pattern of growth, which indicates motility.

In this exercise, you will use several bacterial broth cultures for the hanging drop preparation to determine which is/are motile and which is/are not motile. You will need to observe the hanging drop and look for motility, which is a self-directed movement. Bacterial species that are nonmotile exhibit an erratic **Brownian motion**, which is a phenomenon caused by molecules striking the organisms and displacing them briefly. It lacks the directed movement exhibited by motile cells.

PROCEDURE

I. The Hanging Drop Preparation

1. Thoroughly clean a concave depression slide and a coverslip.
   - Using an applicator stick, place a small amount of petroleum jelly at the corners of the coverslip (FIGURE 11.1).
   - Position the coverslip face up on the desk.
2. Obtain a broth culture of a bacterial species, and place two or three loopfuls in the center of the coverslip.
3. Invert the concave depression slide, and lower it onto the inoculated coverslip, pressing gently so that the petroleum jelly seals the corner of the coverslip to the slide.
   - Quickly invert the slide so that the drop hangs into the concave depression, as shown in FIGURE 11.2.
4. Observe the slide with the low-power (10×) objective (FIGURE 11.3), and lower the light considerably by adjusting the iris diaphragm. Then, locate the edge of the drop using the coarse and fine adjustment knobs on the microscope.
   - Now switch to high power (40×) objective, and locate the edge of the drop again using the fine adjustment knob on the microscope. Careful focusing and reduction of the light to achieve contrast are essential to success.

FIGURE 11.1 Petroleum jelly is placed on each corner of a coverslip.

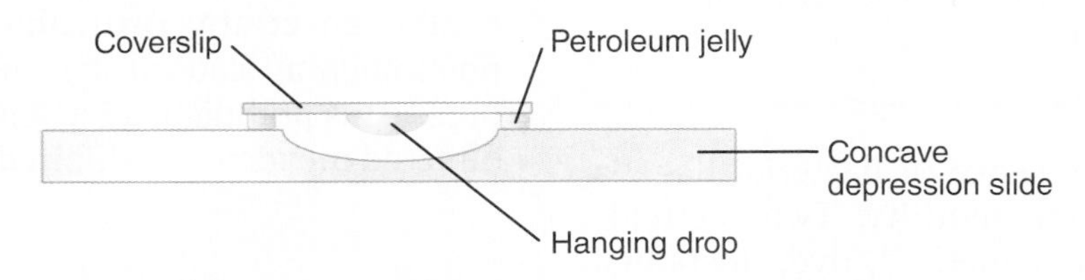

FIGURE 11.2 The final orientation for the hanging drop preparation.

- Locate organisms within the fluid, and note whether the motion of the cells has direction, or exhibits erratic Brownian movement.
- Note in the Results section the patterns of motion displayed by the organisms, where they accumulate, their relative sizes and shapes, any configurations they display, and the speed of their movement. When your observations are complete, enter representations of the cells, showing the direction of motion.
- Since the slides contain live organisms, they should be placed into a beaker of disinfectant after use.
- Rinse the slide with disinfectant before reuse.

5. Repeat steps 1-4 with the other two species of bacteria.
   - Observe these slides for the presence of motility or Brownian movement.
   - Representations of these organisms and their pattern of movement may be entered in the Results section.
   - Conclusions may be drawn on the type of motion observed as evidence of the presence of flagella.

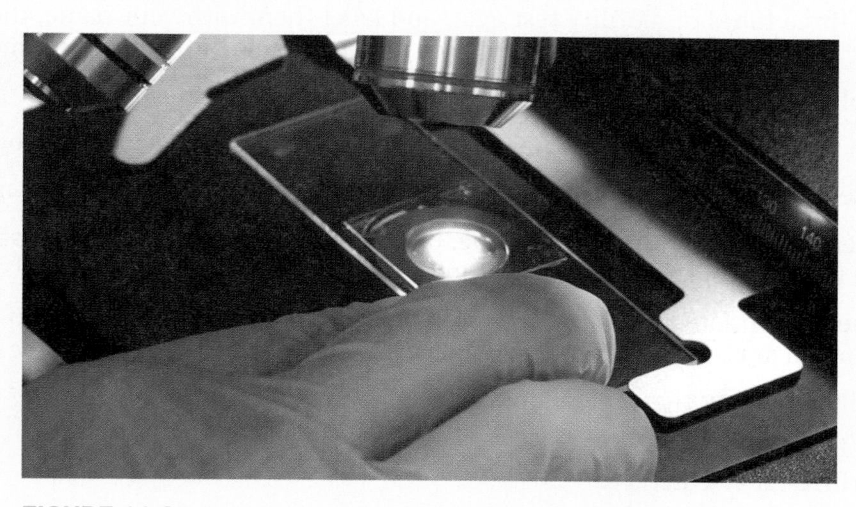

FIGURE 11.3 Observe the slide using the low-power (10×) objective before switching to the high power (40×) objective.

Quick Procedure

Hanging Drop Preparation

1. Prepare coverslip with petroleum jelly adhesive.
2. Add two or three loopfuls of broth culture to coverslip.
3. Invert a depression slide onto the coverslip.
4. Turn the depression slide upright; observe.

When a sample of a motile bacterial species is stabbed with an inoculating needle into a semisolid **motility test agar** (0.4% agar rather than 1,5%), the cells can swim away from the stab. A diffuse "cloud" of growth extending from the stab indicates motility. Nonmotile bacteria will grow only along the stab.

II. Motility Test Agar Preparation

1. Obtain broth cultures of the three bacterial species to be used.
   - Obtain three tubes of motility test agar, and label them with your name, the date, the codes of the three bacterial organisms to be used, and the name of the medium "motility test agar."
   - Each tube will be stabbed with one of the three bacterial species.
2. Obtain an inoculating needle and hold it as you would hold an inoculating loop.
   - Sterilize the needle in the Bunsen burner flame and let it cool for 1-15 seconds.
   - Remove the cap from the culture tube, briefly flame the neck and then aseptically obtain a sample from one of the unknown cultures.
   - Stab the appropriate tube of motility test agar by inserting the needle into the center of the medium and poke the needle down about halfway to the bottom of the tube.
   - Carefully withdraw the needle along the same line. Flame the beck of the tube and replace the cap.
   - Re-sterilize the needle.
3. Stab each of the remaining two tubes of motility test agar with one of the other bacterial species.
4. Incubate the tubes at 37°C for 24 to 48 hours, or as directed by the instructor. At the end of the incubation period, the tubes may be refrigerated to preserve them until observed.
5. Observe the tubes and determine which bacterial species is/are motile. Enter representations of the tubes in the Results section, showing the evidence for motility and, by inference, the presence of flagella.
6. The motility agar tubes should be placed in the appropriate disposal area.

Name: ______________________ Date: __________ Section: __________

Observations of Motile Bacteria

EXERCISE RESULTS **11**

RESULTS

BACTERIAL MOTILITY

I. Hanging Drop Preparation

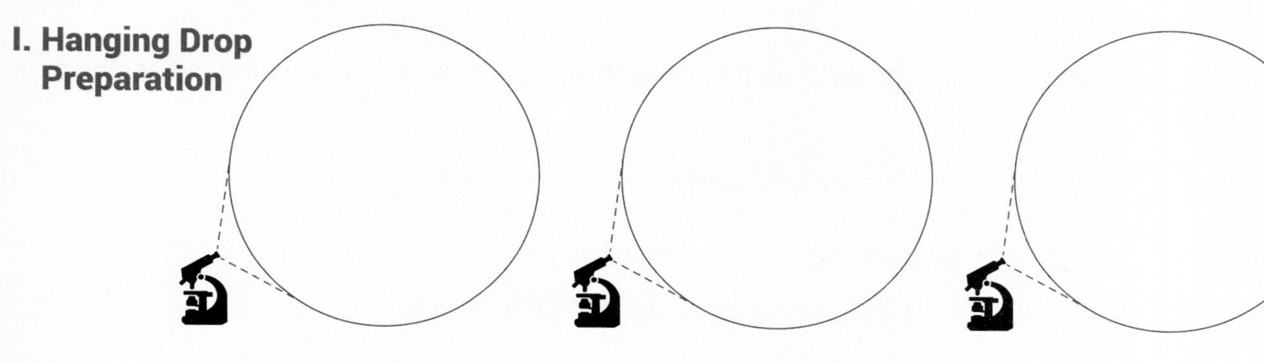

Organism: ______________ ______________ ______________

Total magnification: ______________ ______________ ______________

Motility observed? [ ] yes [ ] no [ ] yes [ ] no [ ] yes [ ] no

II. Motility Test Agar Preparation

Organism

Organism: __________ __________ __________

Observations and Conclusions

____

____

____

____

____

____

Questions

1. In what ways can Brownian motion be distinguished from true bacterial motility in the hanging drop preparation?

   ____

   ____

   ____

2. Explain why a control tube with a nonmotile bacterial species is necessary for reliable determination of bacterial motility by (a) the hang drop preparation and (b) by the motility test agar preparation.

   ____

   ____

   ____

   ____

Bacterial Staining Techniques

PART IV

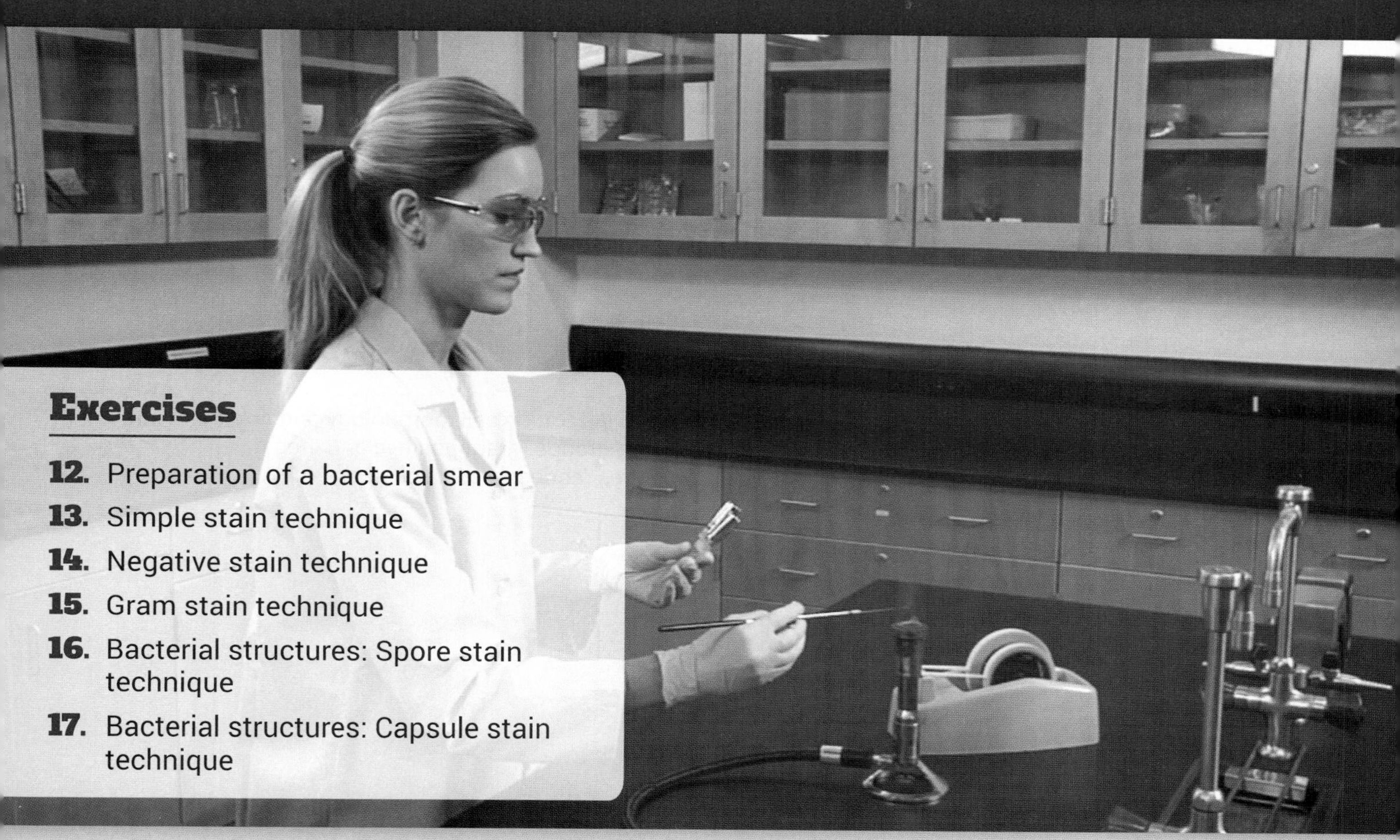

Exercises

12. Preparation of a bacterial smear
13. Simple stain technique
14. Negative stain technique
15. Gram stain technique
16. Bacterial structures: Spore stain technique
17. Bacterial structures: Capsule stain technique

Learning Objectives

When you have completed the exercises in Part IV, you will be able to:

- Prepare a bacterial smear.
- Calculate bacterial cell size and determine bacterial cell shape and arrangement using simple stains.
- Complete a Gram stain to correctly identify gram-positive and gram-negative bacteria.
- Identify (1) bacterial spores and capsules using special differential stains and (2) bacterial motility.

Watch These Videos

- How to Make a Bacterial Smear
- How to Make a Simple Stain Preparation
- How to Make a Negative Stain Preparation
- Preparing a Gram Stain
- Preparing an Endospore Stain
- How to Make a Capsule Stain Preparation
- How to Make a Hanging Drop Preparation

Bacterial cells are not easy to see with brightfield microscopy because (1) the cells are extremely small and (2) most appear colorless against the bright background of the microscope field. Although a light microscope will provide the **resolution** and **magnification** to see them, there still is a contrast issue. This problem is solved by using a colored chemical **stain** that imparts color to bacterial cells or structures (Exercises 12 and 13) or to the background (Exercise 14). Such **simple staining** procedures provide the contrast needed to carry out cell measurements and observations (TABLE IV.1). In addition, **differential staining** procedures, which use two contrasting colored stains, will allow you to separate bacteria into one of two basic groups (Exercise 15) and to visualize bacterial structures such as endospores (Exercise 16) and capsules (Exercise 17). Bacterial cell motility will be examined using a different technique.

TABLE IV.1 Types and Uses for Staining

Type of Staining	Purpose
Simple (Uses one stain)	To measure cell size To determine cell morphology (bacillus, coccus, spiral) To determine cell arrangements
Differential (Uses two contrasting stains)	To measure cell size and determine cell morphology and arrangements To separate bacteria into groups (gram-positive and -negative) To visualize cell structures (endospores, capsules)

Preparation of a Bacterial Smear

EXERCISE **12**

Materials and Equipment

- 24-hour pure cultures (agar slant and/or broths) of selected bacterial species
- Glass slides
- Wax marking pencil or felt marker
- Inoculating loop and/or needle
- Bunsen burner

Watch This Video

- How to Make a Bacterial Smear

Most bacteria have no color, so they generate little contrast in the microscope field. Therefore, to see bacteria with the microscope, it is necessary to apply color to a bacterial smear by using a staining reagent.

The preparation of a bacterial smear is important, as the smears are an essential element of laboratory microbiology. Bacteria are placed on a glass slide and "fixed" with heat, partly to ensure that they remain attached to the glass.

This exercise also will demonstrate the important techniques for handling bacterial cultures. Each is a "pure culture"; it contains only one species of bacterium. The bacteria have been cultivated in a liquid growth medium such as nutrient broth or on a solid growth medium such as nutrient agar. Your instructor will specify the medium used.

How to Make a Bacterial Smear

PREPARATION OF A BACTERIAL SMEAR

> **PURPOSE:** to prepare a bacterial sample for staining.

A **bacterial smear** is a thin layer of bacteria placed on a slide for staining. Preparing the smear requires attention to a number of details that help prevent contamination of the culture and ensure safety to the preparer. Each step in this procedure is important, and each should be followed carefully as a prerequisite to successful work.

Prior to starting, watch **Microbiology Video: How to Make a Bacterial Smear** to familiarize yourself with this procedure.

PROCEDURE

1. Before beginning this exercise, clean your bench top with disinfectant. Then, place a clean sheet of paper toweling or other absorbent material over the area to be used. Should a spill occur, the paper will absorb the liquid and the paper can be disposed of as directed by the instructor.

2. Wash a glass slide with soap, rinse it well, and dry it thoroughly, preferably with a lint-free cloth. Place the cleaned slides on your paper towel.
3. With a wax marking pencil or felt marker, mark off one or two circular areas on the slide that are about the diameter of a dime. This target area is where smears will be placed. One area should be reserved for a **slide label**. The label should include the initials of the bacterial species. Some laboratory workers prefer to use the underside of the slide for marking to prevent wax from mixing with the smear. Others use the top because the wax barrier holds the dye in that area and prevents spreading.
4. Obtain cultures of the bacteria, and place the tubes in a suitable rack on the bench.
5. Sterilize your inoculating loop. If agar slant cultures are used, place a loopful of water into one area of the slide. If broth cultures are used, no additional liquid is necessary.
6. Light the Bunsen burner. The smear is now ready to be prepared.
7. Referring to FIGURE 12.1A, hold the loop firmly, as you would hold a pencil, and

When lighting the Bunsen burner for use in the lab, be sure to face it away from you.

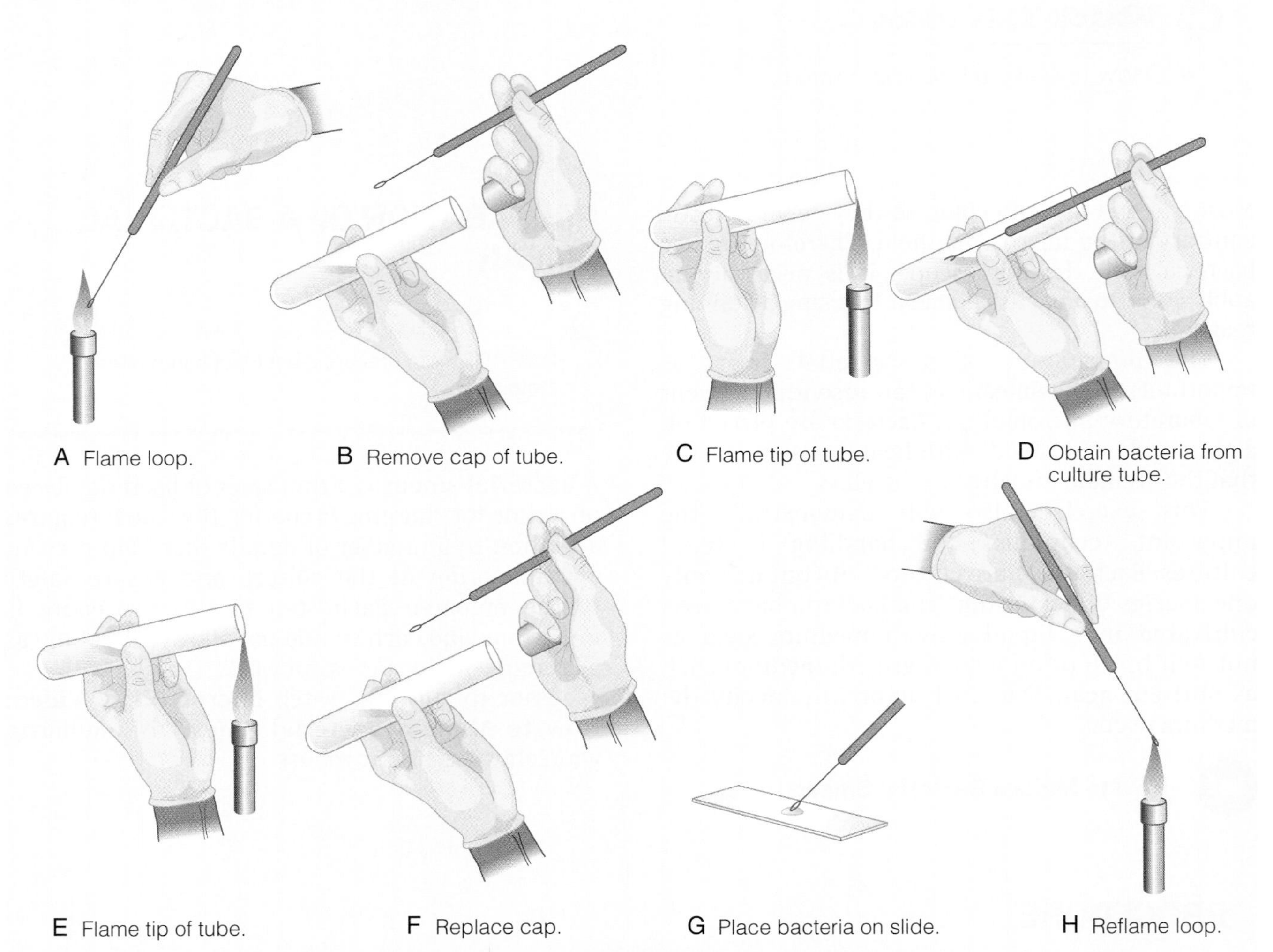

FIGURE 12.1 **Important steps in the preparation of a bacterial smear from a broth culture.** If using a broth culture, the liquid is placed directly on the slide. If a slant culture is used, a loopful of water is placed on the slide before adding the small sample of bacteria.

place it nearly vertically into the blue tip portion of a Bunsen burner flame for a few seconds or until it glows red hot. Allow the loop to cool several seconds.

8. Take the bacterial culture in the opposite hand and remove the cap, as shown in FIGURE 12.1B. Place the tip of the tube in the flame for a few seconds (FIGURE 12.1C). This will burn off any dust or lint and kill any airborne organisms that might happen to fall into the tube.
9. Insert the sterile loop into the tube.
   - If a slant culture is used, touch the loop to the main portion of the slant. Be careful to only remove a tiny amount of the bacteria sample from the surface of the bacterial slant.
   - If a broth culture is used, dip the loop carefully into the liquid (FIGURE 12.1D). You may want to add aseptically a few more loopsful to ensure you can easily find the bacterial cells with the microscope.
10. Remove the loop, reflame the tip of the tube briefly (FIGURE 12.1E), then replace the cap (FIGURE 12.1F), and return the culture to the rack on the desk.
11. For slant cultures, mix the bacteria on the loop in the drop of water on the slide (FIGURE 12.1G and FIGURE 12.2), and swirl the liquid through out the target area. For broth cultures, simply swirl the loopful of bacteria throughout the target area. Complete the procedure by reflaming the loop (FIGURE 12.1H).
12. Prepare other smears in the remaining target areas on the same or separate slides with different bacteria as directed by the instructor. As noted, it is important that you become comfortable with the methods for handling cultures and preparing smears since they are essential elements of laboratory microbiology. All smears on one slide should be prepared before proceeding with the next step.
13. **Air-dry** the smear until all the liquid evaporates. An electric warming tray may be used for this purpose. Use caution to avoid extreme heat because cell distortion and splattering may occur.

Do not blow on the slide or wave the slide in the air during the air drying step.

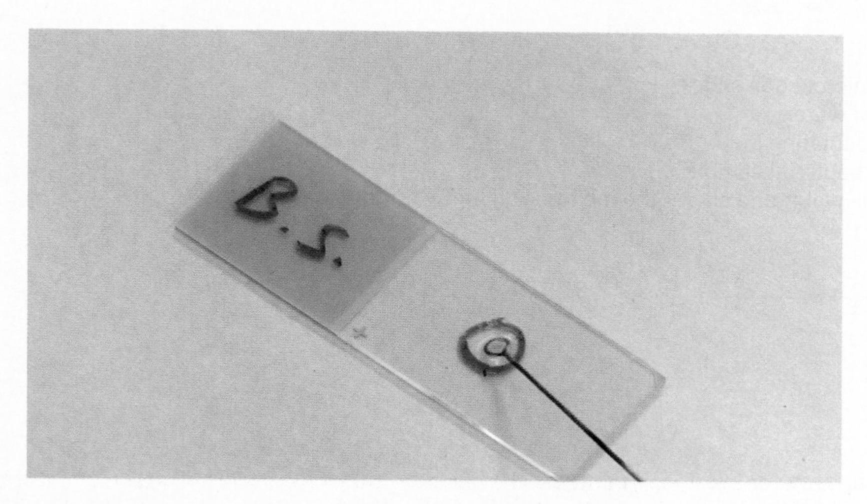

FIGURE 12.2 For slant cultures, mix the bacteria within a drop of water that previously was placed on the slide.

14. **Heat-fix** the smears by quickly passing the slide through the Bunsen burner flame three times (FIGURE 12.3). The heat should contact the underside of the slide. This procedure 1) kills any bacteria that may still be alive, 2) facilitates stain penetration, and 3) fixes cells to the slide so that they do not wash off when stained. The bacterial smear on the slide is ready to be stained.

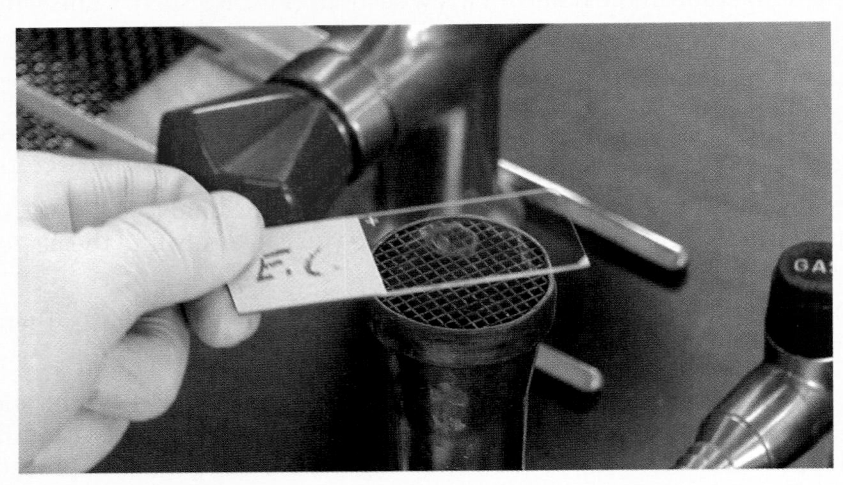

FIGURE 12.3 A slide is heat-fixed by passing it quickly over a flame three times.

Quick Procedure

Smear Preparation: Broth Culture

1. Label a clean microscope slide. Make a circular target area.
2. Sterilize the inoculating loop.
3. Remove a loopful of the bacterial broth.
4. Smear the broth on target area.
5. Air dry.
6. Heat fix.

Smear Preparation: Agar Slant Culture

1. Label a clean microscope slide.
2. Add a loopful of water.
3. Sterilize the inoculating loop.
4. Remove a tiny bacterial sample.
5. Swirl loop in the water and smear the mixture within the target area.
6. Air dry.
7. Heat fix.

Name: ______________________ Date: __________ Section: __________

Preparation of a Bacterial Smear

EXERCISE RESULTS **12**

RESULTS

PREPARATION OF A BACTERIAL SMEAR

Draw a representation of a broth prepared smear and an agar slant prepared smear. Lay the slide smear on top of the print of this lab exercise. A good smear should be quite thin, such that the print should be legible through the slide smear.

Broth smear

Agar slant smear

Observations and Conclusions

____

____

____

Questions

1. Identify the steps taken in the preparation of a bacterial smear to prevent contamination of the culture and the preparer of the smear.

____

____

____

2. For the slant preparation, what will happen if too large drop of water was put on the slide?

3. Identify the three possible outcomes if an air-dried smear were not heat-fixed before proceeding with the staining step.

4. What might happen if a slide were overheated during the heat-fixing step?

5. After a smear has been stained, do you believe a thin, hazy smear or a thick, dense smear will make the best preparation for microscope examination? Explain your reasoning.

Simple Stain Technique

EXERCISE

13

Materials and Equipment

- Basic stains such as crystal violet, methylene blue, and safranin
- Absorbent or blotting paper
- Air-dried, heat-fixed bacterial smears
- Wash bottle
- Brightfield microscope

Watch This Video

- How to Make a Simple Stain Preparation

Once bacteria are stained, they may be observed and studied with respect to their shape, size, and arrangement.

After a bacterial smear is prepared, a **basic stain** is applied. Such a stain carries a positive electrical charge. The negatively charged surface and cytoplasm of bacteria attracts the stain, and staining takes place. This will produce a distinctive contrast between the stained organisms and the bright background of the microscope field. In this exercise, we shall explore the proper methods for simple staining. The technique is called the **simple stain technique** because only a single stain is used. Other staining techniques are explained in Exercises 14, 15, and 16.

How to Make a Simple Stain Preparation

THE SIMPLE STAIN TECHNIQUE

PURPOSE: to determine the size, shape, and arrangement of bacterial cells.

The simple stain technique is a rapid and effective way to prepare a bacterial smear for viewing. It is a one-step procedure in which the smear is covered with stain and allowed to sit undisturbed for a minute or so, during which the bacterial cytoplasm chemically unites with the basic stain. The remaining, unbound stain is then washed away.

PROCEDURE

1. Select air-dried and heat-fixed prepared bacterial smears.
2. At the staining rack or staining tray, cover the smears with the selected basic stain (FIGURE 13.1), and let the preparation remain undisturbed for one minute (FIGURE 13.2). The instructor may modify these directions for individual situations.

3. Using a wash bottle, gently wash off the excess stain with a steady stream of water by holding the slide parallel to the stream. Then, blot the slide with absorbent paper or between the pages of a blotting (bibulous) paper booklet (FIGURE 13.3). Do not wipe the slide. When all the water has evaporated completely, the slide is ready for examination.
4. Prepare your microscope for use.
5. Start by observing the slide with the low power (10×) objective and then the high power (40×) objective. No drawings should be made yet. Now place a drop of immersion oil directly on the smear and switch to the oil immersion (100×) objective, as described in Exercise 10 (and Microbiology Lab Video: Measuring Cell Size) (FIGURE 13.4). After locating the bacterial cells, thoroughly scan the smear, looking for a very thin layer of bacteria apart from other clumps and masses.
6. Determine the shape, size, and arrangement of the bacterial cells. When you are confident of your observations, use a pencil with a sharp point to draw representations of the bacteria under oil immersion in the Results section of this exercise. Usually about three dozen cells per drawing will suffice. Include the full binomial name of the organisms, as well as the total magnification and stain used.
7. When your slide work is completed, the oil may be removed from the slide by gently blotting in the blotting paper booklet. If the slide is to be retained, label with a slide label containing the names of the organisms, the stain, your name, and the date. If it is not going to be retained, wash the slide with soap and water to remove the material. Dry the slide.

If using slide labels, do not put them in your mouth for moistening purposes.

FIGURE 13.1 Use a basic stain, such as crystal violet.

FIGURE 13.2 Stain the smear for approximately one minute.

FIGURE 13.3 A booklet of bibulous (blotting) paper is used to dry the slide after it has been prepared.

FIGURE 13.4 Immersion oil is placed on the smear before positioning the oil immersion (100×) lens.

8. Additional smears may be prepared of tooth-gum scrapings, a sample of yogurt or sour cream, a hay infusion, a yeast suspension, water from a bird bath, a soil sample, or other specimens supplied in the laboratory. Add your drawings in the Results section.
9. At the conclusion of the laboratory period, use lens paper to wipe the microscope objectives; the oil immersion objective should be wiped last.
10. The microscope should be stored with the low-power objective in position, the stage lowered, and the power cord wound neatly around the base or arm.

Quick Procedure

Simple Stain Technique

1. Stain heat-fixed slide for 1 min with basic stain.
2. Wash and dry the slide.
3. Observe with the light microscope.

Name: ________________________________ Date: __________ Section: __________

Simple Stain Technique

EXERCISE RESULTS **13**

RESULTS

SIMPLE STAIN TECHNIQUE

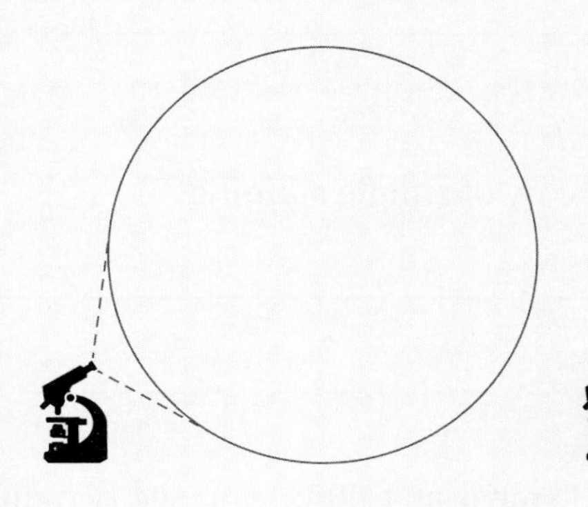

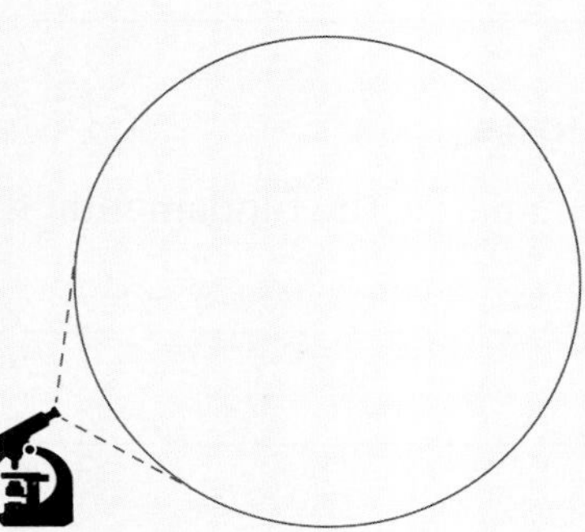

Organism:	____________	____________
Stain:	____________	____________
Total magnification:	____________	____________
Cell shape:	____________	____________
Cell arrangement:	____________	____________

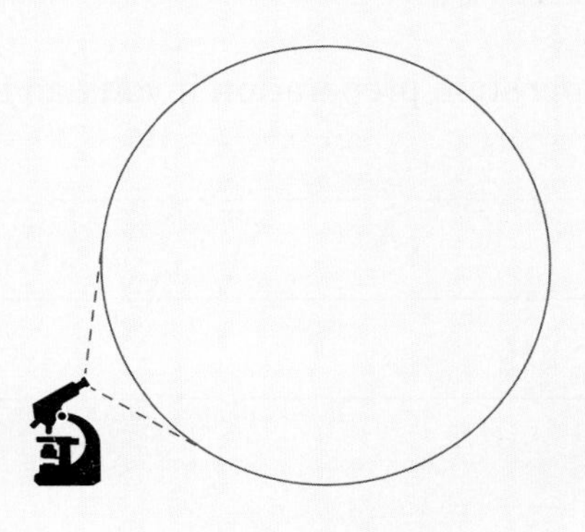

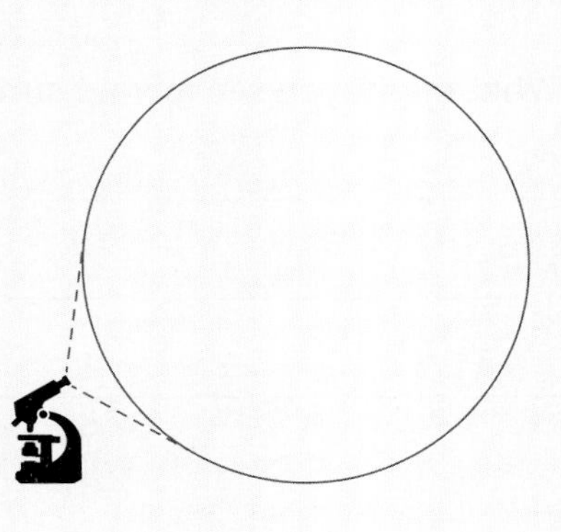

Organism:	____________	____________
Stain:	____________	____________
Total magnification:	____________	____________
Cell shape:	____________	____________
Cell arrangement:	____________	____________

Observations and Conclusions

Questions

1. Summarize the fundamental theory of simple staining.

2. When you observe your stained smear at 1,000×, you see two different-shaped bacteria; some are rod shaped while others are spherical. Explain this result, considering the original source culture was pure.

3. What might you see in your simple stain preparation if you had forgotten to heat fix the smear?

Negative Stain Technique

EXERCISE

14

Materials and Equipment

- Inoculating loop
- Cultures of selected bacterial species
- Nigrosin stain or India ink
- Toothpicks and dental floss
- Glass slides
- Brightfiled microscope

Watch This Video

- How to Make a Negative Stain Preparation

The **negative stain technique** permits one to see unstained bacteria on a stained background and determine bacterial cell shape. **Acidic stains**, such as nigrosin, congo red, and India ink, carry a negative charge and therefore are repelled by the negatively charged bacteria. The technique may be used with samples of bacterial cultures and especially for organisms in dental plaque because the spiral bacteria (spirochetes) in plaque do not stain well with other techniques. Cells appear clear against a dark background. Review **Microbiology Video: How to Make a Negative Stain Preparation** to see this technique.

 **How to Make a Negative Stain Preparation**

Precautions must be taken in this technique. Excessive stain will make the organisms difficult to locate. Also, the stain must not be allowed to become contaminated because it will support bacterial growth.

NEGATIVE STAIN TECHNIQUE

PURPOSE: to accurately measure cell size and determine cell shape in an undistorted manner.

PROCEDURE

1. Wash a slide until it is scrupulously clean. It is important to remove extraneous debris and particles from the slide because these may be mistaken for bacteria. Rinse the slide and dry it, preferably in air or with a lint-free cloth.
2. Aseptically obtain a loopful of bacteria according to the procedure described in Exercise 11 (and Microbiology Video: How to Make a Bacterial Smear). Place it near the end of the slide.
3. Now place a very small drop of acidic stain (nigrosin or India ink) in the drop of bacteria. Using a toothpick, mix the cells with the stain on the slide.

4. Referring to FIGURE 14.1A, at a 30-45° angle, back the short edge of a second (spreader) slide into the mixture.

Be careful not to touch the stained area of the slide, since the bacteria remain alive until the slide has air dried.

5. Spread the mixture in one smooth movement down the length of the slide surface as shown in FIGURE 14.1B. Note that the stain is pulled rather than pushed across the slide to give a light and even distribution. As the far edge is neared, lift the spreader slide to "feather" the edge of the smear. The spreader slide should be flamed briefly to destroy any bacteria that may have contaminated it.

6. Allow the slide to air-dry on the laboratory bench (FIGURE 14.2) or warming tray. Do not heat-fix the slide.

7. Additional slides may be made with different species of bacteria or with a sample of dental plaque. To prepare a slide of plaque, use a clean toothpick or piece of dental floss to gather material from between the junction of teeth and gums. Add this material to a drop of stain on the edge of the slide, then spread as before.

8. Survey the slide with the low-power (10×) lens, looking for tiny pinpoints of light (bacteria). Then, locate the negative stained bacteria under 40×. Add a drop of immersion oil and view the stained smear with the oil immersion (100×) lens. Search for white or clear cells on a dark background as shown in FIGURE 14.3. Scan the slide for several minutes.

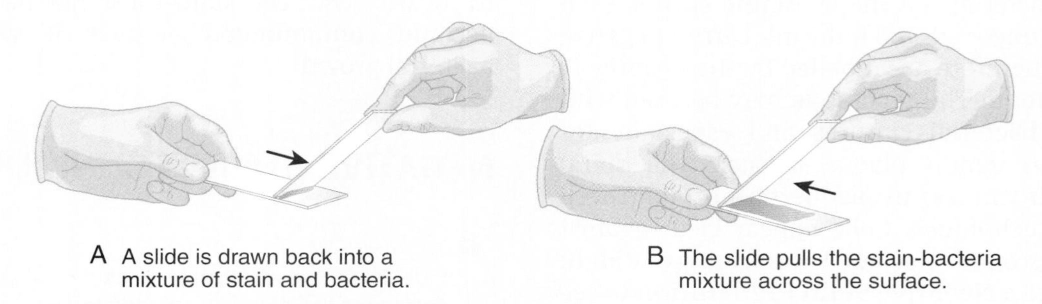

A A slide is drawn back into a mixture of stain and bacteria.

B The slide pulls the stain-bacteria mixture across the surface.

FIGURE 14.1 The procedure for preparing a negative stain.

FIGURE 14.2 Air dry the slide.

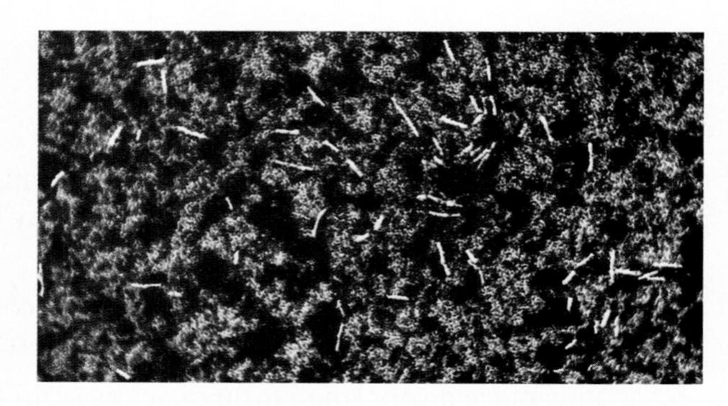

FIGURE 14.3 The appearance of bacteria after negative staining.

9. Prepare drawings in the Results section, being sure to shade the background. Note the size, shape, and arrangement of the cells. Include the full binomial name of known organisms together with the total magnification and dye used. If required, label the slides with a slide label containing your name, the name of the organism, the stain used, and the date. Retain them for later reference.

Quick Procedure

Negative Stain

1. Add a loopful of bacteria near the end of slide.
2. Place small drop of nigrosin or India ink in the drop of bacteria.
3. Smear across face of slide.
4. Air-dry; observe.

Name: ______________________ Date: __________ Section: __________

Negative Stain Technique

EXERCISE RESULTS **14**

RESULTS

NEGATIVE-STAINED BACTERIAL SMEARS

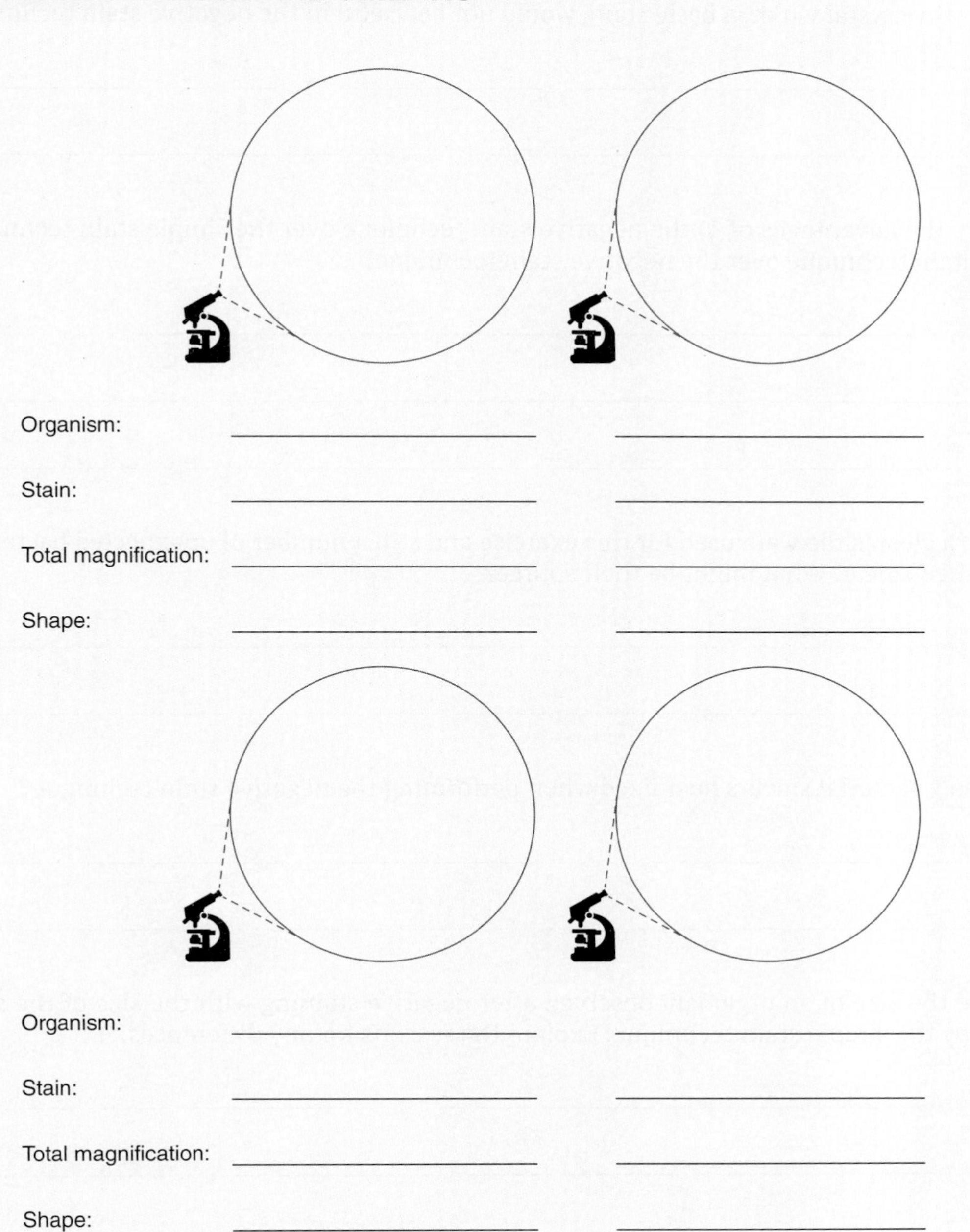

Observations and Conclusions

Questions

1. Explain why crystal violet, a basic stain, would not be useful in the negative stain technique.

2. What are the advantages of 1) the negative stain technique over the simple stain technique and 2) the simple stain technique over the negative stain technique?

3. Suppose a clean slide were used for this exercise and still a number of unexpected bacteria appeared in the finished smear. What might be their source?

4. Why aren't bacterial smears heat fixed when performing the negative stain technique?

5. Compare the size of an organism observed after negative staining with the size of the same organism stained by the simple stain technique. Explain the reasons for any differences.

Gram Stain Technique

EXERCISE

15

Materials and Equipment

- Cultures of selected bacterial species
- Crystal violet
- Gram's iodine
- 95% ethyl alcohol
- Safranin
- Bibulous paper
- Brightfield microscope
- Wax marking pencil or felt marker

Watch This Video

- Preparing a Gram Stain

In 1884, Christian Gram, a Danish physician, discovered that certain bacteria, after being stained with crystal violet and iodine, would lose their color on subsequent treatment with alcohol while other bacteria would retain their color. Bacteria could thus be separated into two groups based on their reaction to this procedure. Those retaining the stain are known as gram-positive bacteria, while those losing it are gram-negative bacteria.

The **Gram stain technique** is perhaps the most important differential staining procedure in microbiology because the great majority of bacteria are either **gram-positive** or **gram-negative**. (Spirochetes and mycobacteria are notable exceptions—they do not respond to the procedure.) Watch **Microbiology Video: Preparing a Gram Stain** to see how the procedure is done.

The differential staining has been elucidated through electron microscopy. The crystal violet stain and iodine form a complex that exits easily from gram-negative bacteria but not from gram-positive bacteria when alcohol is applied. Also, large amounts of peptidoglycan in the cell walls of gram-positive bacteria trap the complex. Gram-negative bacteria have less peptidoglycan and thus fail to trap it.

The Gram stain technique is a four-part procedure in which two dyes, a mordant, and a decolorizing agent are used (TABLE 15.1). It is important to note that the terms "Gram positive" and "Gram negative" do not refer to electrical charges but are merely used for convenience to indicate the retention (positive) or loss (negative) of the crystal violet-iodine complex.

Preparing a Gram Stain

GRAM STAIN TECHNIQUE

PURPOSE: to differentiate between gram-positive and gram-negative bacteria; to determine cell size, shape, and arrangement.

TABLE 15.1 Gram Stain Reagents

Reagent	Function
Crystal violet	Primary stain: stains all bacteria blue to purple
Gram's iodine	Mordant: enhances reaction between cell wall and primary stain
Ethyl alcohol or acetone	Gram-positive bacteria retain the primary stain because of the peptidoglycan and teichoic acid cross-links. Gram-negative bacteria lose the primary stain because of the large amount of lipopolysaccharide in the cell wall.
Safranin	Counterstain: no effect on gram-positive bacteria; stains gram-negative bacteria pink to red

PROCEDURE

1. Prepare air-dried, heat-fixed smears of selected bacterial organisms on clean slides, as outlined in Exercise 11.
2. Flood the smears with **crystal violet** (the **primary stain**), and allow the stain to remain for 1 minute as indicated in FIGURE 15.1A.
    - Rinse with water (FIGURE 15.1B and FIGURE 15.2).
    - Do not blot the slide dry; simply gently shake off the excess water.
3. Flood the smears with **Gram's iodine** solution for 1 minute (FIGURE 15.1C). Rinse with water, but do not blot (FIGURE 15.1D).
4. Decolorize the smears with **95% ethyl alcohol** as follows:
    - Pour a few drops of alcohol on the smears (FIGURE 15.1E), and rock the slide back and forth for 15 seconds.
    - Allow the alcohol to drip off, then repeat this decolorization procedure for another 15-second period. (As a general rule, decolorization should continue until the ethyl alcohol running off is no longer purple. However, this may be difficult to discern when performing the procedure for the first time.)
5. Rinse the slide with a gentle stream of water but do not blot (FIGURE 15.1F). Excessive decolorization may cause gram-positive bacteria to lose their color (appear gram-negative), and insufficient decolorization may allow gram-negative bacteria to retain a purple color (appear gram-positive). At this point, gram-positive bacteria retain the crystal violet iodine complex while gram-negative bacteria lose the complex and would appear transparent if observed with the microscope (FIGURE 15.3).
6. Flood the smears with **safranin** (the **counterstain**), and allow the stain to remain for 30 seconds (FIGURE 15.1G and FIGURE 15.4).
    - Rinse gently with water (FIGURE 15.1H), and blot the slide using the bibulous paper (FIGURE 15.1I).
    - Safranin, a red dye, will stain bacteria that lost the crystal violet-iodine complex during decolorization (i.e., the gram-negative bacteria). It will have no effect on the color of the gram-positive bacteria.

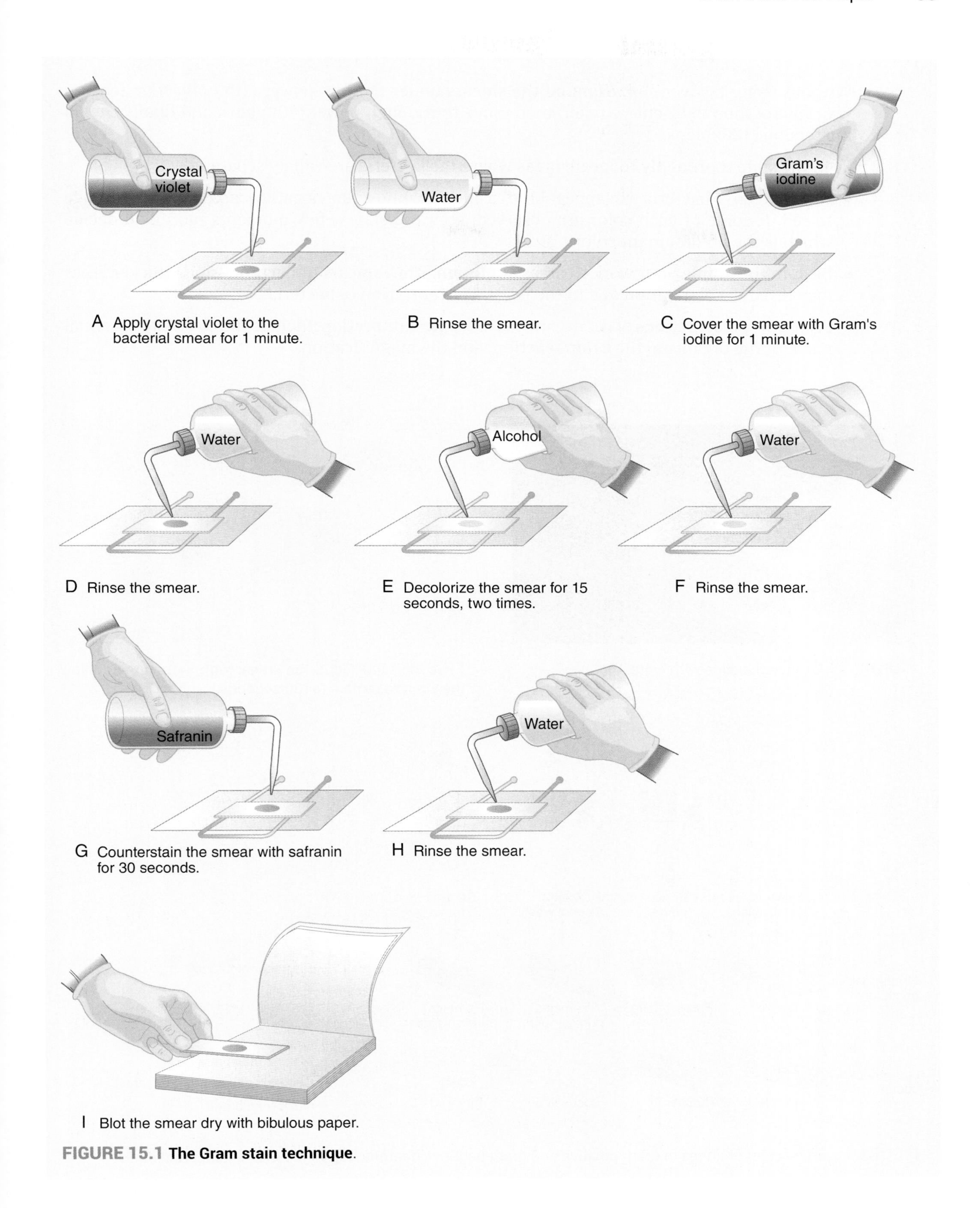

FIGURE 15.1 The Gram stain technique.

7. Without using a cover slip, examine the smears under the low-power (10×) objective lens on your microscope to orient yourself, then move to the high power (40×) lens, and finally the oil immersion (100×) lens.
8. Scan the slide thoroughly to locate areas where the bacteria are widely separated.
   - Gram-positive bacteria will appear blue to purple, while gram-negative bacteria will be orange to red. If some of each color are observed, scan the slide a few moments more and decide which is the predominant color.
   - Be sure to confine your work to areas containing thin smears of bacteria, since thicker areas may resist decolorization due to the heavy concentration of bacteria.
   - Prepare representations of the organisms in the Results section. Include the complete binomial name of the organism, the Gram reaction, and the magnification.

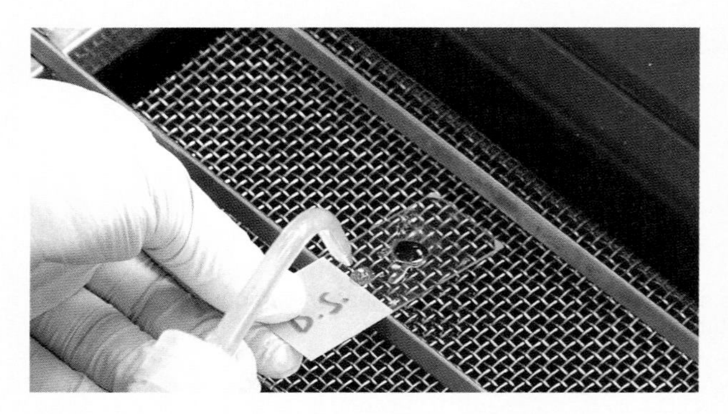

FIGURE 15.2 Rinse the slide with water.

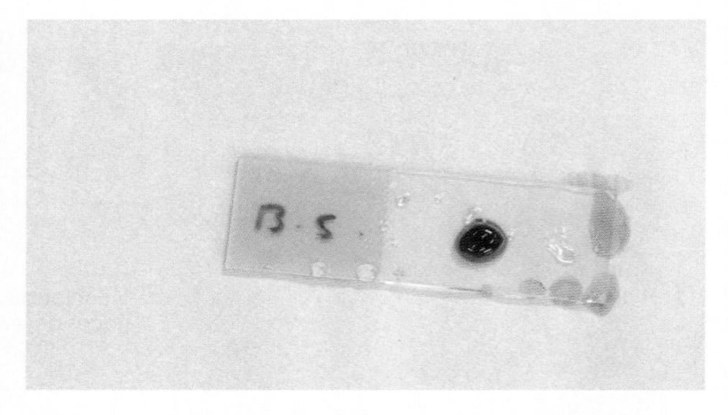

FIGURE 15.4 Flood the smear with safranin and allow the stain to remain for 30 seconds.

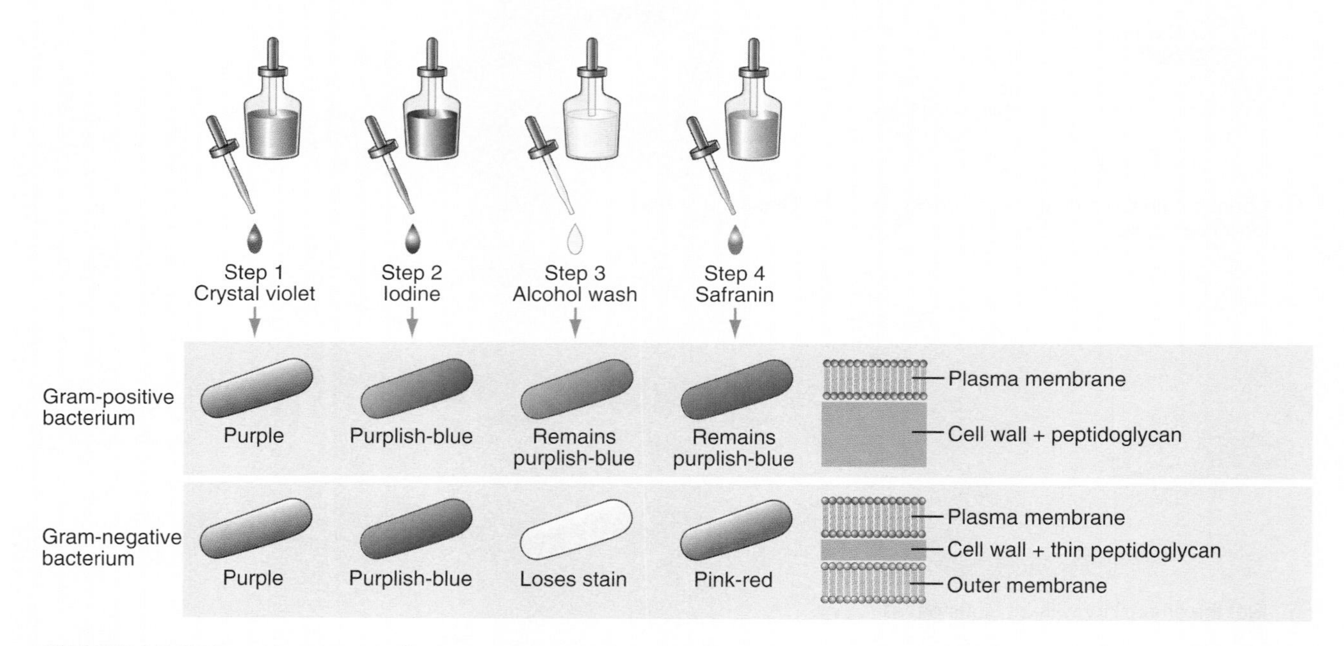

FIGURE 15.3 The color changes in gram-positive and gram-negative bacteria during the Gram stain procedure.

- If the slide will be retained, remove the oil by gently blotting the slide. Otherwise, clean the slide with soap and water. Dry the slide with a paper towel.
- Label the slide with your name, the organisms' names, the date, and the designation "Gram stain."

9. It is valuable to prepare Gram stains of mixtures of gram-positive and gram-negative bacteria to test your ability to perform this technique. If a prepared smear is not provided, add a loopful of bacteria to a loopful of water on a slide. Now, flame the loop, obtain a sample of a second organism, and mix the two organisms together in the water. Additional practice may be obtained by Gram staining samples of dental plaque obtained from the tooth-gum junction. In this case, a mixture of forms and Gram reactions will be observed. Record all drawings in the Results section.

Quick Procedure

Gram Stain

1. Stain heat-fixed slide with crystal violet for 1 min; rinse.
2. Flood with Gram's iodine for 1 min; rinse.
3. Decolorize for two 15-sec periods with alcohol; rinse.
4. Stain with safranin for 30 sec; rinse; dry; observe.

Name: ______________________ Date: __________ Section: __________

Gram Stain Technique

EXERCISE RESULTS **15**

RESULTS

GRAM-STAINED BACTERIAL SMEARS

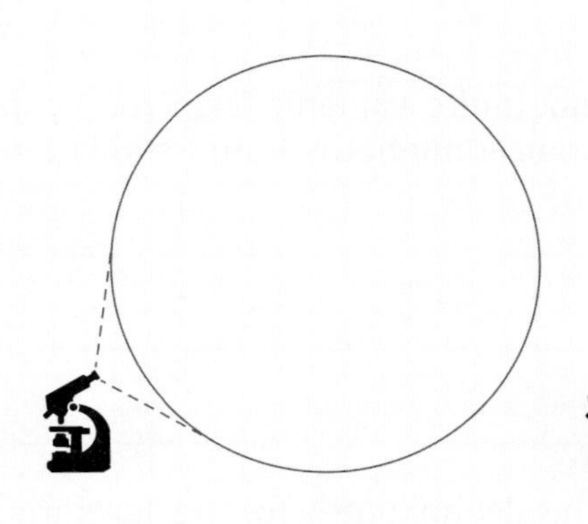

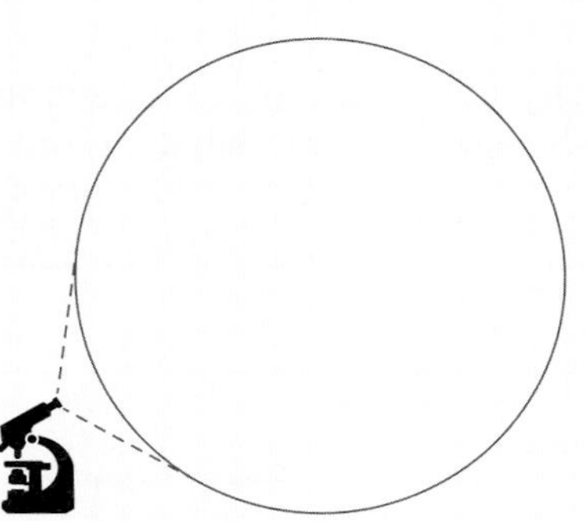

Organism:	______________	______________
Gram reaction:	______________	______________
Total magnification:	______________	______________
Cell shape:	______________	______________
Cell arrangement:	______________	______________

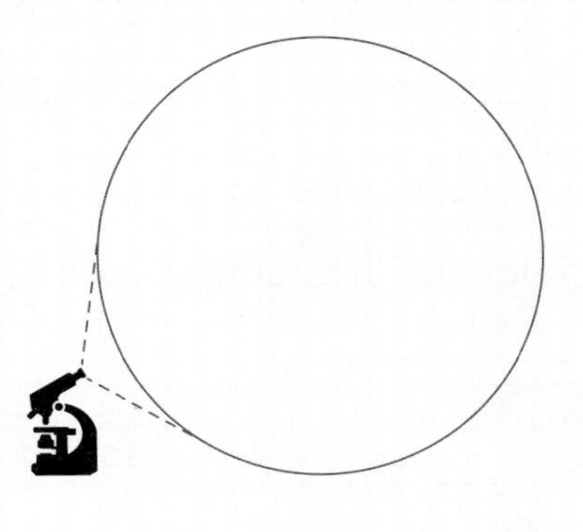

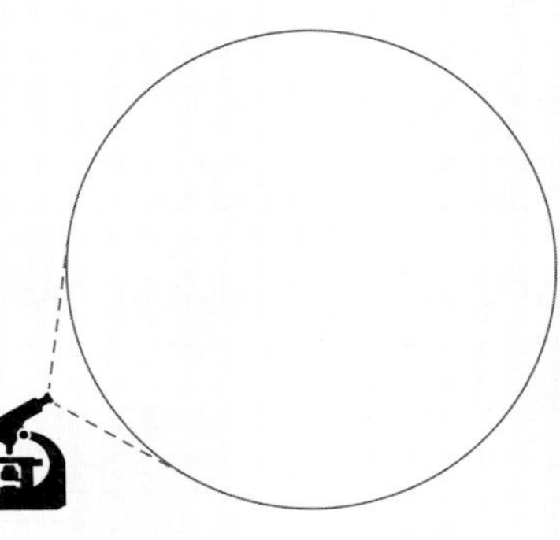

Organism:	______________	______________
Gram reaction:	______________	______________
Total magnification:	______________	______________
Cell shape:	______________	______________
Cell arrangement:	______________	______________

Observations and Conclusions

____

____

____

____

Questions

1. Why is the Gram stain technique more valuable than the simple stain technique in the diagnostic laboratory? Under what circumstances might the simple stain be preferable to the Gram stain technique?

____

____

____

2. Suppose a student reached for the alcohol bottle for the decolorization step and instead took the bottle of distilled water. What would be the color of a gram-positive and a gram-negative bacterium at the conclusion of the procedure? Explain thoroughly.

____

____

____

3. How might the Gram stain technique be of value in establishing the purity of a culture of bacteria?

____

____

4. In your opinion, what is the most critical step in ensuring you will have a good Gram-stained bacterial smear?

____

____

____

5. Describe the procedures needed to properly see Gram-stained organisms with the oil-immersion lens.

____

____

____

Bacterial Structures: Spore Stain Technique

EXERCISE 16

Materials and Equipment

- Cultures of *Bacillus* and/or *Clostridium* species
- Steaming apparatus
- Forceps or clothespin
- 5% malachite green
- Safranin
- Newspaper or paper towels to cover laboratory desk
- Blotting paper
- Brightfield microscope

Watch This Video

- Preparing an Endospore Stain

Endospores are formed by members of a few gram-positive bacterial genera, such as *Bacillus* and *Clostridium*. The spores are extremely resistant structures; resistant to boiling water temperatures for two hours or more. They contain little water and exhibit very few chemical reactions. When the external environment is favorable, the spore's protective layers break down and the vegetative cell emerges to grow and reproduce. Among the notable diseases caused by different species of endospore-forming bacteria are tetanus, botulism, gas gangrene, and anthrax (FIGURE 16.1).

In this exercise, bacterial endospores will be visualized by special staining techniques.

SPORE STAIN TECHNIQUE

PURPOSE: to view and contrast vegetative cells from endospores with the microscope.

Identifying endospore-forming pathogens is important in food and medical microbiology. However, the spores contain numerous protective layers, which cannot be penetrated easily by stain using the simple or Gram stain techniques. It therefore is necessary to apply heat to assist stain penetration. The heat will force the stain into the spores and vegetative cells, and both will become green. By rinsing the slide, the vegetative cells will lose their color but the spores will remain green. Counterstaining with safranin will stain the vegetative cells pink, but have no effect on the spores.

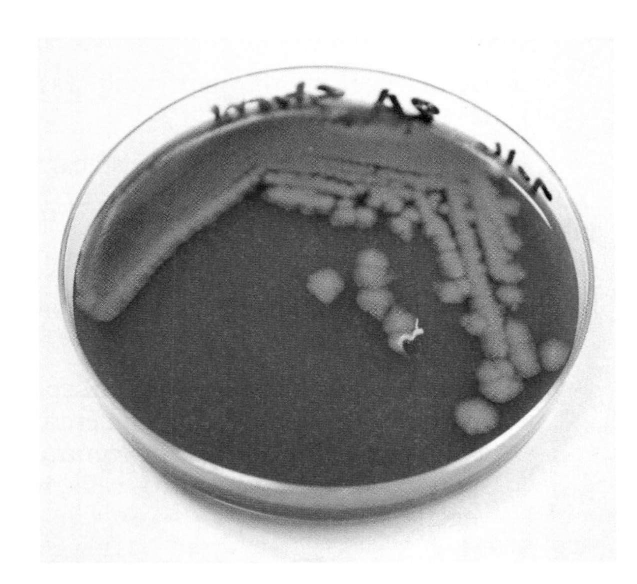

FIGURE 16.1 Colonies of *Bacillus anthracis* growing on blood agar.

PROCEDURE

I. The Schaeffer-Fulton Method

1. Prepare air-dried, heat-fixed smears of *Bacillus* and/or *Clostridium* species as outlined in Exercise 12. The instructor will indicate which organisms are to be used. Normally, a smear will contain both spores and vegetative cells.
2. While the smear is air drying, set up the staining apparatus as follows:
   - Since dripping can occur, newspaper or paper towels should be used to cover the laboratory hood for this procedure.
   - Set up a steaming apparatus consisting of a Bunsen burner, tripod, wire pad, beaker of water, and slide rack of glass rods, as illustrated in FIGURE 16.2A.

The fumes from malachite green are toxic. Always do the staining in the fume hood.

3. Light the flame to begin heating the water.
4. Cut a piece of blotting (bibulous) paper just large enough to cover the smear. (Do not use lens paper for this purpose.)
   - When the water in the beaker has begun boiling, balance the slide over the steam, and cover the smear with the blotting paper. Do not allow the paper to hang over the edge of the slide since this will cause stain to drip.
5. Saturate the blotting paper with **malachite green** stain (FIGURE 16.2A).
   - Allow the slide to remain over the steam for 3 minutes, continually adding stain during this period to keep the paper wet.

The slide will become rather hot when placed over the boiling water. Be sure to use the forceps (or clothespin) to handle the slide.

6. After 3 minutes, use forceps or a clothespin to carefully remove the slide from the steam bath.
   - Gently peel off the paper and wash the slide thoroughly with a gentle stream of water (FIGURE 16.2B).
   - It is not necessary to blot the smear.
7. Flood the smears with **safranin** for 1 minute (FIGURE 16.2C and FIGURE 16.3).
   - Rinse the slide with water (FIGURE 16.2D) and blot it dry (FIGURE 16.2E).
8. Using the microscope, examine the smears under the low-power (10×) lens, then high power (40×) and oil immersion (100×).
   - Scan the slide and note the oval spores stained green and the long vegetative cells stained red-orange. Look carefully to determine whether any spores are still within vegetative cells, and note the spores' position (central, subterminal, or terminal). Young cultures often contain spores within the vegetative cells, while older cultures contain more free spores and fewer vegetative cells.
   - Draw representative spores and vegetative cells in the Results section.

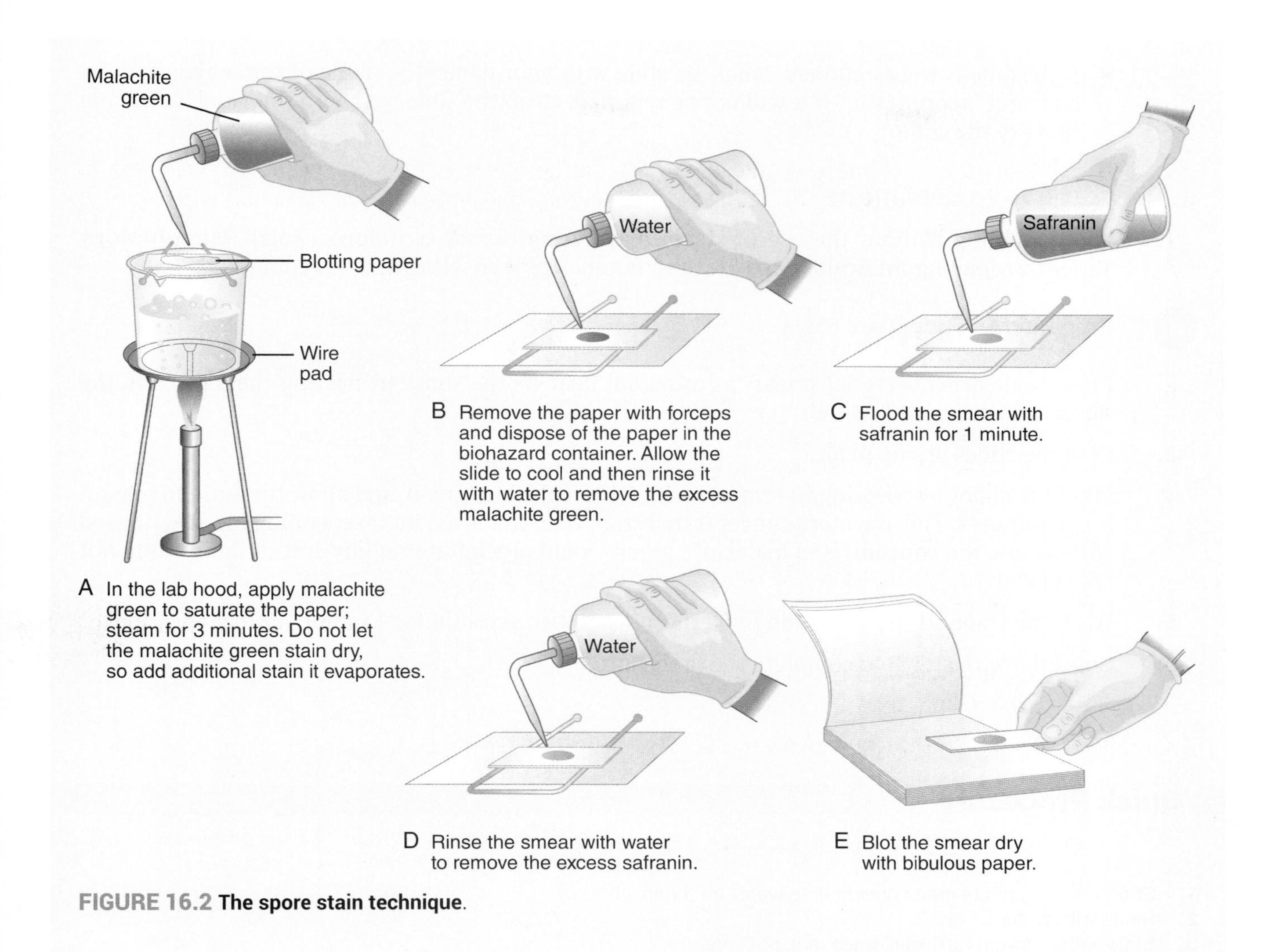

FIGURE 16.2 The spore stain technique.

FIGURE 16.3 Counterstain with safranin for approximately 1 minute.

- If the slide is to be retained, label the slide with your name, the name of the organisms, the date, and "spore stain." If it will not be retained, clean the slide well with soap and water, and then dry the slide.

II. An Alternative Technique

1. Spore staining without the use of steam may be performed as follows (watch **Microbiology Video: Preparing an Endospore Stain** to familiarize yourself with the technique):

Preparing an Endospore Stain

2. Prepare air-dried bacterial smears as usual but heat-fix the slides by passing them through the Bunsen flame about 20 times.
3. Cool the slides briefly in air.
4. Stain the slides by covering the smears with 7.5% malachite green, and allow the stain to remain for 10 minutes. This is a more concentrated stain than that used in the steaming method (if used with steam, the concentrated malachite green would precipitate rapidly and staining could not take place).
5. Wash the stain off the slide, and flood the smears with safranin for 1 minute as in step 7, above.
6. Continue with step 8 to complete the procedure.

Quick Procedure

Spore Stain

1. Stain with malachite green over boiling water for 3 min.
2. Rinse with water.
3. Stain with safranin for 1 min; rinse; dry; observe.

Name: ______________________ Date: __________ Section: __________

Bacterial Structures: Spore Stain Technique

EXERCISE RESULTS **16**

RESULTS

SPORE STAIN TECHNIQUE

Observations and Conclusions

____

____

____

____

____

Questions

1. Why is heat necessary for the successful performance of the spore stain technique using the Schaeffer-Fulton Method?

____

____

____

____

2. Explain how the extremely high resistance of bacterial endospores has an influence on sterilization practices, food microbiology, and disease processes.

____

____

____

Bacterial Structures: Capsule Stain Technique

EXERCISE 17

Materials and Equipment

- Selected encapsulated bacterial species
- Saline solution (0.85% NaCl)
- Nigrosin or India ink
- Crystal violet
- Blotting paper
- Brightfield microscope

Watch This Video

- How to Make a Capsule Stain Preparation

The **capsule** is a layer of polysaccharides and proteins secreted by certain bacteria, including many pathogens. The layer adheres to the cell surface and serves as a buffer between the cell and its external environment. The capsule protects the bacterium against dehydration and traps nutrients from the surrounding environment. It contributes to the establishment of disease by protecting the cell from immune attack by white blood cells. Various species of bacilli and cocci form capsules. When thin and flowing, the layer is called a **slime layer**. The term **glycocalyx** refers to both capsule and slime layer.

In this exercise, bacterial capsules will be visualized by special staining techniques.

Visualization of the bacterial capsule is a two-step procedure involving negative and simple staining. In the first step, the acidic stain stains the background area to outline the capsule. The cells then are stained with crystal violet in the second step. Water and heat should not be used in either step because capsules are easily destroyed by both. Also, it is helpful to use milk cultures of organisms because media containing milk encourage capsule production. Watch **Microbiology Video: How to Make a Capsule Stain Preparation** to familiarize yourself with the technique.

How to Make a Capsule Stain Preparation

CAPSULE STAIN TECHNIQUE

PURPOSE: to detect the presence or absence of a bacterial capsule.

PROCEDURE

1. Prepare a **negative stain** of a selected bacterial organism using nigrosin or India ink and the technique described in Exercise 13 (FIGURE 17.1). Allow the slide to air-dry thoroughly on the laboratory bench or warming tray.
2. Flood the slide with **crystal violet** for 1 minute.
3. Very carefully wash the excess stain from the slide using **saline solution** instead of water (FIGURE 17.2). Saline will help preserve the integrity of the capsule. Remember to be gentle because the slide has not been heat-fixed, and the bacteria may be lost with the stain if too much saline is applied. Blot the slide very gently.

Be very gentle with the slide and avoid excess rinsing!

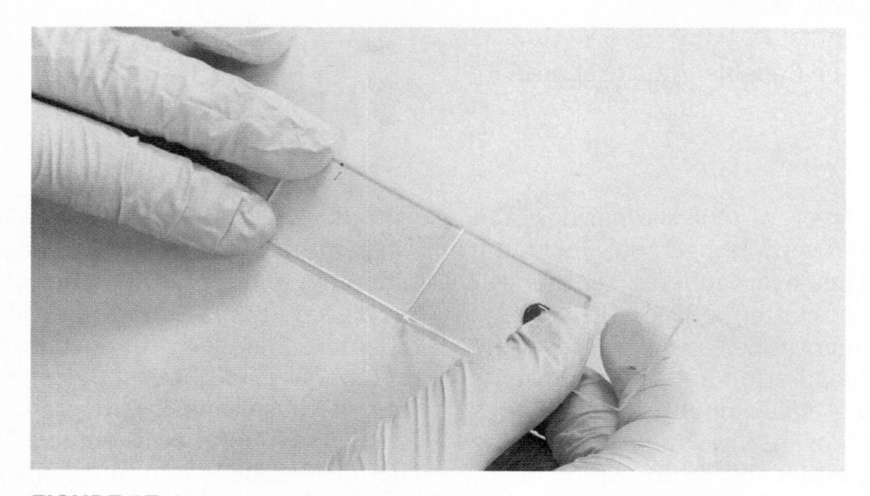

FIGURE 17.1 Prepare a negative stain of a selected bacterial organism.

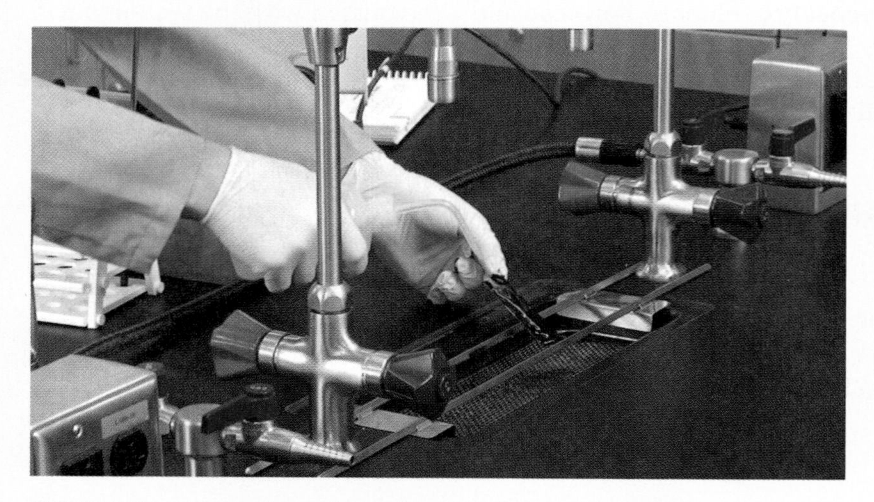

FIGURE 17.2 Use saline solution instead of water to wash the excess stain from the slide.

4. Using the microscope, observe the slide under low power, then high power, and finally oil immersion (**FIGURE 17.3**).

   Search for purple cells surrounded by capsules, which appear as white halos as shown in **FIGURE 17.4**. The background should be stained dark, and the cells and halos will appear as "motheaten" areas within the mat of stain. It may be necessary to repeat this technique several times before a successful capsule stain is observed.

5. Enter representations of encapsulated bacteria in the appropriate spaces in the Results section.
   - Be sure to label the capsules to distinguish them from the cells.
   - If the slide is to be retained, label the slide. Otherwise, place the slide in a beaker of disinfectant.

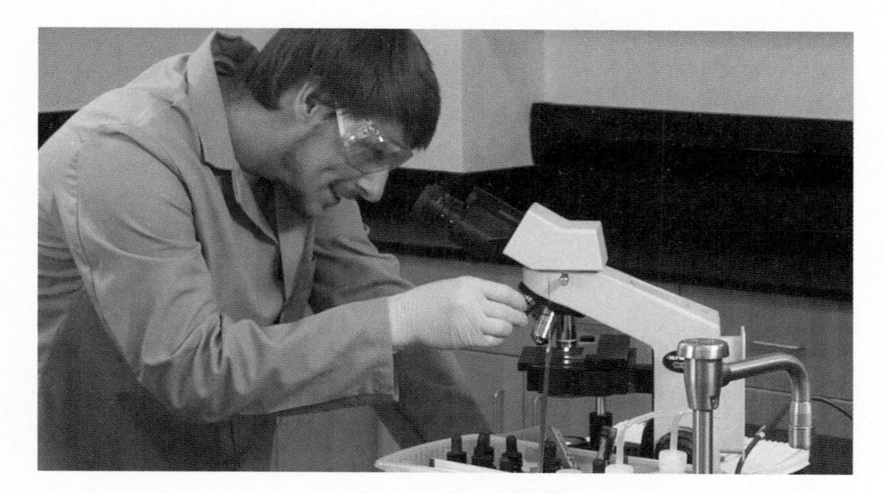

FIGURE 17.3 Observe the slide under the microscope.

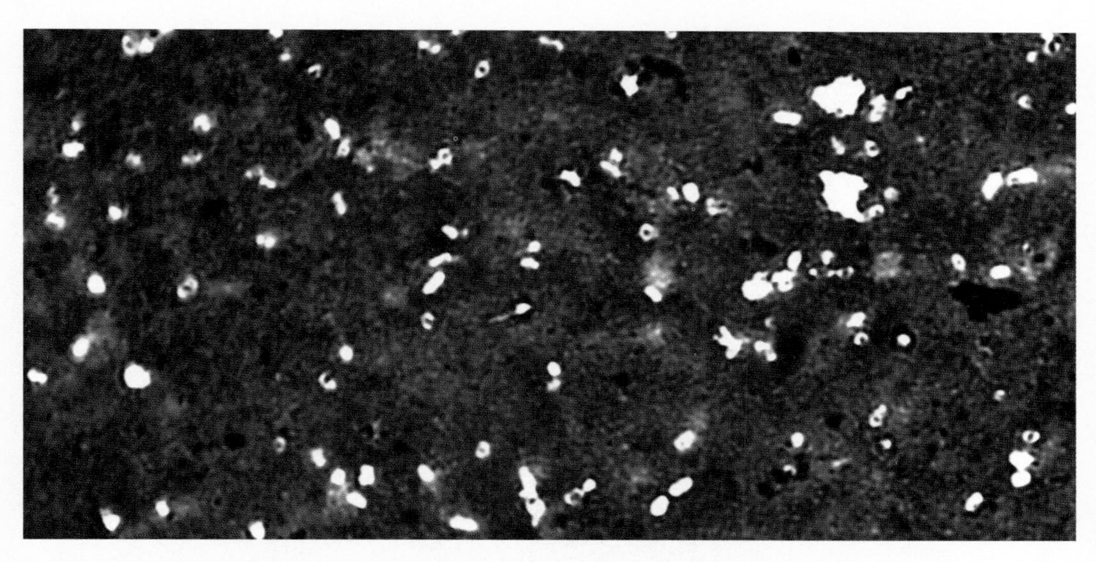

FIGURE 17.4 An image of several bacterial rods stained to show their capsules.

Quick Procedure

Capsule Stain

1. Prepare negative stain of bacterial smear with nigrosin as in Exercise 13.
2. Stain with crystal violet 1 min.
3. Rinse gently with saline solution; dry; observe

Name: ______________________ Date: __________ Section: __________

Bacterial Structures: Capsule Stain Technique

EXERCISE RESULTS **17**

RESULTS

CAPSULE STAIN TECHNIQUE

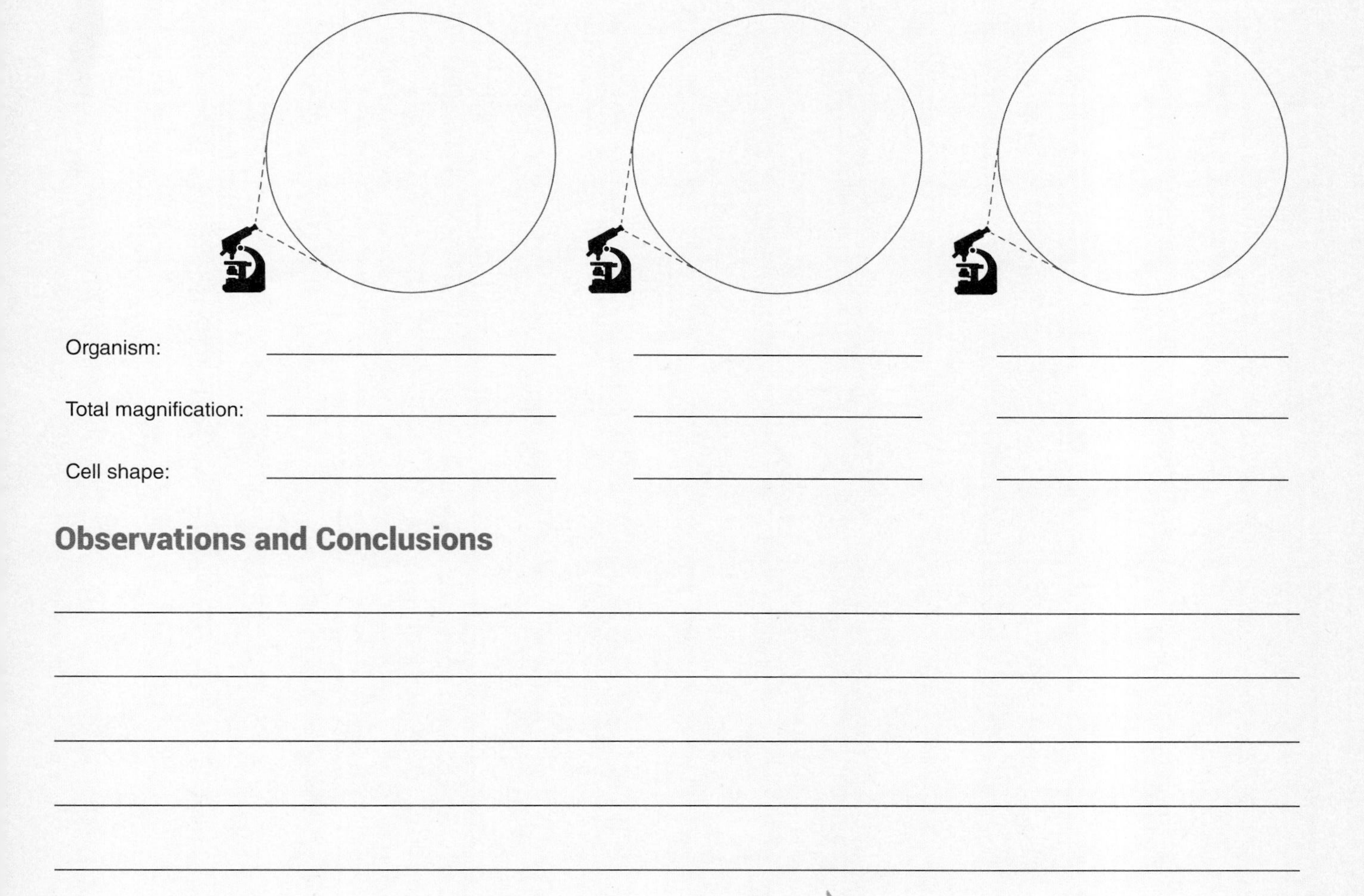

Observations and Conclusions

Questions

1. What mistakes might contribute to the inability to locate any cells on the slide at the conclusion of the capsule stain technique? How can these mistakes be corrected?

Viruses and Eukaryotic Microorganisms

PART V

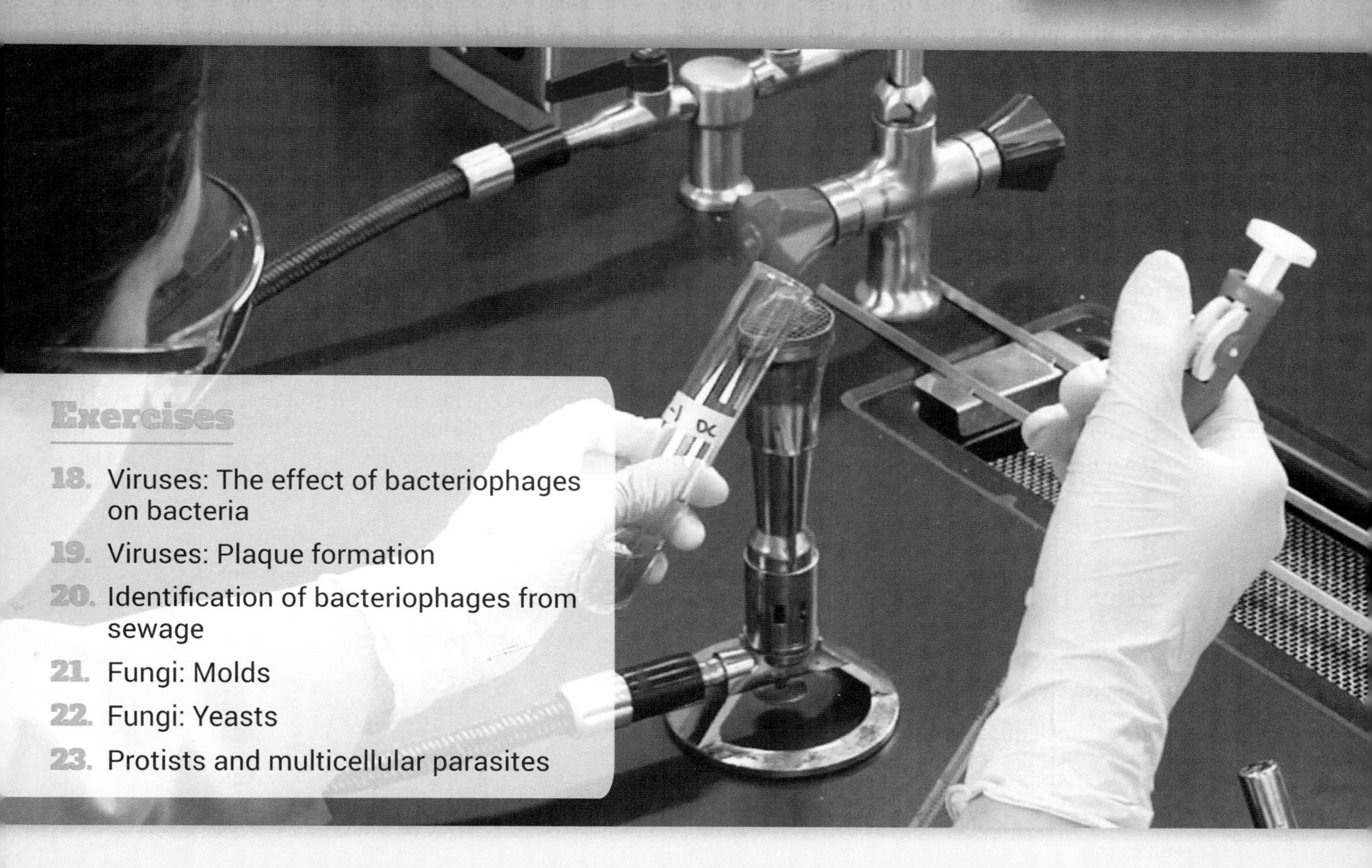

Exercises

18. Viruses: The effect of bacteriophages on bacteria
19. Viruses: Plaque formation
20. Identification of bacteriophages from sewage
21. Fungi: Molds
22. Fungi: Yeasts
23. Protists and multicellular parasites

Learning Objectives

When you have completed the exercises in Part V, you will be able to:

- Distinguish the structure of viruses from other microorganisms.
- Detect the presence of replicating bacteriophages through cell lysis and plaque formation, and enumerate phage numbers.
- Contrast mold and yeast growth and reproduction.
- Test for the presence of yeast metabolism through enzyme activity.
- Compare the characteristics of protists and multicellular parasites.

Certainly the "menagerie" of microorganisms is immense. Besides the unicellular bacteria, which are **prokaryotic organisms**, there are the **viruses**, which are acellular entities, many of which consist of nothing more than genetic information wrapped in a protein coat. The **eukaryotic organisms**, including the unicellular protists (protozoa and algae), and the multicellular fungi, plants, and animals are all similar in having a nuclear envelope that surrounds the DNA and in containing a variety of membrane-enclosed cellular structures that comprise most of the cellular organelles (TABLE V.1).

The viruses are truly different from any of the prokaryotic and eukaryotic microorganisms. Most have very definite geometric shapes. When **bacteriophages** (phages), which are viruses that infect bacteria, find suitable bacteria, the viruses enter the cells and replicate, eventually releasing a substantial number of progeny phage and killing the host cells (Exercise 18). In culture, waves of successive infections and cell lysis produce clear areas called **plaques** that are produced in the bacterial lawn (Exercise 19). Phages can be detected in materials such as sewage by such plaque formation (Exercise 20), which also can be used to estimate bacteriophage numbers.

The **fungi** include the molds and yeasts. These eukaryotic organisms have a unique growth form and produce many products that are beneficial in industrial and commercial industries. The study of the molds and yeasts (Exercises 21 and 22) provides a strong contrast to laboratory studies of the bacteria. Some of the unicellular **protists** and multicellular **flatworms** and **roundworms** (animals) are considered **parasites** because they are pathogenic or parasitic on other organisms, including humans. Most protists are not parasitic and survive on dead and decaying matter in ponds and very damp soil (Exercise 23). The **multicellular parasites** mentioned above are uncommon in developed countries, so they often are studied in the microbiology lab through microscopic observation of prepared slides or perhaps from feces from other animals (Exercise 23).

TABLE V.1 Basic Characteristics of Microorganisms

Characteristic	Viruses	Prokaryotes	Eukaryotes
Contain genetic information	Yes	Yes	Yes
Presence of a cell nucleus	No	No	Yes
Presence of membrane-enclosed organelles	No	No	Yes
Capable of independent reproduction or replication	No	Yes (most)	Yes

Viruses: The Effect of Bacteriophages on Bacteria

EXERCISE 18

Materials and Equipment

- Inoculating loop and/or inoculating needle
- Suspension of bacteriophage T4
- 24-hour broth culture of *Escherichia coli* strain B
- 24-hour broth culture of another strain of *Escherichia coli*
- Tubes of nutrient broth
- Tubes of phenol red lactose broth

Viruses are obligate intracellular parasites that replicate within the cytoplasm of a host organism. Those viruses capable of replicating within bacteria are called **bacteriophages**, or **phages**.

Viruses are ultramicroscopic entities, and an electron microscope is needed to see them clearly. Many viruses appear as 20-sided geometrical figures, while others are helical, and others have complex shapes. The shape is determined by the viral genes and is an important characteristic in viral classification.

When viruses replicate within their host cells, they rely upon the metabolic machinery of the cell for the necessary functions of replication. The most dramatic type of phage infection is the **lytic pathway**, in which the host bacterium disintegrates as a result of phage replication. Should the viruses replicate within and be released from bacteria growing on the surface of a nutrient agar plate, clear areas of bacterial destruction will result.

The interaction between a bacteriophage and its host bacterium can be an important tool in the identification of bacterial strains because bacteriophages are highly specific for their bacterial hosts.

THE EFFECT OF BACTERIOPHAGES ON BACTERIA

PURPOSE: to detect the presence of replicating phages in susceptible bacteria.

Virulent bacteriophages are those that destroy (lyse) their bacterial hosts at the conclusion of the replication process. The viral nucleic acid enters the bacterial cytoplasm and after about 15 minutes, the viral genes have encoded enzymes and other proteins for the construction of new viruses. Several hundred bacteriophages are produced per bacterium, and as they are released, the bacteria undergo lysis and die. The new bacteriophages infect other bacterial cells and infections continue until all the bacteria have been destroyed. In a broth culture, a cloudy suspension of bacteria would become relatively clear.

PROCEDURE

1. Obtain three tubes of nutrient broth, a tube of the phage suspension, and broth cultures of *E. coli* B and a second strain of *E. coli*.
   - Also obtain three tubes of **phenol red lactose broth.** This medium contains a pH indicator called phenol red and an inverted Durham tube. The indicator in the tube turns yellow when acid is produced, and a bubble is trapped in the inverted tube when carbon dioxide gas is produced.
2. Inoculate one of the nutrient broth tubes with a loopful of *E. coli* B and a second broth tube with the other strain of *E. coli.*
   - Leave the third tube as an uninoculated control. (At the instructor's suggestion, a different species of bacterium may also be included, using a fourth tube of nutrient broth.)
   - Inoculate two tubes of phenol red lactose broth in the same way and leave the third tube as a control.
   - Label each tube with your name, the date, the name of the medium, and the contents. (Another control tube of each medium inoculated with *E. coli* B but no phage may be included at the instructor's suggestion.)
3. Inoculate all six tubes with a loopful of a suspension of bacteriophage T4.
4. Incubate all the tubes at 37°C for 24 to 48 hours. If the tubes cannot be examined immediately, they may be refrigerated until the next laboratory session.
5. Examine the three nutrient broth tubes.
   - Look for evidence of growth in the tubes, indicated by cloudiness in the tube.
   - It may be necessary to vigorously tap the bottom of the tube to suspend any bacteria that have sedimented.
   - The control tube should be clear with no evidence of bacterial growth.
   - Note the tubes in which growth has or has not occurred, and draw your conclusions on the effect of the bacteriophage T4 on its host bacterium.
   - Note whether the phage had any effect on the other strain of *E. coli* tested, and place your conclusions in the Results section.
6. Examine the three tubes of phenol red lactose broth. Normally, *Escherichia coli* will grow in this medium and produce both acid and gas. As noted previously, acid is displayed by a yellowing of the medium, and gas is indicated by a gas bubble trapped in the inverted tube.
   - As you examine the tubes, note in the Results section whether *Escherichia coli* has multiplied and grown in the tube.
   - If no acid or gas is present and the tube remains the original color, then *E. coli* has not grown.
   - Alternately, if acid and gas have been produced, the *E. coli* have multiplied.
   - From your observations, conclude whether the bacteriophage destroyed the host organism.

Name: ______________________ Date: __________ Section: __________

Viruses: The Effect of Bacteriophages on Bacteria

EXERCISE RESULTS 18

RESULTS

THE EFFECT OF BACTERIOPHAGES ON BACTERIA

Contents:	______	______	______	______
Growth:	______	______	______	______

Contents: ______ ______ ______ ______

Growth: ______ ______ ______ ______

Observations and Conclusions

______

______

______

______

______

______

______

Questions

1. Since phages only infect and destroy bacterial cells, might it be possible to use viruses to treat and cure bacterial diseases in humans?

______

______

______

______

Viruses: Plaque Formation

EXERCISE 19

Materials and Equipment

- Suspension of bacteriophage T4
- 24-hour broth culture of *Escherichia coli* strain B
- 24-hour broth culture of another strain of *Escherichia coli*
- Plates of nutrient agar, or materials for their preparation
- Tubes of 9.9 ml sterile saline
- Mechanical pipetters and pipettes

PLAQUE FORMATION

PURPOSE: to detect and visualize areas of bacterial lysis caused by phage infection in a bacterial lawn.

When bacteriophages lyse their bacterial hosts on an agar surface, a clear circular area is seen amid the confluent lawn of bacterial growth. This area is called a **plaque**. It indicates that bacteriophages have destroyed their specific bacterial host.

PROCEDURE

1. Obtain two plates of nutrient agar, or pour two plates at the direction of the instructor.
2. After the agar has hardened, inoculate one plate with *E. coli* B, and one plate with another strain of *E. coli*.
3. To prepare an even **lawn of bacteria**, streaking should be performed with a sterile swab.
   - Dip a sterile swab into a broth culture of *E. coli*, and prepare a lawn of bacteria by swabbing the surface of the agar in three directions, as shown in FIGURE 19.1. Be sure to cover all parts. Then make a final swab around the perimeter.
   - A separate swab should be used for each plate, and precautions should be taken to avoid airborne contamination by lifting the lid only enough to permit access.
   - When the swabbing is complete, the swab should be deposited in the beaker of disinfectant as directed by the instructor.
4. After inoculation, the plates should be labeled with your name, the date, the medium, and the name of the bacteria used. They may be divided into sectors if the instructor recommends.

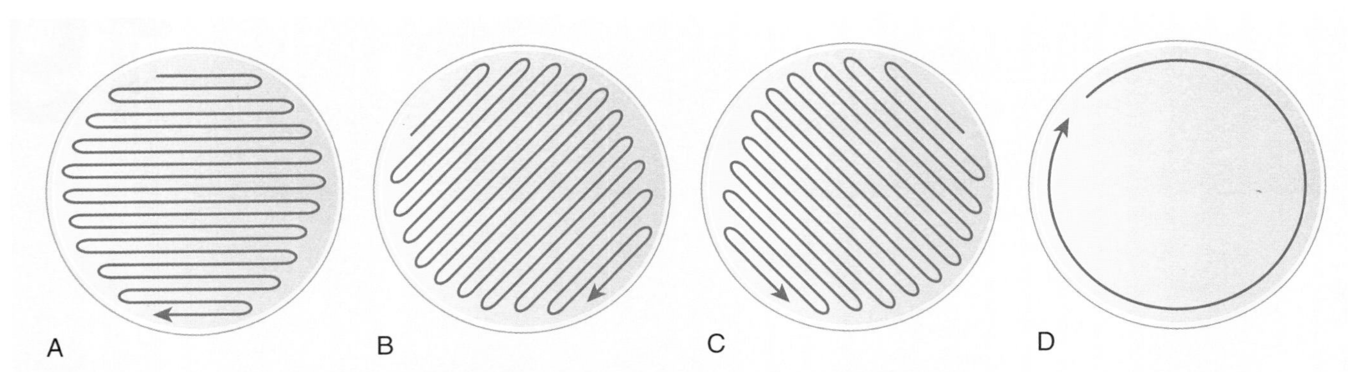

FIGURE 19.1 The "3+1" procedure for making a lawn of bacteria on a plate of medium.

5. Obtain a loopful of phage suspension from the phage culture and make one streak across the center of the inoculated plate, being careful not to dig into the surface of the agar.
   - Inoculate both plates in this manner.
   - At the instructor's direction, various dilutions of the phage suspensions may be made and samples of various dilutions may be inoculated to different sectors of the plate.
6. Incubate all plates at 37°C for 24 to 48 hours.
   - Observe the plates for clear lines in an otherwise uniform lawn of bacterial growth. These lines indicate regions where bacteriophages have replicated within the bacteria and destroyed them.
   - Note the plate in which a clearing has occurred and compare this plate with the other containing a nonspecific bacterial host.
   - Consider and explain the specificity of the bacteriophage with its host organism.
   - Results may be expressed in the appropriate diagrams in the Results section, and conclusions concerning virus–host specificity should be drawn.
7. It is valuable to make dilutions of the phage suspension to determine the number of phage particles in a sample.

Name: ______________________ Date: __________ Section: __________

Viruses: Plaque Formation

EXERCISE RESULTS 19

RESULTS

PLAQUE FORMATION

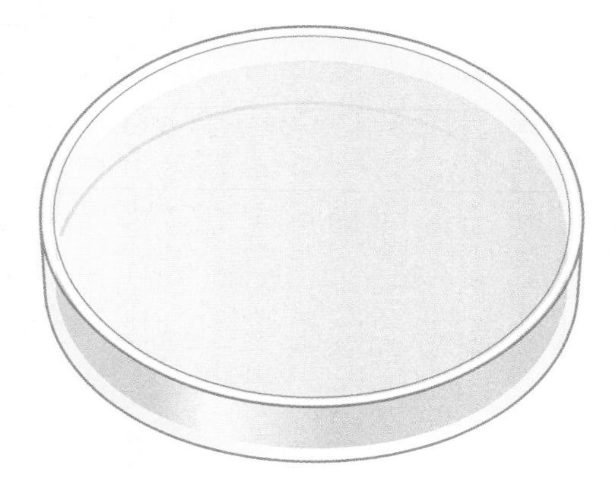

Bacterium: ______________________ ______________________

Observations and Conclusions

____

____

____

____

____

Questions

1. Which factors might account for the specificity of certain viruses for certain bacteria?

2. Suppose a nutrient broth tube was inoculated with nothing but a loopful of phage and it became cloudy after several hours of incubation. What might be a possible explanation?

Identification of Bacteriophages from Sewage

EXERCISE **20**

Materials and Equipment

- Sample of raw sewage from a local sewage treatment plant
- Inoculating loops
- Plates of nutrient agar
- Filtering apparatus with 0.45 μm membrane filter
- Nutrient agar plates, or materials for their preparation
- 24-hour broth culture of *Escherichia coli* strain B
- Disposal gloves

The interaction between a bacteriophage and its host bacterium can be an important tool in the identification of water quality. Polluted water often contains fecal bacteria and these potential pathogens can be identified by the presence of bacteriophages in the polluted water.

IDENTIFICATION OF BACTERIOPHAGES FROM SEWAGE

PURPOSE: to detect the presence of phage particles (coliphages) from raw sewage.

Because *Escherichia coli* is commonly found in sewage and polluted water, its bacteriophages (**coliphages**) also are present. Identifying ***E. coli* bacteriophages** (**coliphages**) present in sewage is a relatively easy exercise requiring that all bacteria be removed from the sewage and the liquid tested. Bacterial cell lysis (Exercise 18) and/or plaque formation (Exercise 19) are used for the identification.

A sewage treatment plant should be visited to obtain a small sample of raw sewage. Approximately 100 ml is needed.

PROCEDURE

1. The sewage fluid should be filtered through a membrane filter apparatus such as that described in Exercise 92. A membrane filter having a pore size of 0.45 μm is used to trap bacteria in the sewage fluid. The material emerging from the filter (the bacteria-free filtrate) should contain only bacteriophages.

Use disposal gloves to handle raw sewage carefully because the sample may contain human pathogens.

2. To detect the presence of coliphages, follow the procedures in Exercises 18 and 19.
    - For example, loopfuls of the phage filtrate may be inoculated to tubes containing *E. coli* strain B and incubated to determine whether *E. coli* cells were infected and destroyed (lysed) by coliphage.
    - Alternately, loopfuls of the filtrate can be streaked onto plates containing a lawn of *E. coli* strain B. The filtrate also can be diluted, and samples used for streaking. After incubation, the plates can be examined to determine if *E. coli* cells were infected and destroyed (lysed) by coliphage
3. Record your results in the appropriate portion of the Results section and indicate whether the identification of bacteriophages was successful.

Name: ______________________ Date: __________ Section: __________

Identification of Bacteriophages from Sewage

EXERCISE RESULTS 20

IDENTIFICATION OF COLIPHAGES FROM SEWAGE

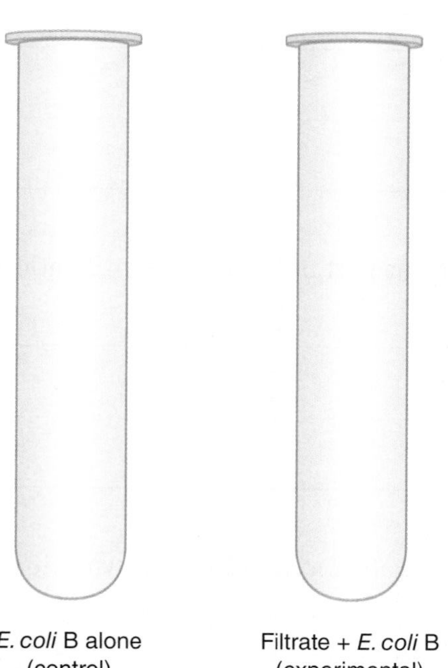

E. coli B alone (control)

Filtrate + *E. coli* B (experimental)

Plaques from sewage phages

Observations and Conclusions

____

____

____

____

____

Questions

1. Besides raw sewage, what other environments or sources might provide possible alternatives for obtaining coliphages specific for *E. coli*? Hint: Where are *E. coli* cells normally found in the human body?

___

___

___

2. How could coliphages be used to determine if polluted recreational waters, such as a lake or stream, is a threat to wildlife and human health?

___

___

___

3. Why must a control sample (*E. coli* B alone) be included as part of the tube identification?

___

___

___

4. On an agar plate containing plaques, you observe a small *E. coli* colony growing in the middle of a plaque. Explain why this colony is present.

___

___

___

Fungi: Molds

EXERCISE

21

Materials and Equipment

- Selected mold cultures and/or prepared slides
- Plates of Sabouraud dextrose agar or potato dextrose agar
- Deep tubes of Sabouraud dextrose agar or potato dextrose agar
- Sterile dropper pipettes, straws, applicator sticks
- Cotton balls, glass rods or pipe cleaners, petroleum jelly
- Hydrogen peroxide
- Inoculating needle
- Petri dishes
- Microscope
- Disinfectant

The microbial world includes a broad variety of eukaryotic organisms having considerably more detail and structural complexity than bacteria. Fungi are among these organisms.

FUNGI

PURPOSE: to identify the structures and features that are characteristic of molds.

Fungi are a diverse group of over 100,000 described species of molds and yeasts. **Molds** grow as long, tangled filaments of cells, each filament known as a **hypha** (pl. **hyphae**). The mass of filaments called a **mycelium** (pl. **mycelia**) soon becomes visible to the unaided eye as it spreads across the surface of its growth medium. Spores eventually form on specialized hyphae that form the reproductive structures called **fruiting bodies**, and the mold takes on the color of the spore pigments (FIGURE 21.1).

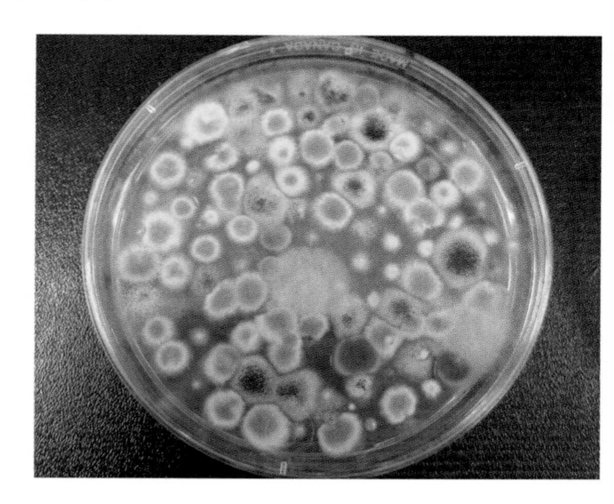

FIGURE 21.1 This image shows an example of several mold species growing as colonies on a nutrient agar plate. The colors are due to the pigments produced by the spores.

PROCEDURE

The Study of Molds

1. Mold Cultivation: Sabouraud dextrose agar (or potato dextrose agar) contains the acidity and added carbohydrate favored by molds. It is a preferred medium for cultivating fungi.
   - Prepare a plate of the medium according to the instructor's directions.
   - Label the plate with your name, the date, the name of the medium, and the designation "mold isolation."
2. Mold Isolation: There are a number of methods available to discover the variety of molds in the environment. Here are a few suggestions:
   - Remove the lid from a plate of Sabouraud dextrose agar (or potato dextrose agar) and allow the agar surface to be exposed to the air in the lab for 15 to 30 minutes. Alternatively, leave the plate open to the air outdoors for 15 to 30 minutes.
   - Use a sterile cotton swab to collect dust or other material from the surface of an object. Gently streak the swab over the agar surface.
   - Allow the exhaust from a vacuum cleaner to spray on the agar surface. (This can be particularly interesting for revealing how fungal spores are broadcast.)
   - Rub a piece of blue cheese on the agar surface to deposit fungal spores for cultivation.
3. The Agar Plug Technique: Another way to inoculate a plate with a sample of fungus is to use the agar plug technique as illustrated in FIGURE 21.2. To perform this technique, begin with a plate of medium that supports the growth of molds.
   - Use a sterile plastic straw to remove a plug of agar medium from the plate (FIGURE 21.2A and B). The plug of agar can be pushed out of the straw with a long sterile applicator stick placed inside the straw.

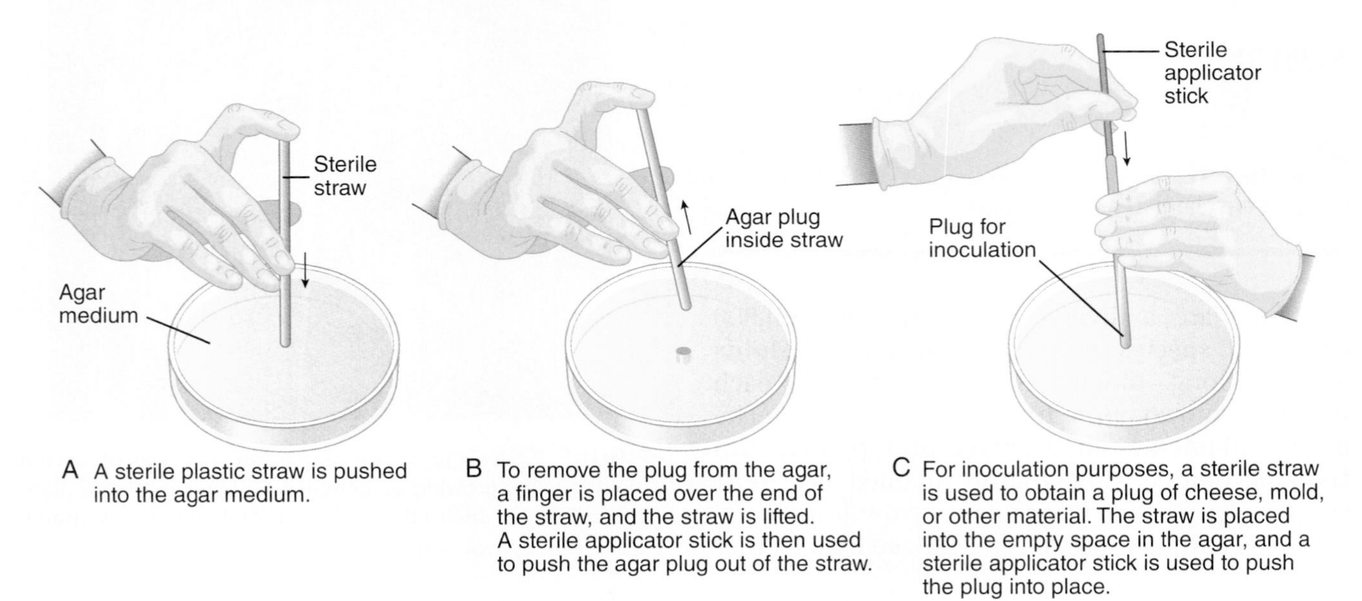

A A sterile plastic straw is pushed into the agar medium.

B To remove the plug from the agar, a finger is placed over the end of the straw, and the straw is lifted. A sterile applicator stick is then used to push the agar plug out of the straw.

C For inoculation purposes, a sterile straw is used to obtain a plug of cheese, mold, or other material. The straw is placed into the empty space in the agar, and a sterile applicator stick is used to push the plug into place.

FIGURE 21.2 The agar plug technique.

- Now use a sterile straw to obtain a plug of fungus or a plug of blue cheese and insert the plug into the empty space, again using a sterile applicator stick inside the straw (FIGURE 21.2C). The plate should be incubated at room temperature for about one week.

4. Viewing the Mycelium: Different molds may be examined by viewing the mycelium on the plate directly with the low power (10×) objective of the microscope. The underside of the plate should be used.
   - The edge of the colony may be observed as long as the medium is not too thick.
   - Representations of the molds in culture should be presented in the appropriate space in the Results section.
   - Wet mount preparations may be made from the colonies as described next.

Do not open the plates because spores are easily dispersed in the air and could contaminate other growth media in the lab.

5. Wet Mounts: Microscopic observations of molds may be made by preparing **wet mounts**. To prepare a wet mount, place a large drop of water on a clean slide.
   - Using an inoculating needle, pick off a small amount of mycelium from the mold culture and deposit it in the drop of water.
   - Next, tease apart the hyphae of the mycelium, and cover the preparation with a coverslip. Two or three preparations may be made next to one another on a slide, but they should not be allowed to dry out.
6. Observations: Observe the molds under the low (10×) and high power (40×) objectives, and choose the lens that gives the most detail and clarity. Note the size, shape, and unique characteristics of the hyphae. Locate cellular features such as nuclei, cross walls, granules, and vacuoles. Watch for various types of spores, and note the complexity or simplicity of the hyphae.
   - Place labeled drawings of each mold in the appropriate space in the Results section.
   - Use oil immersion only at the instructor's direction.
   - When your work is complete, the wet mounts should be wrapped in paper toweling and saturated with disinfectant before being discarded in the designated container.
   - Prepared slides also may be available for observations of molds.
7. Mold Slide Culture: A slide culture is a preparation that permits the observer to view a mold microscopically while it is growing on a glass slide. The preparation is set up as shown in FIGURE 21.3 and as follows:
   - Prepare ridges on a slide with petroleum jelly, preferably from a syringe (FIGURE 21.3A).
   - Set a clean coverslip atop the ridges (FIGURE 21.3B).
   - Obtain a tube of melted Sabouraud dextrose agar (or potato dextrose agar) and inoculate it with a piece of mold from a culture. Mix the contents well.
   - Using a sterile dropper pipette, obtain a small amount of the liquid inoculated agar and place it into the area under the coverslip (FIGURE 21.3C). The space should be half-filled.
   - Place the slide culture on a bent glass rod or pipe cleaner within a Petri dish, and add a wet cotton ball or paper towel to provide moisture (FIGURE 21.3D).
   - Set the dish aside to incubate for several days.

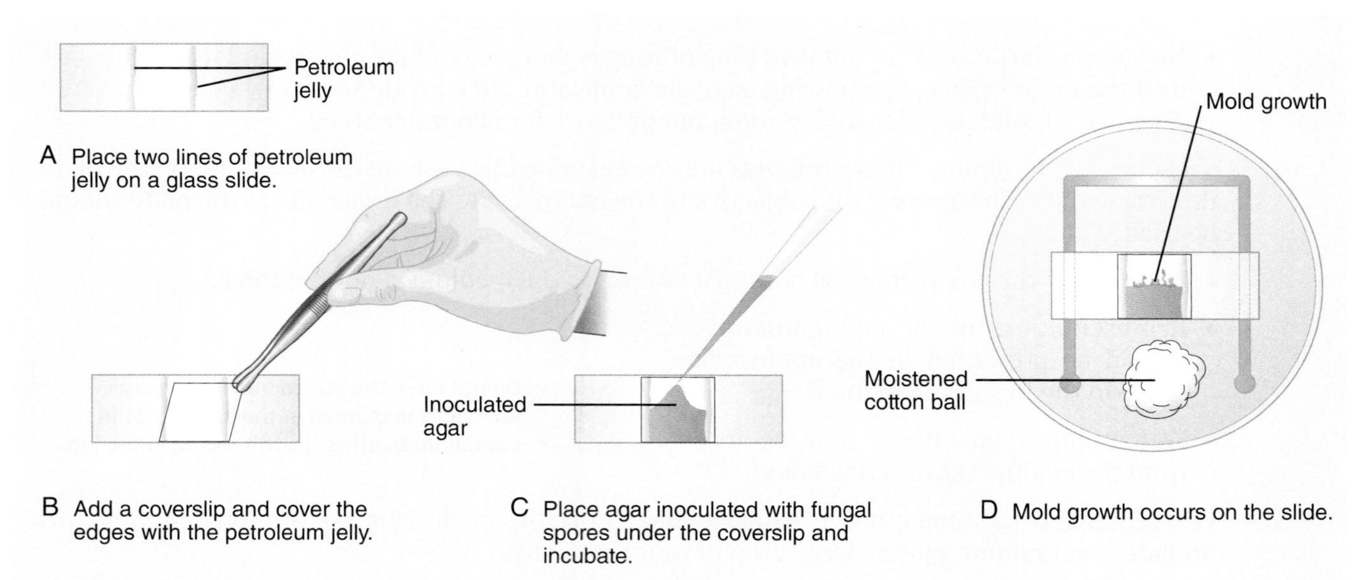

FIGURE 21.3 Preparation of a slide culture for studying molds.

- When growth is deemed optimal, microscopic observations of the mold may be made by placing the slide under the microscope. You will note the complete features of the mold, and you will be able to study all its parts without disturbing it.

Name: ______________________ Date: __________ Section: __________

Fungi: Molds

EXERCISE RESULTS 21

RESULTS

FUNGI

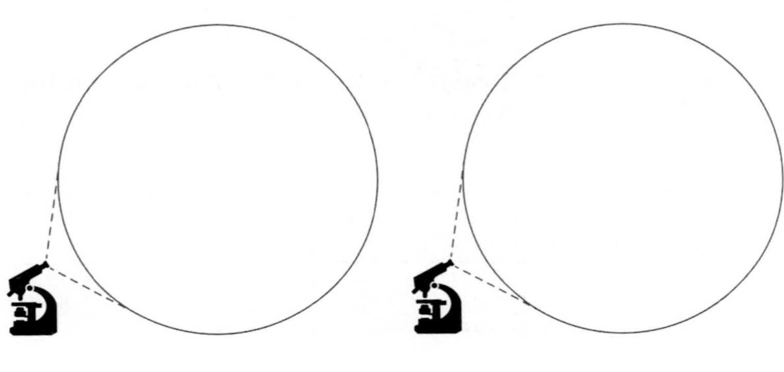

Organism: ______________ ______________

Magnif: ______________ ______________

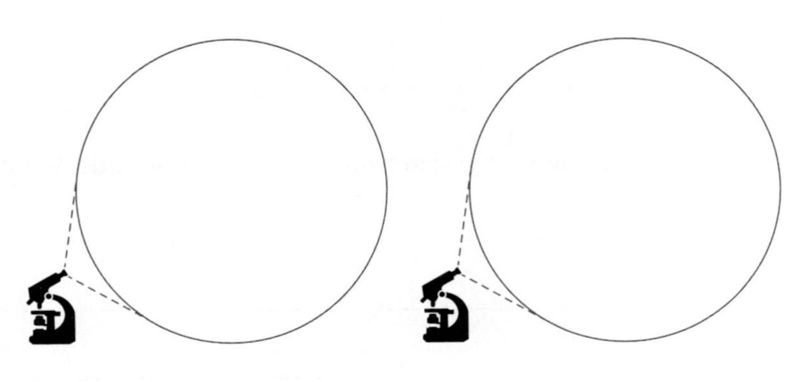

Organism: ______________ ______________

Magnif: ______________ ______________

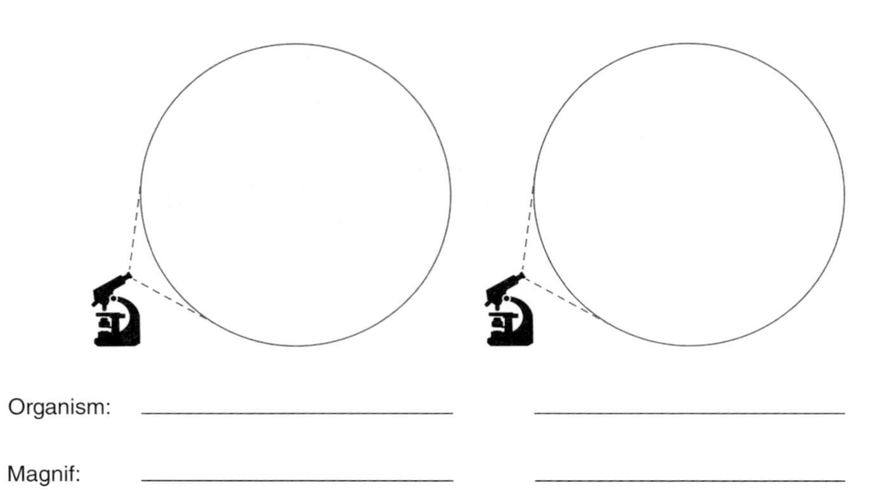

Questions

1. Why is Sabouraud dextrose agar of greater value than nutrient agar for the isolation of molds from the environment?

2. What conditions might prevent a successful slide culture of a mold?

3. Explain the advantages of a wet mount preparation over a prepared slide. What disadvantages are there?

4. It is noted that certain molds come close but do not overgrow one another on Sabouraud dextrose agar. How might this be explained?

EXERCISE 22

Fungi: Yeasts

Materials and Equipment

- Yeast cultures (*Saccharomyces cerevisiae*)
- Plates of Sabouraud dextrose agar or potato dextrose agar
- Deep tubes of Sabouraud dextrose agar or potato dextrose agar
- Sterile disposal plastic pipettes, straws, applicator sticks, swabs
- Slides and coverslips
- Cotton balls, glass rods or pipe cleaners, petroleum jelly
- Methylene blue, malachite green, safranin (optional)
- Hydrogen peroxide
- 5% glucose solution
- 5% lactose solution
- 10-ml and 25-ml test tubes
- Pencil
- Petri dishes
- Microscope

The microbial world includes a broad variety of eukaryotic organisms having considerably more detail and structural complexity than bacteria. Fungi (yeasts) are among these organisms. Yeasts are present throughout the environment and are found on plants, humans, and other animals. Many of the organisms are harmless **saprobes**, but several species are human **pathogens**.

FUNGI: YEASTS

PURPOSE: to identify the structures and features that are characteristic of the yeasts.

Yeasts are unicellular microorganisms, colonies of which on agar resemble bacterial colonies. Yeast cells are eukaryotic with a distinct nucleus, nuclear membrane, and other organelles. The colonies usually appear cream-colored or another light color. They multiply asexually via budding, a process in which a new cell forms at the periphery of the parent cell, then grows and breaks free to assume an independent existence. The yeasts used in this exercise—*Saccharomyces cerevisiae*—are a rare cause of human infection, such as thrush and vulvovaginitis.

PROCEDURE

The Study of Yeasts

1. Microscope Observations: To study the morphological characteristics of yeasts, place a drop of *Saccharomyces cerevisiae* or commercial yeast suspension on a slide.
   - Add a coverslip, and locate the preparation under the low- (10×) and high-power (40×) objectives. The yeasts will be seen floating about. They are clear, oval organisms resembling air bubbles, but with internal detail.
   - Observe the nucleus, vacuoles, and other cytoplasmic details of the cells.
   - Watch for "buds" at the surface of the cells. These are young cells formed by mitosis and destined to break free of the parent cell.
   - Place drawings of the yeasts in the appropriate space in the Results section.
2. Staining Yeasts: Yeasts may be stained in various ways. For example, you may add a drop of methylene blue to a drop of yeast cells to stain the cells and increase contrast. In addition, an air-dried, heat-fixed preparation of yeasts may be simple-stained, as outlined in Exercise 13. This will provide a permanent slide.
3. Yeast Isolation: To isolate and cultivate yeasts, use a plate of Sabouraud dextrose agar (or potato dextrose agar). Here are two sources for possible yeasts:
   - Obtain yeasts from the surface of the tongue by swabbing the tongue with a sterile swab and applying the swab to the plate.
   - The hazy surface of a piece of fresh fruit such as an apple (or a sample of unpreserved apple cider) is another possible source of yeasts; still another is a yeast cake. Use sterile swabs to collect yeasts from these surfaces.
   - After several days, the yeast will form white or cream-colored colonies, and the plate will have a "yeasty" odor similar to that of beer.
   - Slides may be made from the yeast colonies to verify the presence of yeasts.
4. Catalase Production: Yeasts are known producers of catalase. Catalase is an enzyme that breaks down hydrogen peroxide to form water and oxygen:

   $$2H_2O_2 \xrightarrow{catalase} 2H_2O + O_2$$

   Yeast catalase may be demonstrated as follows:
   - Obtain a dropper pipette.
   - Withdraw a small volume of hydrogen peroxide into the pipette.
   - Expel one drop of hydrogen peroxide onto one or a few yeast colonies growing on agar. Almost immediately, the hydrogen peroxide will start to break down and oxygen bubbles will rise in the tube.
5. Yeast Enzymes: Yeasts contain many of the enzymes used in the metabolism of glucose and other carbohydrates. The presence of these enzymes is demonstrated by the following procedure:
   - Obtain a small 10-ml test tube and place in it 9 ml of 5% glucose solution and 1 ml of yeast suspension. This should fill the tube to the brim (add water if necessary).
   - Obtain a larger 25-ml test tube and invert it over the small tube.
   - Using a pencil, push the small tube high up into the larger tube until the surface of the small tube touches the bottom of the large tube.

- Quickly invert the large tube. At this point the large tube will be upright and the small tube will be upside down.
- Set the tube aside in a warm environment, and observe the contents of the small tube at regular intervals. Refer to Step 6 to complete this exercise.
- As the minutes pass, you will see the bubbles in the small tube gradually increase in size. Through the chemical process of respiration, yeast enzymes are breaking down the glucose and releasing gas, mainly carbon dioxide gas. The process is identical to that happening in the early stages of fermentation where yeasts metabolize the glucose in grape juice. The exercise may be modified to study the effects of yeasts on different carbohydrates (for example, sucrose, lactose, maltose) or on different concentrations of the same carbohydrate, (for example, 1%, 5%, 10%). It also is possible to study the effects of an inhibitor substance on the chemical process. The effects of temperature, pH or other factors may also be measured by altering the procedure to suit your needs.

6. Lactose Metabolism: It is interesting to note that many humans lack the enzyme lactase and are unable to digest lactose. These individuals suffer from a condition called **lactose intolerance**. Fortunately, they can add a commercially prepared lactase product such as Lactaid® or Dairy Ease® to dairy foods and digest the lactose this way.

   The procedure explained in Steps 5 and 6 will be used in this exercise.

   - Obtain a small 10-ml test tube and fill it with 9 ml of 5% lactose solution and 1 ml of yeast suspension. This should fill the tube up to the brim (add water if necessary).
   - Prepare a second tube of lactose and yeast, but this time add two drops of commercial Lactaid® or Dairy Ease®.
   - Invert two larger 25-ml test tubes over the two smaller tubes and push the small tubes high up into the larger tubes until they touch the bottoms of the large tubes.
   - Quickly invert the large tubes so the small tubes are upside down.
   - Set the tubes aside in a warm environment, and observe the contents at regular intervals.
   - As the minutes pass, you will note the development of gas bubbles in one tube but not the other. As described on the previous page, gas bubbles mean that carbohydrate metabolism is taking place. In this instance, the lactose is being broken down and the glucose resulting from its breakdown is being metabolized. In which tube is this metabolism taking place? Why?
   - Enter your observations in the Results section and suggest ways of modifying the exercise to discover other properties of lactase activity.

Name: ______________________ Date: __________ Section: __________

Fungi: Yeasts

EXERCISE RESULTS **22**

RESULTS

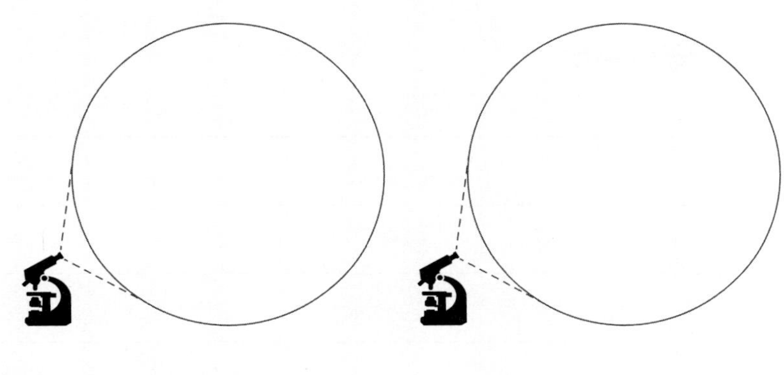

Organism: ______________ ______________

Magnif: ______________ ______________

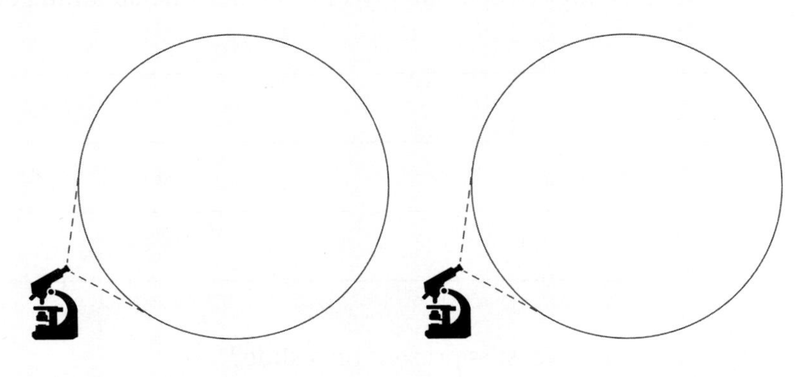

Organism: ______________ ______________

Magnif: ______________ ______________

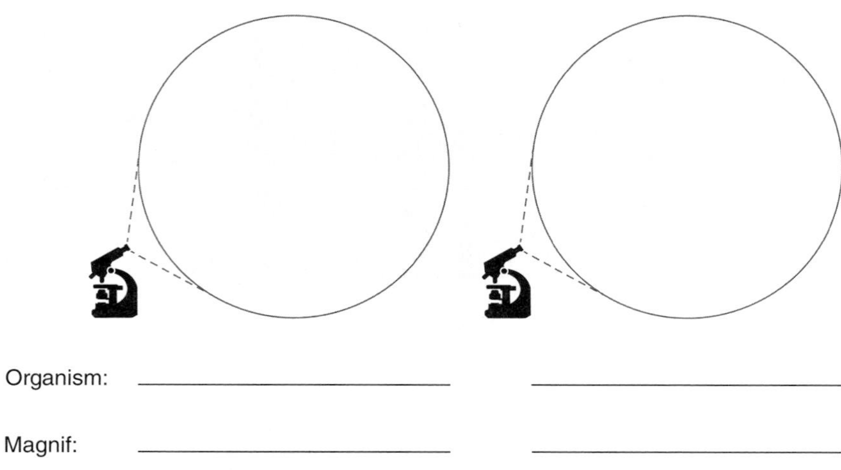

Organism: ______________ ______________

Magnif: ______________ ______________

Observations and Conclusions

Questions

1. Describe the colonial morphology of *Saccharomyces cerevisiae* on Sabouraud dextrose agar.

2. How can you easily determine if yeast is present on a slide?

3. Can yeasts metabolize lactose?

Protists and Multicellular Parasites

EXERCISE 23

Materials and Equipment

- Samples of pond and aquarium water
- Fresh hay infusions
- 10% methylcellulose
- Microscope
- Prepared slides of protists and multicellular parasites
- Dilute acid, India ink, yeasts
- Thread
- Salt and sugar solutions

The microbial world includes a broad variety of eukaryotic organisms having considerably more detail and structural complexity than bacteria. **Protists** and **multicellular parasites** are among these organisms. Many of the organisms are harmless **saprobes**, but several species are human **pathogens**. This exercise will give you an opportunity to work with the eukaryotic microorganisms and increase your familiarity with them.

Multicellular parasites are the **flatworms** and **roundworms** that commonly inhabit the natural environment and infect millions of people on a global basis. In the strict sense, they are animals, but they are studied in microbiology because of their small size and ability to cause infectious disease. The study of these pathogens and multicellular parasites constitutes the discipline of **parasitology**.

PROTISTS AND MULTICELLULAR PARASITES

PURPOSE: to become familiar with some features characteristic of the protists and multicellular parasites.

The protists include over 200,000 species. The organisms are unicellular and microscopic, and most are free-living species, although a few species (informally called the protozoa) cause serious human diseases such as sleeping sickness, trichomoniasis, and malaria (FIGURE 23.1).

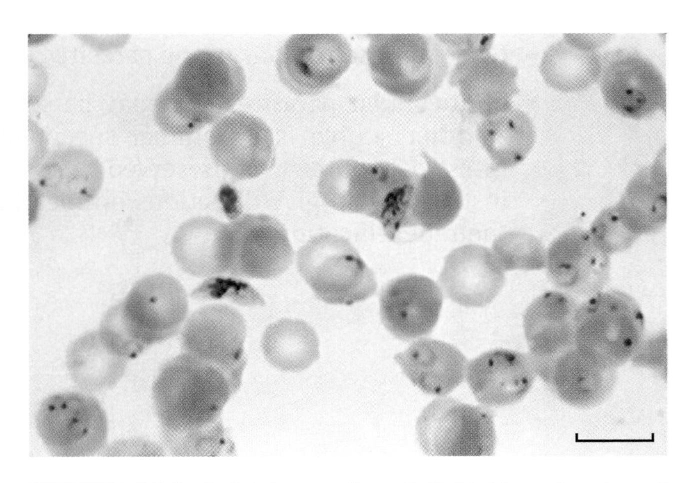

FIGURE 23.1 *P. falciparum* (small light blue structures) is one of the five *Plasmodium* species that cause malaria. (Bar = 10 µm.).

Courtesy of CDC/Dr. Greene; Steven Glenn, Laboratory & Consultation Division.

PROCEDURE

1. Microscope Observations: Protists may be observed from cultures, samples of pond or aquarium water, or a fresh hay infusion. Plant material from a pond or aquarium is another good source for these eukaryotic microbes. When using plant material, rub a piece of the plant on the slide to dislodge the protists.
   - A drop of 10% methylcellulose may be added to provide a viscous environment and slow the movement of protistan cells.
   - Alternately, a thin fragment of lens paper will provide threads to trap the protists for observation.
   - Observe the organisms for size, shape, cellular features, locomotor organelles (e.g., flagella, cilia, pseudopods), and other distinctive characteristics.
   - Drawings should be prepared in the appropriate spaces, with labels pointing to the important characteristics. Prepared slides may be used for this exercise.
2. Estimating Size: To estimate the size of protistan cells, follow the procedure outlined in Exercise 10.
3. Protist Behavior: Various types of behavior may be observed in the protists.
   - A drop of dilute acid may be placed at the edge of the slide to see how the protists respond. India ink may be used in a second demonstration.
   - It also is possible to mix a sample of yeasts with the protists to observe the feeding mechanism.
   - Another way to study protist behavior is to dip a piece of thread in a salt or sugar solution and place the thread in the drop of protists before adding the coverslip. Threads dipped in sugar solution and ammonia solution should elicit different behaviors.
4. Multicellular Parasites: Prepared slides will be available to study the multicellular parasites. Using whichever magnification is best for the specimen, observe their shape, size, and distinctive features. Anatomical structures should be visible in the parasite. Note the salient features used in its identification.
   - Place representations of the parasites in the appropriate space of the Results section.
   - Multicellular parasites also may be sought in fresh feces from a dog, cat, or other animal. Another source is an animal specimen used in anatomy laboratories. For example, the intestinal contents of preserved rats and cats may contain parasites suitable for study. The instructor will give directions on locating the parasites. Parasitology manuals will assist in their identification.

Name: ______________________ Date: __________ Section: __________

Protists and Multicellular Parasites

EXERCISE RESULTS 23

RESULTS

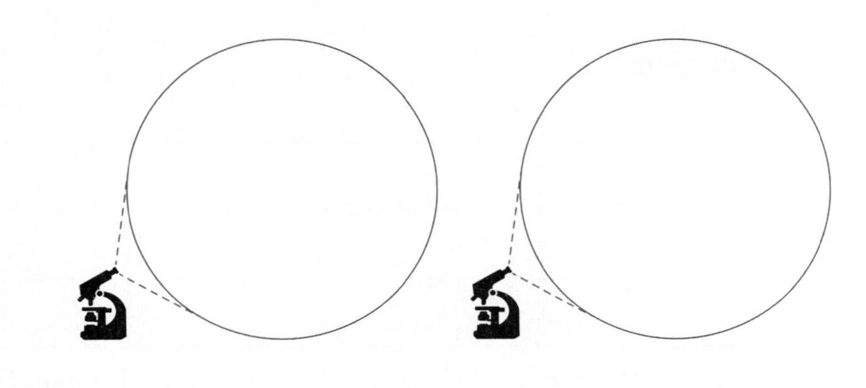

Organism: ______________ ______________

Magnif: ______________ ______________

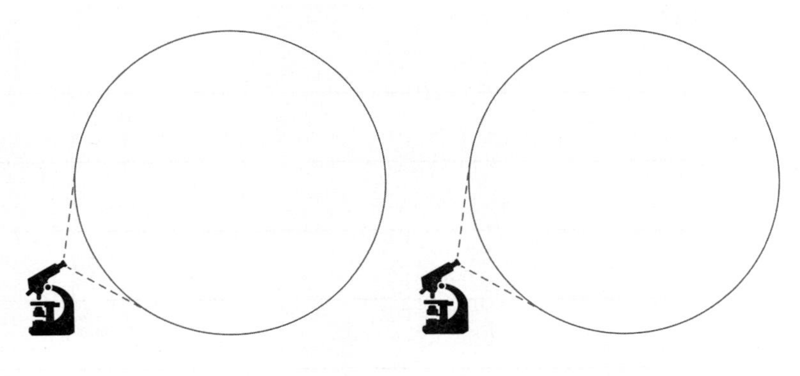

Organism: ______________ ______________

Magnif: ______________ ______________

Organism: ______________ ______________

Magnif: ______________ ______________

Observations and Conclusions

Questions

1. How do the fungi, protists, and multicellular parasites compare with the bacteria in size and morphological features?

2. Why are different behaviors elicited from thread dipped in sugar solution and thread dipped in ammonia solution?

Control of Microorganisms

PART VI

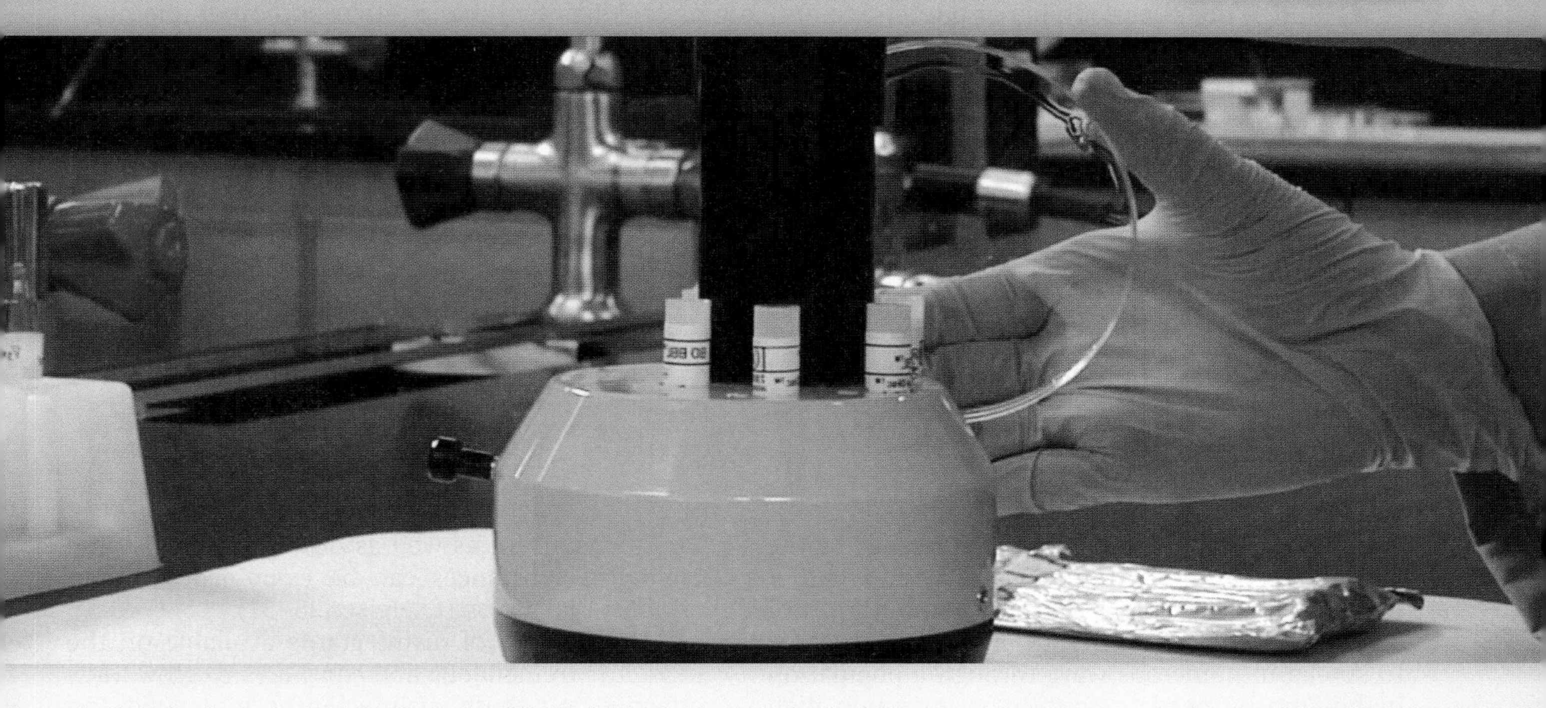

Exercises

24. The effect of physical agents on bacteria: Heat tolerance
25. The effect of physical agents on bacteria: Ultraviolet light
26. The effect of physical agents on bacteria: Drying
27. The effect of physical agents on bacteria: Other physical agents
    - Incineration
    - Osmotic pressure
    - Cold temperature
    - Autoclaving
28. The effect of chemical agents on bacteria: Antiseptics and disinfectants
29. The effect of chemical agents on bacteria: Metals
30. The effect of chemical agents on bacteria: Other chemical substances
31. The effect of chemical agents on bacteria: Lysozyme
32. Evaluation of disinfectants and antiseptics: Effectiveness of disinfectants on inanimate objects
33. Evaluation of disinfectants and antiseptics: Effectiveness of disinfectants and hand washing on the skin surface
34. Evaluation of disinfectants and antiseptics: Mouthwashes and oral bacteria
35. Evaluation of disinfectants and antiseptics: The effect of antibiotics on bacteria

Learning Objectives

When you have completed the exercises in Part VI, you will be able to:

- Assess the effects of heat, UV light, drying, and other physical agents on bacteria.
- Judge the susceptibility of bacteria to antiseptics and disinfectants, metals, and other chemical substances, including lysozyme.
- Evaluate the effectiveness of disinfectants on contaminated inanimate objects, on the skin surface, and as mouthwashes.
- Complete a standardized disk susceptibility test (Kirby-Bauer method) to evaluate the sensitivity of bacteria to antibiotics.

Watch These Videos

- How to Prepare a Disk Diffusion Assay
- How to Prepare the Kirby-Bauer Test

We often hear about and become concerned by the latest outbreak of an infectious disease, whether it be the Ebola virus, influenza, or an outbreak of salmonellosis. The ability of infectious diseases to spread in numerous ways through a population highlights the need to continually maintain high levels of sanitation and strengthen public health measures to ensure potentially infectious agents are controlled or destroyed.

Looking at bacterial diseases, physical agents are one way to keep infectious agents under control. Physical agents include, among others, moist heat, ultraviolet radiation, and drying. Experiments can be designed to effectively show how such physical agents limit microbial growth and the opportunity to spread into a community or population (Exercises 24–27).

In many cases, the more common and familiar agents of control are chemical compounds. Supermarkets are full of products, many good and some unnecessary, to keep microorganisms under control. Chemical agents, such as antiseptics and disinfectants, are key to any public health initiative or good sanitation practice when it comes to a potential for spread of an infectious disease. These chemicals, as well as other commercial and biological substances can be examined for their antibacterial action (Exercises 28–31).

The number of disinfectants available on the market is tremendous and continues to grow. How effective are these disinfectants? It depends on whether they are designed for inanimate objects, as handwashing agents, or as mouthwashes (Exercises 32–34). The relative effectiveness of disinfectants can be determined by selecting one of several evaluation procedures.

Historically, besides improving the standards of public health, the most influential event that has led to control or elimination of infections is the development of antibiotics. By definition, **antibiotics** are chemical inhibitors produced by microorganisms. Many other therapeutic agents have been developed and synthesized by chemists for use in the body and are referred to as **chemotherapeutic agents**. Here, we will lump them all under the umbrella of antibiotics. The effectiveness of such agents on inhibiting bacterial growth can be evaluated using the standardized disk susceptibility test, often referred to as the Kirby-Bauer method (Exercise 35).

The Effect of Physical Agents on Bacteria: Heat Tolerance

EXERCISE 24

Materials and Equipment

- Broth cultures of four selected bacteria
- Nutrient agar plates
- Wax pencil
- Inoculation loop
- Thermometers and tubes of water
- Water bath

Physical agents have broad application for controlling and destroying microorganisms. **Pressurized steam**, for example, is used in the autoclave to sterilize bacteriological media and an assortment of medical supplies and equipment.

In this exercise, the effects of heat will be tested on various species of bacteria. Organisms will be exposed to a physical agent of destruction, and their survival will be determined by their ability to grow in bacteriological media after the exposure has been completed.

HEAT TOLERANCE

PURPOSE: to determine the heat tolerance of different species of bacteria.

Heated water is a moist heat method commonly used for destroying microorganisms. The principal effect on the microorganisms is **denaturation**, a process that involves altering a protein's structure. Many enzymes and structural proteins of the cell are inactivated by this method. Death follows rapidly.

Bacteria vary in their susceptibility to heat, with bacterial spores displaying unusually high resistance. In this section, various bacteria in nutrient broth will be exposed to water at different temperatures. Survival will be determined by streaking the heat-treated bacteria on nutrient agar plates and determining if growth has occurred after 48 hours.

PROCEDURE

1. Set up a 40°C water bath on the laboratory bench, using a beaker one-third filled with water.
   - A hot plate or ringstand apparatus may be used, as shown in FIGURE 24.1.
   - A thermometer should be placed in a tube of water in the bath to determine temperature.

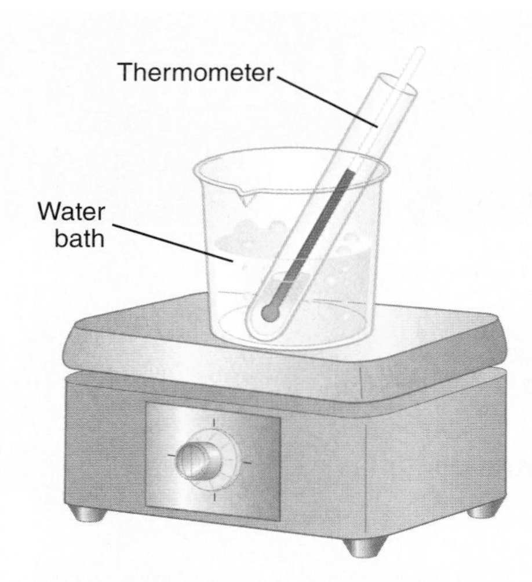

FIGURE 24.1 Setup for an apparatus to test the effect of moist heat on bacteria.

2. Obtain four different broth cultures of bacteria that are coded by a letter or number (or both).
3. Obtain five nutrient agar plates, and on the bottom side, use a wax pencil to divide the plate into four sectors to accommodate each of the four bacteria.
   - Label each sector on one plate with the code for each bacterium.
   - Repeat this procedure for the other four plates.
   - Also include your name.
4. The bacterial broths will be exposed to temperatures for 40°, 60°, 80°, and 100°C.
   - Mark one plate (on lid) as "40," another as "60," the third as "80," and the fourth as "100."
   - Mark the top of the fifth plate "control."
5. Using aseptic technique, make a single-line streak from each bacterial broth onto its respective sector of the "control" plate.
6. When the water bath has reached 40°C, place the four bacterial broth tubes in the bath.
   - Note the time and leave the tubes in the bath for 10 minutes.
   - Watch the thermometer and monitor the temperature so that it does not go above 40°C.

Do not let the water bath temperature go above 40°C.

7. After the 10-minute incubation, remove the tubes and aseptically inoculate a loopful of each bacterial broth as a single-line streak onto its respective sector of the "40" plate.
8. Raise the water bath temperature to 60°C and repeat steps 6 and 7, streaking a loopful onto the respective sectors of the "60" plate.
9. Raise the water bath temperature to 80°C and repeat steps 6 and 7, streaking a loopful onto the respective sectors of the "80" plate.

10. Finally, raise the water bath temperature to 100°C and repeat steps 6 and 7, streaking a loopful onto the respective sectors of the "100" plate.

11. Incubate the five plates at 37°C in the inverted position for 24 to 48 hours.
    - If the plates will not be observed until the following week, move them to the refrigerator after 48 hours.

12. Observe the five plates for the presence of growth. Growth indicates survival of the bacteria at the temperature tested.
    - Enter the codes for the bacterial broths (or species if provided by your instructor) in TABLE 24.1 of the Results section.
    - If the growth at a specific temperature is about equal to the control, indicate the abundant growth as (+++).
    - If the growth is moderate (not as abundant as the control), indicate moderate growth as (++).
    - If there is minimal growth compared to the control, indicate slight growth as (+).
    - If there was no growth, mark the box with a (–).

13. From the data in Table 24.1, compare the heat tolerance for the four different bacteria tested.
    - Determine the minimal amount of heat necessary for complete destruction of the bacteria and record your results in TABLE 24.2 in the Results section.

Name: ________________________ Date: __________ Section: __________

The Effect of Physical Agents on Bacteria: Heat Tolerance

EXERCISE RESULTS 24

RESULTS

HEAT TOLERANCE

TABLE 24.1 Growth of Bacteria after Heat Exposure

Temperature (°C)	Organisms Tested			
	1.	2.	3.	4.
40				
60				
80				
100				
Control				

+++ = abundant growth; ++ = moderate growth; + = slight growth; - = no growth.

TABLE 24.2 Minimum Heating Time for Complete Destruction

Organism	1.	2.	3.	4.
Time				

Observations and Conclusions

Questions

1. What was the point of including a "control" in this exercise?

2. How do higher temperatures affect the bacteria?

3. How does moist heat specifically kill organisms?

The Effect of Physical Agents on Bacteria: Ultraviolet Light

EXERCISE 25

Materials and Equipment

- Ultraviolet light at about 265 nm
- Broth cultures of three selected bacteria, including one sporeformer
- Inoculation loops
- Sterile cotton swabs
- Nutrient agar plates

Physical agents have broad application for controlling and destroying microorganisms. **Ultraviolet (UV) light** is used to decontaminate the air in operating rooms and food service operations.

In this exercise, the effects of ultraviolet light will be tested on various species of bacteria. Organisms will be exposed to ultraviolet light, and their survival will be determined by their ability to grow in bacteriological media after the exposure has been completed.

ULTRAVIOLET LIGHT

PURPOSE: to determine the lethal effect of UV light on bacterial growth.

UV light having a wavelength of about 265 nm affects the chromosomal DNA of microorganisms by inducing adjacent thymine and cytosine bases to bind together in the molecule, forming dimers. If the microorganism is unable to repair the DNA damage, this binding prevents DNA replication during binary fission.

In this section, the lethal effect of UV light will be explored at various exposure times. This section also demonstrates the poor penetration of ultraviolet light and the effects of plastic shielding.

PROCEDURE

1. Obtain or prepare three plates of nutrient agar for use in this exercise.
   - On the bottom side, use a wax marking pencil to divide the plate into three sectors, to accommodate three organisms.
   - Label the plate accordingly.
   - Three sterile cotton swabs will be needed.
   - The instructor will assign an exposure time to the ultraviolet (UV) light, such as 30 seconds, 1 minute, 5 minutes, or 10 minutes.

2. Different sectors will be inoculated with different bacteria. Inoculate each sector by aseptically removing a loopful of the broth culture of bacteria and drawing a single streak across the center of the sector as in Exercise 24.

Be careful not to look directly into the UV light, since damage to the eye may occur.

3. Remove the cover from one plate and expose the bacteria to the UV light for the assigned amount of time.
   - Leaving the cover on the plate, expose the bacteria in the second plate for an equal amount of time. The distance to the light should be about 6 inches.
   - The third plate is a control plate and will receive no exposure to UV light.
   - After the exposure period, incubate all three plates in the inverted position at 37°C for 24 to 48 hours.
   - Move the plates to the refrigerator if they will not be observed until the following week.
4. Examine the three plates and compare the two exposed plates to the control plate.
   - Determine whether a reduction of growth took place. Reduction or absence of growth demonstrates the lethal effect of UV light.
   - Compare the covered and uncovered plates to one another, and note whether the plastic lid shielded the bacteria from the light.
   - In the Results section, enter labeled diagrams of the plates, demonstrating the effect of ultraviolet light, and explain your observations.
5. If various time periods of exposure were used, it may be possible to determine the minimum time required for complete destruction of the bacteria. Refer to the uncovered plate where direct exposure to the UV light took place.
   - In TABLE 25.1 of the Results section, enter the names of the bacteria and various exposure periods used. Enter a (+) for presence of growth and a (-) for absence of growth.
   - From the data, attempt to determine the minimum amount of exposure time lethal to each organism, and enter this data in TABLE 25.2.
6. An alternative method of testing various time periods of exposure to UV light is as follows:
   - Place a card over an inoculated plate of nutrient agar (inoculated as in Exercise 66), then slide the card back so that one-quarter of the plate is exposed (FIGURE 25.1).
   - Permit the UV light to irradiate the surface for 1 minute.

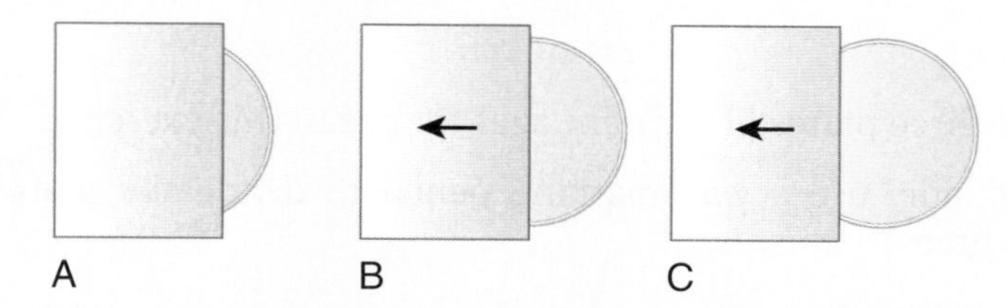

FIGURE 25.1 A method for exposing an inoculated agar plate to a gradation of ultraviolet light over a period of minutes.

- Slide the card back so that one-half of the plate is exposed, and irradiate for another minute.
- Slide the card back to expose three-quarters of the plate, and place the plate under the UV light for another minute.
- Remove the card and expose the entire plate for a final minute before incubating it. The effect will be to expose the plate for a gradation of time periods from one to five minutes.
- Theoretically, the survival of the bacteria exposed to the UV light should vary according to the time of exposure and be reflected in the gradation of growth.

Quick Procedure

Effect of UV Light on Bacteria

1. Inoculate plates of test medium with bacteria.
2. Expose plates to UV light for a specified period.
3. Incubate.
4. Determine whether bacteria survive and grow.

Name: ______________________ Date: __________ Section: __________

The Effect of Physical Agents on Bacteria: Ultraviolet Light

EXERCISE RESULTS 25

RESULTS

ULTRAVIOLET LIGHT

Nutrient Agar Plates

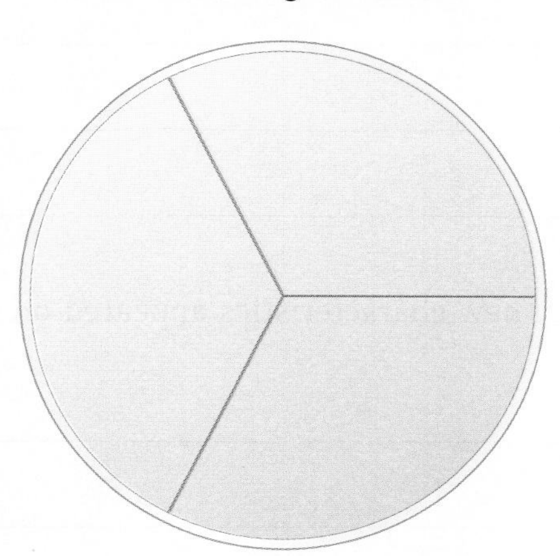

TABLE 25.1	Growth of Bacteria after Direct Exposure to UV Light (Lid off)		
Irradiation Time	Organisms Tested		
	1.	2.	3.
sec			
min			
min			
min			

TABLE 25.2	Minimum Exposure Time for Complete Destruction		
Organism	1.	2.	3.
Time			

Observations and Conclusions

____

____

____

____

____

Questions

1. Why is it a poor idea to sterilize empty Petri dishes with UV light?

____

____

____

2. Suppose a bacterial colony with new characteristics appeared on a plate treated with UV light. What might be the explanation?

____

____

____

The Effect of Physical Agents on Bacteria: Drying

EXERCISE 26

Materials and Equipment

- Broth cultures of three selected bacteria, including a sporeformer
- Sterile screw-capped test tubes
- Sterile cotton swabs and disinfectant
- Tubes of nutrient broth
- Mechanical pipetter, 5-ml pipettes

Physical agents have broad application for controlling and destroying microorganisms. In this exercise, the effects of drying will be tested on various species of bacteria. Organisms will be exposed to a drying method, and their survival will be determined by their ability to grow in bacteriological media after the exposure has been completed.

DRYING

PURPOSE: to determine the resistance of bacterial species to the effects of drying.

Drying is a useful preservation method for foods and other items because water is a prerequisite for life. Most bacteria die in dry environments unless they form spores that are resistant to drying or structures such as capsules that retard drying.

This exercise will demonstrate the effects of drying by placing bacteria in a dry environment for various time periods, then adding nutrient broth to cultivate any remaining live bacteria.

PROCEDURE

1. Select three sterile screw-capped test tubes, three sterile cotton swabs, and broth cultures of three organisms to be tested.
   - Label one test tube for each organism.
   - Include your name, the date, the organism, and the drying time.
   - The instructor will assign a drying time to be used.
2. Inoculate each empty sterile test tube as follows:
   - Dip a sterile cotton swab into the broth culture of bacteria, remove the swab, and rub the swab lightly along the bottom of the test tube.
   - Be careful to avoid depositing a large drop of the broth.
   - One test tube should be used for each organism.

3. Incubate the inoculated tubes at 37°C for the time assigned by the instructor. A minimum of several hours or days may be necessary.
4. After the drying period, pipette 3 ml of sterile nutrient broth into each tube so that it mixes with the residue of dried bacteria. A mechanical pipetter should be used even though the broth is sterile.
   - Re-incubate the tubes at 37°C for 24 to 48 hours, and then refrigerate them until the next laboratory period.
5. Observe the tubes for the presence or absence of growth in the nutrient broth.
   - If the organisms survived the drying period, they will grow in the broth during the second incubation period, and the broth will be cloudy (turbid).
   - If the broth is clear, this indicates that the bacteria failed to survive the drying.
   - In TABLE 26.1, enter the names of the organisms and drying time used.
   - Enter a (+) to indicate growth in the broth tube and a (−) to indicate absence of growth.
   - If different drying times were used, determine the minimum amount of time required for the destruction of each organism by drying.
   - Enter your observations in TABLE 26.2.

Name: ______________________ Date: __________ Section: __________

The Effect of Physical Agents on Bacteria: Drying

EXERCISE RESULTS **26**

RESULTS

DRYING

TABLE 26.1	Growth of Bacteria after Drying Period		
Drying Time	Organisms Tested		
	1.	2.	3.
hr			
hr			
hr			
hr			

TABLE 26.2	Minimum Drying Time for Complete Destruction		
Organism	1.	2.	3.
Time			

Observations and Conclusions

____

____

____

____

____

Questions

1. Name several foods that rely on their dry environment for preservation from bacterial contamination.

2. Which of the organisms tested in Exercises 24, 25, and 26 showed the greatest resistance to heat, UV light, and drying? How can these results be explained?

The Effect of Physical Agents on Bacteria: Other Physical Agents

EXERCISE 27

Materials and Equipment

- Inoculating loop and/or inoculating needle
- Nutrient agar plate
- Broth cultures of selected bacteria
- Autoclave

Physical agents have broad application for controlling and destroying microorganisms. In this exercise, the effects of incineration, osmotic pressure, cold temperature, and autoclaving will be tested on various species of bacteria. Organisms will be exposed to a physical agent, and bacterial cell survival will be determined by their ability to grow in bacteriological media after the exposure has been completed.

OTHER PHYSICAL AGENTS

PURPOSE: to test the effects of other physical agents on bacterial growth and survival.

PROCEDURE

1. Bacteria are susceptible to the effects of **incineration**. To test the effect of incineration on a test bacterium, obtain a nutrient agar plate and divide it in half.
   - On the plate bottom, label the two sides "pretreatment" and "posttreatment."
   - Secure a loopful of a test bacterium and make a single streak across the pretreatment side of the plate.
   - Flame the loop and cool it.
   - Streak the loop across the posttreatment side.
   - Incubate the plate, then note whether growth occurred on both sides of the plate.
   - In the Results section, note your conclusions on the value of incineration as a sterilizing agent.
2. To test the effect of **osmotic pressure** on bacteria, obtain tubes of nutrient broth containing added sodium chloride (NaCl) in amounts of 1%, 5%, and 10%. A fourth tube of normal nutrient broth should also be used.
   - Inoculate the four tubes with a loopful of a test bacterium, and incubate the tubes.
   - At the conclusion of the incubation period, evaluate the growth in the various broth environments by their turbidity and draw your conclusions regarding the effect of osmotic pressure as a preservative measure in foods.

3. **Cold temperature** can have an inhibitory effect on the growth of bacteria.
   - Inoculate two nutrient broth tubes with a test bacterial species, and place one tube in the refrigerator (4°C) and another in the 37°C incubator as a control.
   - At regular intervals, suspend the sedimented bacteria by tapping the bottoms of the tubes, then compare the turbidity of the growth in the tubes as an indication of the degree of bacterial growth.
   - Enter your observations in the Results section.
4. A variation of step 3 can be used to study the effectiveness of **autoclaving**.
   - After inoculating two tubes of nutrient broth with bacteria, one tube is autoclaved and the other is left untreated as a control.
   - Both tubes are then incubated at 37°C.
   - Note the presence of growth.

Name: ______________________ Date: __________ Section: __________

The Effect of Physical Agents on Bacteria: Other Physical Agents

EXERCISE RESULTS 27

RESULTS

OTHER PHYSICAL AGENTS

Observations and Conclusions

____

____

____

____

____

Questions

1. Compare and contrast the effects of incineration, osmotic pressure, cold temperature, and autoclaving.

____

____

____

____

____

2. Name several foods that rely on cold temperature for preservation from bacterial contamination.

____

____

____

3. Name several foods that rely on osmotic pressure for preservation from bacterial contamination.

____

____

____

4. Which physical agent had the greatest effect on the bacteria?

____

____

The Effect of Chemical Agents on Bacteria: Antiseptics and Disinfectants

EXERCISE

28

Materials and Equipment

- Two selected bacterial species
- Plates of nutrient agar or materials
- Beakers of various disinfectants and antiseptics, such as iodine, Merthiolate (thiomersal), formaldehyde, Lysol, 70% ethyl alcohol, phenol, and crystal violet
- Sterile paper disks and cotton swabs
- Forceps in beakers of ethyl alcohol
- Wax marking pencil
- Millimeter rulers

Watch This Video

- How to Prepare a Disk Diffusion Assay

Many **chemical agents** exert an inhibitory effect on bacteria. Among the best known are the antiseptics and disinfectants used in medicine, industry, and the home.

The paper disk method will be employed to test the effect of chemical agents on bacteria. In this procedure, chemical substances are adsorbed onto disks of special paper, which are then applied to an inoculated growth medium. Although convenient for demonstrating antibacterial activity, the method does not take into account such factors as the diffusion rate of the chemical or the effect of the growth medium. Also, the bacterial concentration and chemical concentration are not standardized. The results must therefore be interpreted with these factors in mind.

ANTISEPTICS AND DISINFECTANTS

PURPOSE: to determine the susceptibility of bacteria to antiseptics and disinfectants.

A broad assortment of antiseptics and disinfectants is available for the control and destruction of bacteria (FIGURE 28.1). **Antiseptics** are commonly used on the skin surface, while **disinfectants** are applied to lifeless objects such as tabletops, equipment, and instruments.

In this exercise, saturated paper disks will be used to demonstrate the effect of chemical agents on two species of bacteria (watch **Microbiology Video: How to Prepare a Disk Diffusion Assay** to familiarize yourself with the technique). The bacteria will be streaked onto a nutrient agar medium, and various antiseptics and disinfectants will be applied to the surface of the medium. The destruction of bacteria will be noted by the appearance of a clear **zone of inhibition** where the chemical agent has been applied.

How to Prepare a Disk Diffusion Assay

FIGURE 28.1 Your instructor may have a broad array of disinfectants for this exercise.

PROCEDURE

1. Obtain or prepare two plates of nutrient agar, label the bottom sides with your name and the date, and designate one plate for each of two bacteria to be tested.
   - The depth of the agar should be about 4 millimeters.
   - Obtain broth cultures of two species of bacteria and two sterile cotton swabs.
2. Aseptically dip a swab into the broth culture, press it to the inside of the tube to express the excess fluid, and inoculate one of the nutrient plates by making a **lawn of bacteria.**
   - The swabbing should be performed at three different angles, and a final sweep should be made around the rim of the plate as shown in Figure 19.1 in Exercise 19.
   - Discard the swab in the beaker of disinfectant after use.
   - Inoculate the second plate with the second organism in the same manner.
3. With the lids in place, allow the plates to sit at room temperature for several minutes so that the excess moisture is absorbed into the agar.
   - Use a wax marking pencil to mark the bottom side and divide each plate into four sectors to test four chemical agents.
   - Label the bottom side of each sector with the name of the agent tested.

Chemical disinfectants could be caustic to the skin. Be sure to handle them with care.

4. Lightly flame the forceps and obtain a sterile disk.
   - Impregnate the disk with a chemical agent by touching it gently to the surface of the chemical, as shown in FIGURE 28.2.
   - Dip half the disk into the solution, and allow the chemical to adsorb onto the disk (FIGURE 28.3). This will avoid excessive amounts of chemical agent on the disk.
   - Place the disk on the agar of the appropriate sector, and press lightly to achieve firm contact with the agar surface.

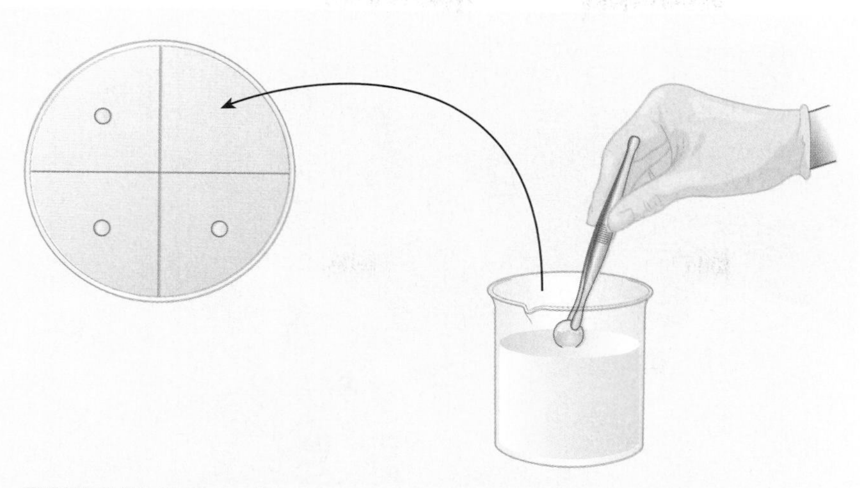

FIGURE 28.2 Method for impregnating the paper disk with the chemical agent.

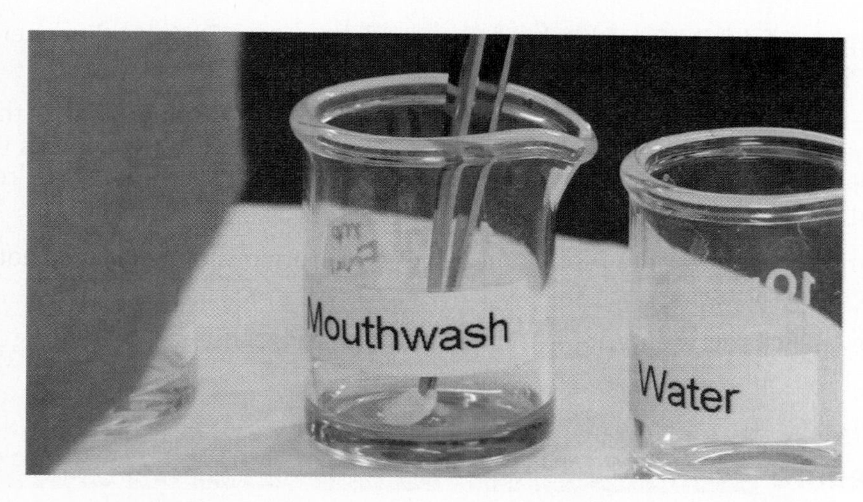

FIGURE 28.3 Only dip half the disk into the solution, and allow the chemical to adsorb onto the disk.

- Saturate three other disks with three other chemicals for the other three sectors.
- An alternative method of saturating disks is to place them on a sheet of paper toweling and add drops of the chemical agent onto the disks. The paper toweling will absorb the excess chemical so it is not carried over to the surface of the plate.

5. Repeat the above procedure for the second plate.
6. Incubate the plates in the inverted position at 37°C for 24 to 48 hours, then refrigerate until observed.
7. Examine the nutrient agar plates for evidence of clear halos or zones of inhibition surrounding the disks (FIGURE 28.4).
   - A zone indicates that the organisms are susceptible to the chemical agent. The word "susceptible" should be noted in TABLE 28.1 in the Results section.

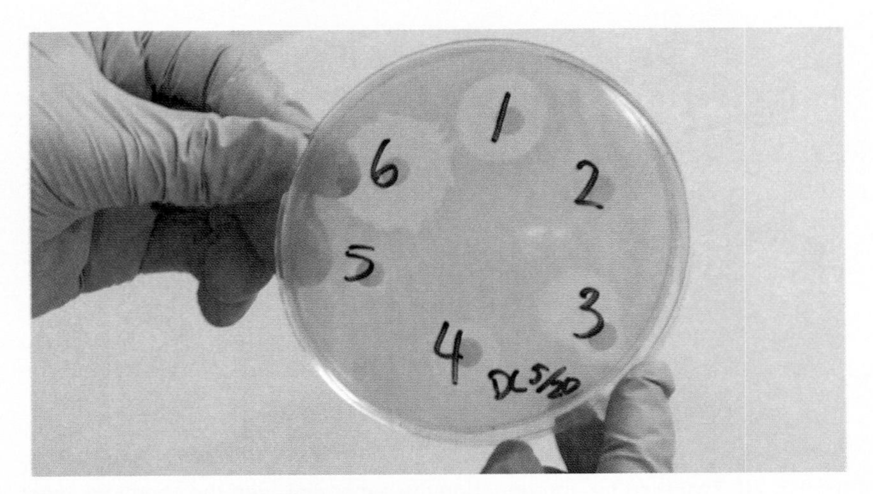

FIGURE 28.4 Zones of inhibition will surround the disks.

- The degree of inhibition may be determined by measuring the diameter of the zone in millimeters, including the disk.
- Use the underside of the plate for this procedure, and record your results in the table, together with the names of organisms and chemical agents used. If growth appears up to the edge of the disk, the organism is resistant to the chemical agent, and the word "resistant" may be entered in the chart.
- Compare the activity of the various chemicals on the organisms tested, and note the differences in your observations.
- Include labeled representations of the plates with the results.

Quick Procedure

Effect of Chemical Agents on Bacteria: Antiseptics and Disinfectants

1. Inoculate a nutrient agar plate with bacteria.
2. Dip sterile disks into various chemical agents.
3. Place disks on the agar surface.
4. Incubate.
5. Observe and measure zones of inhibition.

Name: ______________________ Date: __________ Section: __________

The Effect of Chemical Agents on Bacteria: Antiseptics and Disinfectants

EXERCISE RESULTS 28

RESULTS

ANTISEPTICS AND DISINFECTANTS

TABLE 28.1 Susceptibility or Resistance of Bacteria on Nutrient Agar

Antiseptic or Disinfectant	Organisms Tested			
	1.		2.	
	Zone Diameter	Sus. or Resist.	Zone Diameter	Sus. or Resist.

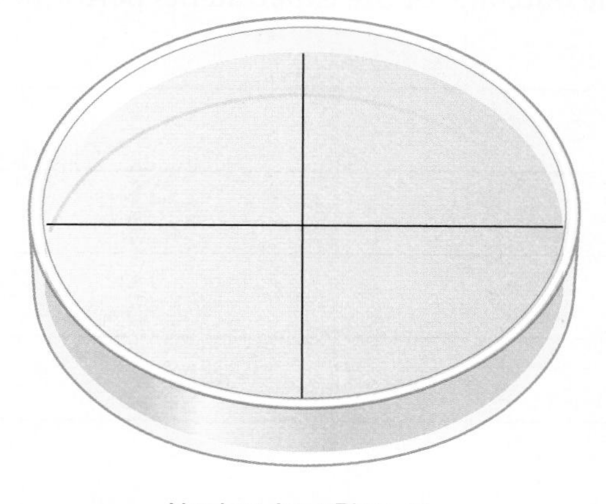

Nutrient Agar Plate #1

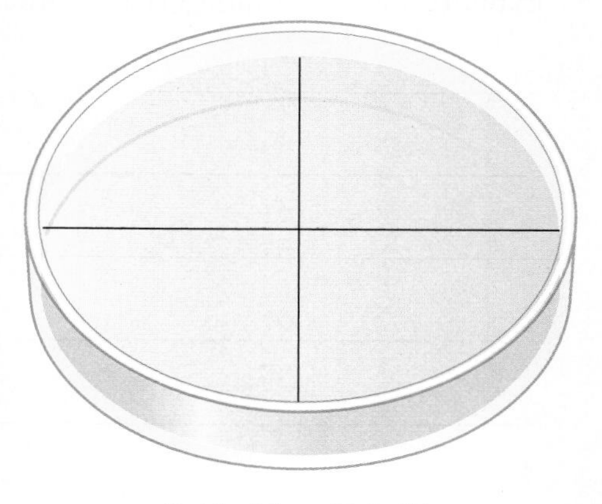

Nutrient Agar Plate #2

Organism: ______________________ ______________________

Observations and Conclusions

Questions

1. List several organic materials other than blood that may affect the disinfecting ability of a chemical agent under practical conditions.

2. What might be the result of dipping paper disks into the chemical tested rather than lightly touching the disks to the surface?

3. Explain several variables that influence the outcome of the experiments performed in this exercise.

The Effect of Chemical Agents on Bacteria: Metals

EXERCISE 29

Materials and Equipment

- Inoculating loop
- Plates of nutrient agar or materials
- Selected bacterial species
- Forceps in beakers of ethyl alcohol
- Samples (disks) of various metals, such as copper, zinc, aluminum, lead, and silver

Many **chemical agents** exert an inhibitory effect on bacteria. Metals may be employed as disinfecting agents because they combine with microbial proteins. Mercury chloride, silver nitrate, and copper sulfate are among the useful compounds of heavy metals.

METALS

PURPOSE: to examine the effects of metals on bacterial growth.

Some metals exert an **oligodynamic activity** on bacteria. This means that a small amount of the metal has a significant effect on the growth of the organism. This effect is generally due to interference with enzyme activity in the metabolizing cell. The following procedure demonstrates the phenomenon of oligodynamic action. It is performed in basically the same way as Exercise 28.

PROCEDURE

1. Prepare two nutrient agar plates and label one for each of the organisms tested.
   - Carefully inoculate each plate by the "3 + 1" method explained in Exercise 19.
2. Clean the metals as follows:
   - First wash the metal disks in hot soap and water.
   - Rinse and, using sterile forceps, dip the metal in alcohol.
   - Allow the alcohol to evaporate.

3. Using sterile forceps, place samples of different metals in different sectors of the agar, pressing gently to ensure firm contact.
    - Do not turn the plates over. Treat both plates similarly.
    - Incubate the plates in the upright position for 24 to 48 hours at 37°C and then refrigerate until observed.
4. Examine the plates for **zones of inhibition** as explained in Exercise 28.
    - Measure the diameters of the zones, and enter your results in TABLE 29.1.
    - Note which metals were inhibitory to which organisms in your observations.
    - Include diagrams of the plates.

Name: ______________________ Date: __________ Section: __________

The Effect of Chemical Agents on Bacteria: Metals

EXERCISE RESULTS 29

RESULTS

METALS

TABLE 29.1 Susceptibility or Resistance of Bacteria

Metal	Organisms Tested			
	1.		2.	
	Zone Diameter	Sus. or Resist.	Zone Diameter	Sus. or Resist.

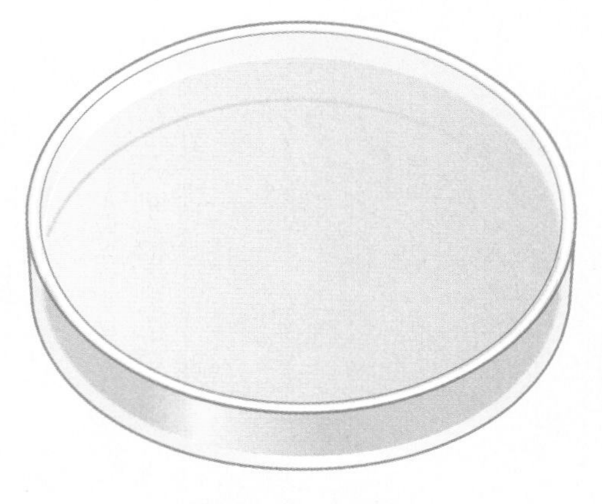

Nutrient Agar Plate

Nutrient Agar Plate

Organism: ______________________ ______________________

Observations and Conclusions

____

____

____

____

____

Questions

1. How do metals interfere with the growth of bacterial organisms?

   ____

   ____

   ____

2. Explain several variables that influence the outcome of the experiments performed in this exercise.

   ____

   ____

   ____

The Effect of Chemical Agents on Bacteria: Other Chemical Substances

EXERCISE 30

Materials and Equipment

- Plates of nutrient agar or materials
- Selected bacterial species
- Forceps in beakers of ethyl alcohol
- Samples of miscellaneous products, such as mouthwashes, hand sanitizers, toothpaste, various foods, body substances, and other substances as noted in the procedure

Many **chemical agents** exert an inhibitory effect on bacteria. Several antimicrobial substances of a chemical nature exist in foods and in the body.

The paper disk method will be employed to test the effect of chemical agents on bacteria. In this procedure, chemical substances are adsorbed onto disks of special paper, which are then applied to an inoculated growth medium. Although convenient for demonstrating antibacterial activity, the method does not take into account such factors as the diffusion rate of the chemical or the effect of the growth medium. Also, the bacterial concentration and chemical concentration are not standardized. The results must therefore be interpreted with these factors in mind.

OTHER CHEMICAL SUBSTANCES

PURPOSE: to test the effectiveness of other chemical agents to inhibit bacterial growth.

Many chemical agents may inhibit the growth of bacteria. Certain of these are pharmaceutical substances; others occur in foods; still others are produced by the body. The inhibitory effects of these miscellaneous agents may be tested by using the basic procedures outlined in Exercises 28 and 29. It is important that observations be made after a brief incubation period of 24 hours or less because the inhibition often is overcome by rapidly multiplying bacteria.

PROCEDURE

1. Prepare inoculated plates of bacteria in nutrient agar using the "3 + 1" method outlined in Exercise 19. The materials to be examined for antimicrobial activity will be noted by the instructor, as explained below.
2. Mouthwashes and other liquid pharmaceutical products may be tested by dipping sterile paper disks into the product and placing the disks on the surface of the agar.
   - Semisolid products such as toothpaste or hand sanitizers may be tested by placing a small sample of the product onto the surface of the agar as described in Exercise 28.

3. The chemicals in newsprint generally keep the pages of a newspaper free of bacteria. This concept may be tested by placing a small square of fresh, untouched newspaper onto the agar surface. Plants are also a source of antimicrobial chemicals. To demonstrate the effect of plant antimicrobials, cut a square from a plant leaf, and place it onto the surface of the inoculated agar.
4. Tears contain the inhibitory enzyme **lysozyme**, which kills gram-positive bacteria. To obtain tears, rub a piece of fresh onion under the eye or squirt lemon in the eye. Now collect the tear drops on a sterile disk and place it on an inoculated agar plate for testing.
5. Foods such as garlic, horseradish, mustard, onion, and whole cloves may be tested for antimicrobial substances by placing small pieces on inoculated agar plates.
6. The acidity in orange juice, lemon juice, sauerkraut juice, and cranberry juice contributes to the growth inhibition of many microorganisms. This phenomenon may be tested by soaking paper disks in the juice and applying them to the inoculated agar surface.
7. At various times, newspaper and journal reports relate the effectiveness of **alcoholic beverages** as antimicrobial agents. The paper disk method may be used to validate these reports. Samples of different beers and wines are tested by soaking paper disks in the sample and placing the disks on the inoculated agar surface. It might be valuable to test white and red wines, light and regular beers, and various types of spirits.
8. All plates should be incubated at 37°C and should be observed for **zones of inhibition** in 24 hours or less. Refrigeration may be used to preserve the plates until they are to be observed. Enter results in TABLE 30.1. Include a diagram of a plate.

Name: ______________________ Date: __________ Section: __________

The Effect of Chemical Agents on Bacteria: Other Chemical Substances

EXERCISE RESULTS 30

RESULTS

OTHER CHEMICAL SUBSTANCES

TABLE 30.1 Susceptibility or Resistance of Bacteria

Chemical Substance	Organisms Tested			
	1.		2.	
	Zone Diameter	Sus. or Resist.	Zone Diameter	Sus. or Resist.

Nutrient Agar Plate

Nutrient Agar Plate

Organism: ______________________ ______________________

Observations and Conclusions

Questions

1. Explain several variables that influence the outcome of the experiments performed in this exercise.

2. List several miscellaneous products, other than those mentioned in this exercise, that may exert an inhibitory effect on bacterial growth. Include why you believe each one will exert an inhibitory effect.

3. Would using another body fluid produce the same results as using tears? If so, which ones?

The Effect of Chemical Agents on Bacteria: Lysozyme

EXERCISE **31**

Materials and Equipment

- Nutrient agar plates or materials for their preparation
- Cultures of *Micrococcus luteus* and/or other bacteria
- Sterile straws, cotton swabs, and applicator sticks
- Beakers of disinfectant
- Fresh egg
- Pipettes and mechanical pipetters
- 99-ml water blank

Many **chemical agents** exert an inhibitory effect on bacteria. Several antimicrobial substances of a chemical nature exist in foods and in the body.

LYSOZOME

PURPOSE: to examine the effects of lysozyme on bacterial growth.

Lysozyme is an antibacterial substance found in human tears and saliva as well as respiratory secretions. It breaks cell wall linkages of gram-positive bacteria, causing the cells to swell and burst through osmotic lysis. Lysozyme is also found in the white of an egg, where it provides natural protection against staphylococci, micrococci, and other gram-positive bacteria. In this exercise, various dilutions of egg white lysozyme are tested for their antibacterial qualities.

PROCEDURE

1. Prepare or pour a plate of nutrient agar, filling the culture dish about two-thirds full.
   - Permit the agar to harden thoroughly and label the plate on the bottom side with your name, the date, and the name of the test organism used. *Micrococcus luteus* is recommended for this exercise due to its high susceptibility to lysozyme.
   - For comparison, other plates may be used with gram-positive and gram-negative bacteria (e.g., *Escherichia coli*).
2. Dip a sterile swab into a culture of *M. luteus* and swab the plate in the "3 + 1" pattern described in Exercise 19, being sure to cover the entire agar surface. If other organisms and plates are to be used, they may be swabbed in this manner.
3. Using a straw, carefully remove four plugs of the inoculated agar from each of the four sectors as shown in Figure 21.1.

- Use the applicator stick to push the plugs out of the straw into the beaker of disinfectant.
- Deposit the straw and the stick into the disinfectant when you have finished with them. Removing the plugs forms a number of wells in the agar.

4. Over a clean beaker, carefully crack open a fresh egg and separate the yolk from the white. Do this by moving the yolk back and forth between the egg shells while allowing the white to fall into the beaker.
5. Using a mechanical pipetter, pipette 1 ml of the egg white into 99 ml of sterile water to make a 1% concentration of egg white.
   - Additional concentrations of 0.5% and 0.1% should also be prepared. To prepare a 0.5% concentration, mix 5 ml of the 1% concentration with 5 ml of sterile water. For a 0.1% concentration, mix 1 ml of the 1% concentration with 9 ml of sterile water.
   - Label the plate bottom with the concentration(s) used.
6. Using fresh sterile pipettes, fill various wells in the inoculated agar with samples of the undiluted and diluted egg white prepared above.
   - Do not overfill the wells, and be sure to label the plate with the concentrations used.
   - Incubate the plate at room temperature in the upright position for approximately 48 hours or longer.
7. Observe the plate for the presence of clear rings of inhibition surrounding the wells. The rings indicate that lysozyme in the egg white has killed the test bacterium.
   - Record your observations in TABLE 31.1 of the Results section and draw an illustration of the plate indicating the areas of inhibition.
   - If various bacteria have been used, draw conclusions on the effect of lysozyme on different organisms.
   - Note any variations in the size of the ring owing to the concentration of lysozyme used, including whether the higher or lower concentration caused more inhibition.

Quick Procedure

Effect of Lysozymes on Bacteria

1. Inoculate plate of nutrient agar with bacteria.
2. Remove plugs of agar to form wells.
3. Add various concentrations of egg white to the wells.
4. Incubate.
5. Observe for inhibition of agar.

Name: ______________________ Date: __________ Section: __________

The Effect of Chemical Agents on Bacteria: Lysozyme

EXERCISE RESULTS 31

RESULTS

EFFECT OF LYSOZYME ON BACTERIA

TABLE 31.1	Susceptibility or Resistance of Bacteria			
Concentration of Lysozyme	Organisms Tested			
	1.		2.	
	Zone Diameter	Sus. or Resist.	Zone Diameter	Sus. or Resist.

Plate #1

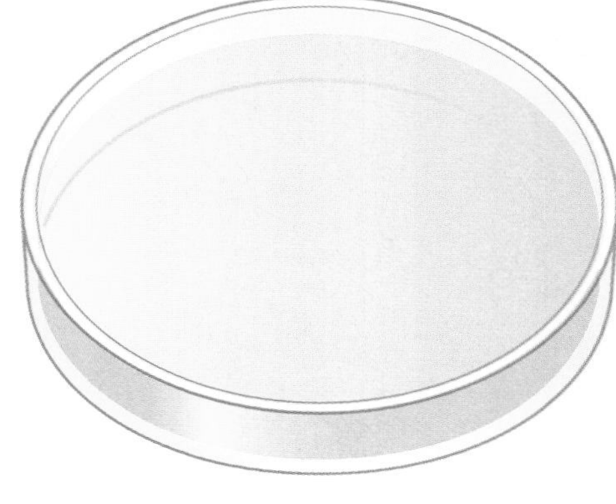

Plate #2

Organism: ______________________ ______________________

Observations and Conclusions

____

____

____

____

____

Questions

1. List several variables that influence the outcome of the experiment performed in this exercise.

____

____

____

2. Suggest ways to modify this exercise to learn more properties of lysozyme.

____

____

____

Evaluation of Disinfectants and Antiseptics: Effectiveness of Disinfectants on Inanimate Objects

EXERCISE 32

Materials and Equipment

- Sterile glass rods 3 inches in length, with rounded edges
- Beakers of two different disinfectants
- Nutrient agar plates, or materials for their preparation

Disinfection is essential for controlling the spread of microorganisms. A good **disinfectant** should kill microorganisms on a surface, but it also must have low toxicity to humans, have a substantial shelf life, be useful in dilute form, and be inexpensive and easy to obtain. Numerous other factors also contribute to the value of a good disinfectant.

Currently, the United States Department of Agriculture (USDA) lists over 8,000 disinfectants for hospital use and thousands more for general use. In this exercise, we shall evaluate the activity of a number of disinfectants on a thermometerlike object.

EFFECTIVENESS OF DISINFECTANTS ON INANIMATE OBJECTS

PURPOSE: to test the effectiveness of disinfectants on objects contaminated with oral bacteria.

Disinfectants may be used to kill microorganisms on inanimate objects **(fomites)** such as instruments and thermometers. It is common practice in the home, for example, to disinfect a thermometer before storing it for later use. In this section, we shall evaluate the effectiveness of disinfectants by using glass rods to simulate thermometers. The bacteria used for the test will be those commonly found in the mouth.

PROCEDURE

1. Prepare a plate of nutrient agar and label it on the bottom side with your name, the date, and the designation "thermometer disinfection."
   - Divide the plate into three equal sectors. One sector will be used as a control, and the other two will be used to evaluate the effectiveness of two disinfectants.

Be certain that the glass rods you use in this exercise have been sterilized.

2. Select three sterile glass rods and place them under your tongue to simulate the use of thermometers. Allow them to remain for 3 minutes.
3. Take one glass rod and streak it three times on the surface of the control sector of the plate. This will represent the untreated sample.

4. Place each of the remaining two rods in a beaker containing different disinfectants. Allow them to stand in the disinfectants for 5 minutes.
5. Wash each rod under a gentle stream of water, and streak the remaining sectors of the plate three times, as you streaked the control.
   - Incubate the plate in the inverted position at 37°C for 24 to 48 hours and then refrigerate until observed.
6. Observe the plate for the presence of bacteria in the three sectors.
   - Compare the bacterial growth from the untreated and disinfected glass rods.
   - Note the number of colonies in each sector rather than the amount of growth, since a single colony may have overgrown the area.
   - Enter a labeled representation of the plate in the Results section, and record your observations on the effectiveness of the two disinfectants tested.

Name: ______________________ Date: __________ Section: __________

Evaluation of Disinfectants and Antiseptics: Effectiveness of Disinfectants on Inanimate Objects

EXERCISE RESULTS 32

RESULTS

EFFECTIVENESS OF DISINFECTANTS ON INANIMATE OBJECTS

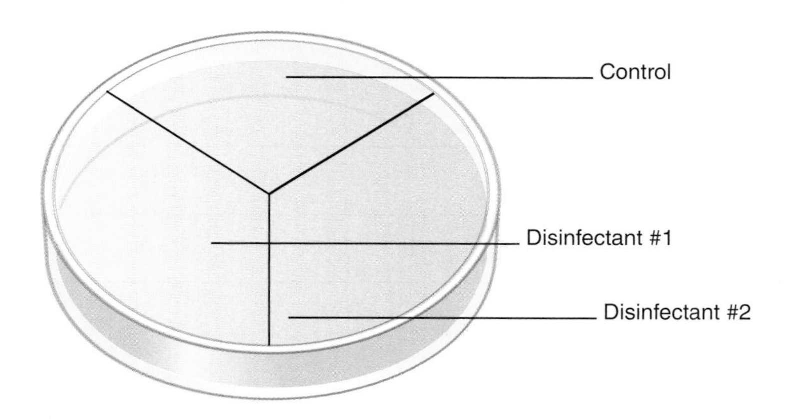

Nutrient Agar Plate

Observations and Conclusions

____

____

____

____

____

Questions

1. What are the advantages of testing a disinfectant on glass rods placed in the mouth rather than in a broth culture of bacteria?

2. Why would a large mass of bacteria in a sector of the plates not necessarily imply poor disinfection by the chemical agent?

Evaluation of Disinfectants and Antiseptics: Effectiveness of Disinfectants and Hand Washing on the Skin Surface

EXERCISE 33

Materials and Equipment

- Plates of nutrient agar, or materials for their preparation
- Beakers of disinfectants

Disinfection is essential for controlling the spread of microorganisms. A good **disinfectant** should kill microorganisms on a surface, but it also must have low toxicity to humans, have a substantial shelf life, be useful in dilute form, and be inexpensive and easy to obtain. Numerous other factors also contribute to the value of a good disinfectant.

In this exercise, we shall evaluate the activity of a number of disinfectants on the skin surface.

EFFECTIVENESS OF DISINFECTANTS AND HAND WASHING ON THE SKIN SURFACE

PURPOSE: to test the effectiveness of disinfectants used on the skin surface.

The fingertips may be important mechanical vectors of microorganisms, especially when the individual is a carrier of such diseases as the common cold, hepatitis A, or amoebiasis. Employees in food service and medical professions especially are aware of the need to use disinfectants to prevent the spread of disease. In this exercise, disinfectants will be tested for their ability to control microorganisms on the fingertips by observing the bacterial population before and after treatment.

PROCEDURE

1. Obtain or prepare one plate of nutrient agar and label it on the bottom side with your name, the date, and designation "skin disinfection."
   - Divide the plate in half.
   - Make a **five-finger impression** by touching the fingers of one hand simultaneously to one-half of the plate. This represents the control area.

2. Dip the fingertips into a beaker of disinfectant for 10 seconds, then place them under a gentle stream of water to remove the excess disinfectant.
   - Allow the fingers to air-dry.
3. Touch the fingertips to the second half of the plate, making another five-finger impression.
   - Incubate the plate at 37°C for 24 to 48 hours in the inverted position and then refrigerate until observed.
4. Observe the plate and note the number and type of bacterial colonies from the control and disinfected fingertips.
   - Place a labeled representation in the Results section, and record your observations on the effectiveness of the disinfectant.
5. This exercise also can be used to show the effectiveness of hand washing on the skin.
   - Prepare a plate of nutrient agar as in step 1.
   - Make a five-finger impression on the control side, then proceed to the sink area and wash your hands by any of several methods for the experimental side. You may try a "normal" wash; or a "vigorous" wash; or a wash with a disinfectant soap or a hand sanitizer; or a wash with a brush (such as a nail brush); or any other experimental condition suggested.
   - Incubate the plate as in step 3, and observe the number and type of bacteria in the control and experimental sides.

Quick Procedure

Effect of Disinfectants on Skin Surface

1. Make a 5-finger impression on plate of nutrient agar.
2. Disinfect fingers for 10 sec.
3. Make a second 5-finger impression.
4. Incubate.
5. Evaluate effect of disinfectant.

Name: ______________________ Date: __________ Section: __________

Evaluation of Disinfectants and Antiseptics: Effectiveness of Disinfectants and Hand Washing on the Skin Surface

EXERCISE RESULTS 33

RESULTS

EFFECTIVENESS OF DISINFECTANTS AND HAND WASHING ON THE SKIN SURFACE

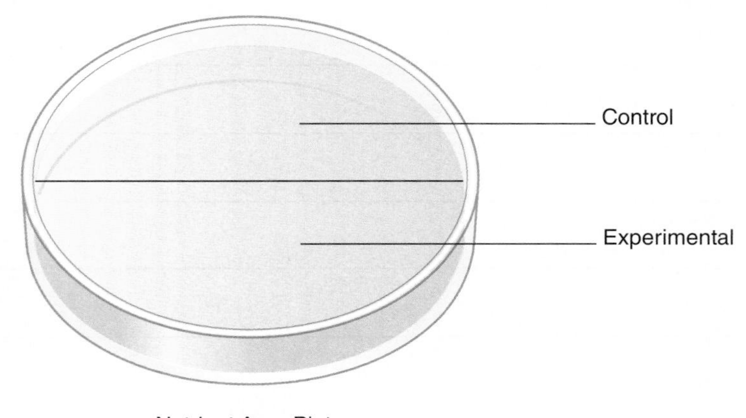

Nutrient Agar Plate

Nutrient Agar Plate

Observations and Conclusions

____

____

____

____

____

Questions

1. Why would a large mass of bacteria in a sector of the plates not necessarily imply poor disinfection by the chemical agent?

2. Would it be possible to sterilize your hands? Support your answer.

Evaluation of Disinfectants and Antiseptics: Mouthwashes and Oral Bacteria

EXERCISE 34

Materials and Equipment

- Sterile tubes for collecting saliva
- Sterile saline and swabs
- Sterile water
- Sterile pipettes and mechanical pipetters
- Nutrient agar plates, or materials for their preparation
- Commercial mouthwashes
- Paraffin (optional)
- Disinfectant
- Sterile forceps, straws, applicator sticks, filter paper disks

Disinfection is essential for controlling the spread of microorganisms. A good **disinfectant** should kill microorganisms on a surface, but it also must have low toxicity to humans, have a substantial shelf life, be useful in dilute form, and be inexpensive and easy to obtain. Numerous other factors also contribute to the value of a good disinfectant.

MOUTHWASHES AND ORAL BACTERIA

PURPOSE: to test the effectiveness of mouthwashes on killing oral bacteria.

In the advertising media, many mouthwashes claim to "kill the germs that cause bad breath." This claim can be tested in the laboratory by exposing oral bacteria from saliva to a series of mouthwashes and determining the killing effects of these mouthwashes.

PROCEDURE

1. Obtain a plate of nutrient agar or prepare one at the direction of the instructor.
   - On the bottom, label the plate with your name, the date, and the designation "mouthwash."
2. Expectorate up to 10 ml of saliva into an empty sterile test tube (chewing a piece of paraffin will help salivation).
   - Fill a second tube with 5 ml of saline, and using a mechanical pipetter, pipette 5 ml of the saliva into the saline.
   - Mix the contents of the tube thoroughly. This tube will be the source of oral bacteria for the exercise.
3. Dip a sterile cotton swab into the diluted saliva and swab the plate from side to side, covering the entire area of the plate.
   - Dip the swab into the saliva a second time and repeat the swabbing, but at a different angle.
   - Dip the swab and repeat the swabbing a third time at a third angle. This heavy inoculation is necessary because the number of oral bacteria growing on a nutrient agar plate is usually not high.

An alternative is to "lick" a plate of sterile nutrient agar. It is sterile, so there is no harm in licking it with your tongue.

   - Discard the swab in a beaker of disinfectant.
4. Obtain a sterile straw and remove a plug of nutrient agar by pushing the straw into the agar and holding the mouth end of the straw with the finger as described in Exercise 21.
   - Push the plug out of the straw with a long sterile applicator stick.
   - Remove as many plugs of agar as mouthwashes to be tested.
5. Using a sterile pipette, place a sample of a selected mouthwash into one of the wells in the agar.
   - Using fresh pipettes, fill the remaining wells with different mouthwashes.
   - Label the tops of the plate with the mouthwash names and place an identifying mark to line up the top and bottom of the plate.
   - An alternative way of performing the test is explained in Step 8.
6. Incubate the plate in the upright position at 37°C for 48 hours or as directed by the instructor.
7. At the end of the incubation period, examine the plate for evidence of bacterial inhibition near the disks.
   - Bacterial inhibition is indicated by a clear **zone of inhibition** surrounding the well.
   - If no zone of inhibition is present, you may assume that the bacteria resisted the effects of the mouthwash under the experimental conditions tested.
   - Measure the diameters of the zones (including the well) in millimeters to further express and quantify your results.
   - In TABLE 34.1, enter the name of the mouthwash tested and indicate whether the bacteria were "susceptible" or "resistant" to the effects of the mouthwash.

- Include the diameter of the zone of inhibition if determined.
- Include a statement of your observations.

8. An alternative way of testing the effect of mouthwashes is as follows:

- Inoculate agar plates as in Steps 1 to 3.
- Lightly flame a forceps and use it to obtain a sterile disk.
- Impregnate the disk with a sample of a selected mouthwash (let the mouthwash adsorb onto the disk) and place the disk onto the agar surface (see Exercise 28).
- Repeat with as many different mouthwashes as available.
- Incubate the plate at 37°C for 48 hours in the inverted position or as instructed.
- Observe the plates as explained in Step 7.

Name: ______________________ Date: __________ Section: __________

Evaluation of Disinfectants and Antiseptics: Mouthwashes and Oral Bacteria

EXERCISE RESULTS 34

RESULTS

MOUTHWASHES AND ORAL BACTERIA

TABLE 34.1 Susceptibility or Resistance of Bacteria

Mouthwash	Zone Diameter	Susceptible or Resistant

Nutrient Agar Plate

Observations and Conclusions

Questions

1. Are there other factors that could contribute to the results observed in this Exercise? Explain.

2. Does the price of a mouthwash correlate to its capacity to kill bacteria?

3. What decisions will you make about your personal use of mouthwash based on this Exercise?

Evaluation of Disinfectants and Antiseptics: The Effect of Antibiotics on Bacteria

EXERCISE 35

Materials and Equipment

- Broth cultures of two selected bacterial species
- Plates of Mueller-Hinton agar, or materials for their preparation
- Sterile cotton swabs and beakers of disinfectant
- Antibiotic disks
- Forceps in beakers of 70% ethyl alcohol
- Inoculating loop
- Millimeter rulers

Watch This Video

- How to Prepare the Kirby-Bauer Test

It is valuable for a physician to have laboratory data indicating whether a bacterium is susceptible or resistant to various antibiotics. This data, together with knowledge of possible side effects, patient allergies, and other information, will be used in prescribing an appropriate antibiotic. The laboratory test should be rapid, accurate, and relatively inexpensive to perform.

THE EFFECT OF ANTIBIOTICS ON BACTERIA

PURPOSE: to evaluate the sensitivity of bacterial species to antibiotics.

In most laboratories, the effect of antibiotics on bacteria is determined by a standardized **Disk Susceptibility Test** The procedure also is referred to as the **Kirby-Bauer method,** after its developers (watch **Microbiology Video: How to Prepare the Kirby-Bauer Test** to familiarize yourself with the technique). Bacteria are streaked on Mueller-Hinton agar, which gives reproducible results and does not inhibit sulfonamides. Paper disks containing known amounts of the antibiotic are added, and after incubation, the plates are observed for clear **zones of inhibition** surrounding the disks. By referring to a standardized table (see Appendix A), the effect of the antimicrobial drug on the organism may be ascertained. A physician may be reasonably certain that if the stated dose of antibiotic exists in the tissues, then bacterial inhibition will take place.

How to Prepare the Kirby-Bauer Test

PROCEDURE

1. Obtain or prepare two plates of **Mueller-Hinton agar**.
   - Label them on the bottom side with your name, the date, and the name of the procedure, and designate one plate for each of two organisms to be tested.
   - The depth of the agar should be about 4 millimeters.
   - Obtain broth cultures of the two organisms and two sterile cotton swabs.
2. Dip a swab into a broth culture of bacteria, press it to the inside of the tube to express the excess fluid, and inoculate one of the Mueller-Hinton plates by making a lawn of bacteria.
   - Perform the swabbing at three different angles, and make a final sweep around the rim of the plate as shown in Figure 19.1 in Exercise 19.
   - Discard the swab in the beaker of disinfectant after use.
   - Inoculate the second plate with the second organism in like manner.

After swabbing plates, always place the swabs into disinfectant to avoid contaminating nearby materials.

3. With lids on, allow the plates to sit at room temperature for about 10 minutes so that the excess moisture is absorbed.
4. Using sterile, flamed forceps or an automatic dispenser (FIGURE 35.1), apply a series of antibiotic disks to the surfaces of the plates as directed by the instructor.
   - The disks should be about 3 cm from each other to prevent overlapping of antibiotic, and at least 2 cm from the edge of the plate (FIGURE 35.2).
   - If the forceps is used, it should be flamed lightly after applying each disk.
   - Apply light pressure to the disks using a forceps or inoculating loop to ensure firm contact.
5. Incubate the plates in the inverted position at 37°C for 16 to 18 hours, and then refrigerate them until observed. The short incubation time is necessary because bacteria may overcome the inhibition, and bacterial over-growth may occur during extended periods.

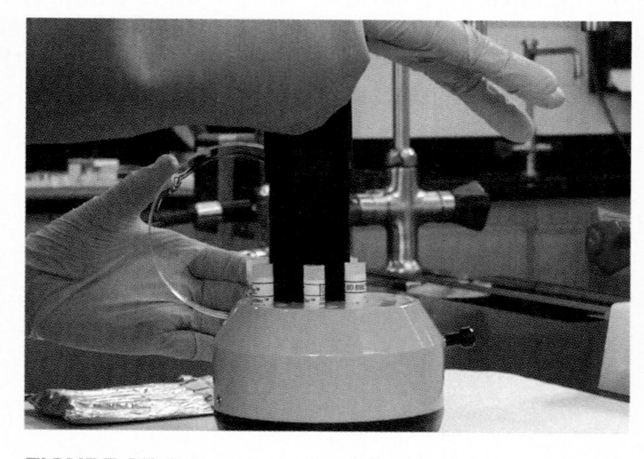

FIGURE 35.1 An automatic disk dispenser.

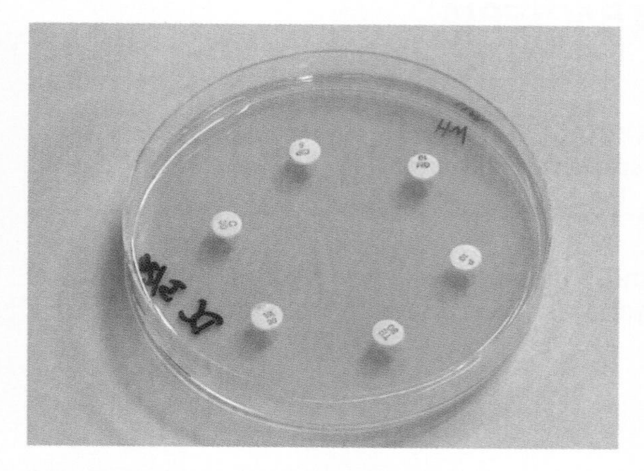

FIGURE 35.2 Antibiotic disks on the plate surface.

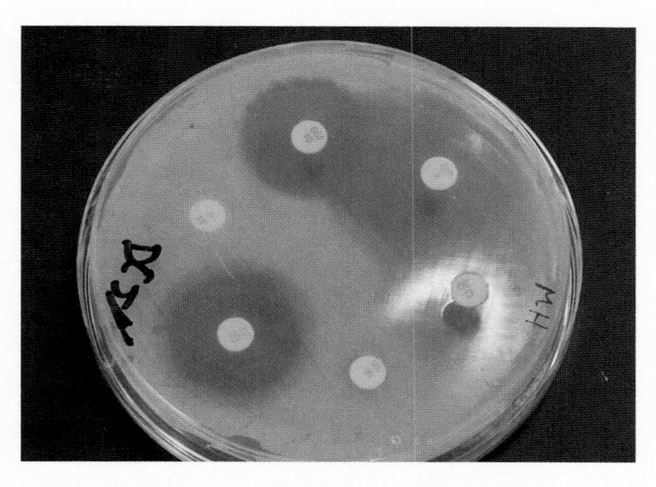

FIGURE 35.3 Zones of inhibition on antibiotic susceptibility plates.

6. Examine the plates for the presence of clear zones of inhibition surrounding the disks, as shown in FIGURE 35.3. These zones represent areas where bacterial growth was inhibited by the antibiotic.
   - With a millimeter ruler, measure the diameter of the zone, including the disk. Measurements should be made to the nearest millimeter.
   - Use the bottom of the plate for your measurements. Enter the determinations in TABLE 35.1 in the Results section.
7. Referring to **Appendix A**, determine whether the organism was susceptible to the antibiotic (S), showed intermediate susceptibility (I), or was resistant (R).
   - Enter these determinations in Table 35.1 next to the diameters.
   - Also, note whether tiny colonies appear within the zone of inhibition.
   - Enter diagrams of the plates in the spaces provided.

Quick Procedure

Antibiotic Susceptibility Test

1. Inoculate a plate of test medium with bacteria.
2. Place antibiotic disks on the surface.
3. Incubate.
4. Observe and measure zones of inhibition.
5. Determine S, I, or R from Appendix A.

Name: ______________________ Date: __________ Section: __________

Evaluation of Disinfectants and Antiseptics: The Effect of Antibiotics on Bacteria

EXERCISE RESULTS 35

RESULTS

TABLE 35.1 Susceptibility or Resistance of Bacteria

Antibiotic	Organisms Tested			
	1.		2.	
	Zone Diameter	S, I, or R	Zone Diameter	S, I, or R

Observations and Conclusions

____

____

____

____

____

Questions

1. List several factors that must be standardized to enhance the usefulness of the antibiotic susceptibility test by the Kirby-Bauer method.

2. Which antibiotics may be broad-spectrum antibiotics and which may be narrow-spectrum antibiotics, as determined from the results in this exercise?

3. How might one explain the appearance of colonies of bacteria within a zone of inhibition?

4. Why is it more advisable to use a swab instead of a loop for preparing a lawn of bacteria on an agar plate?

5. Explain why a shorter incubation time of 16 to 18 hours is preferred in this test to the usual 24 to 48 hours.

Estimating Virus and Bacterial Cell Numbers and Measuring Population Growth

PART VII

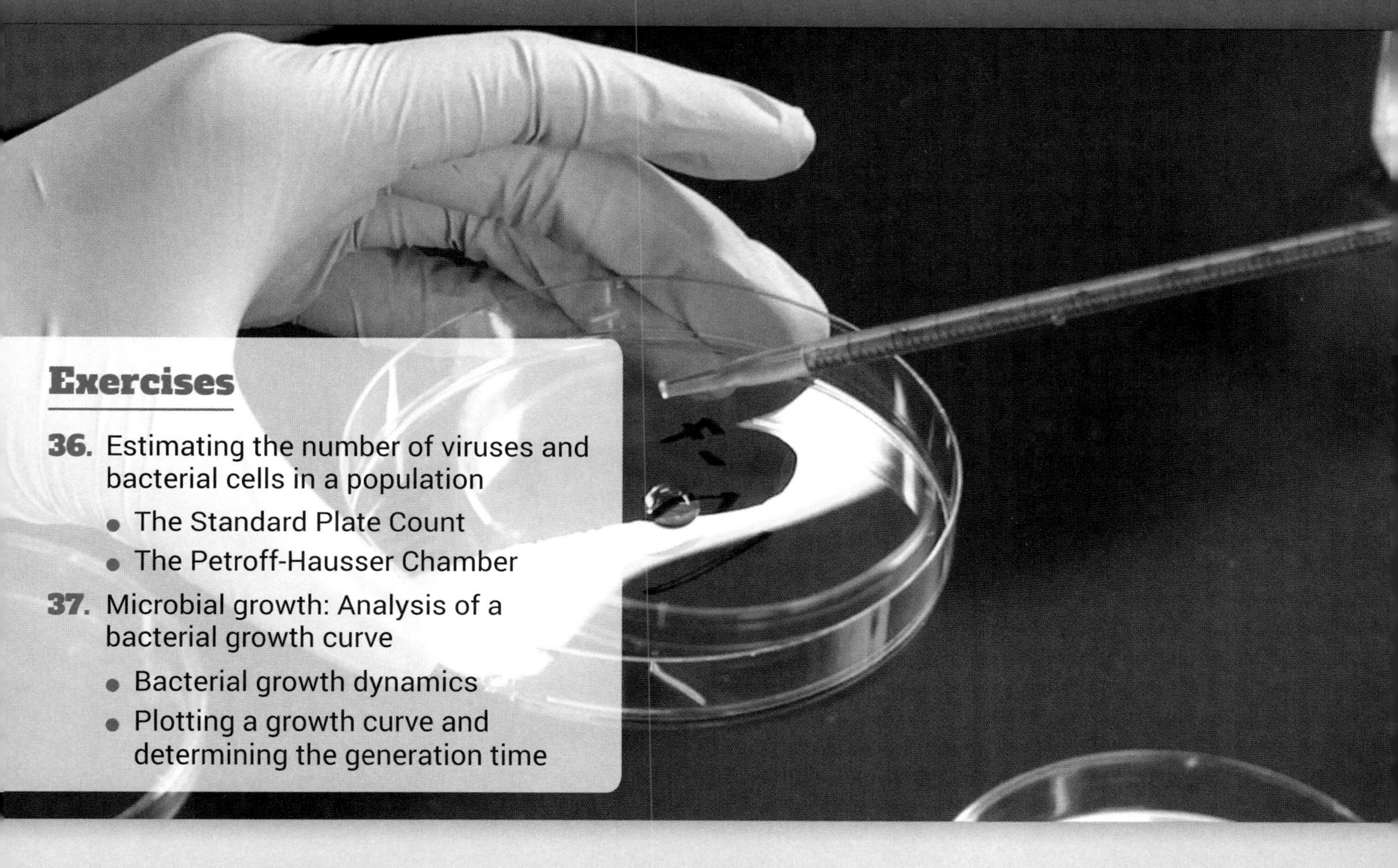

Exercises

36. Estimating the number of viruses and bacterial cells in a population
- The Standard Plate Count
- The Petroff-Hausser Chamber

37. Microbial growth: Analysis of a bacterial growth curve
- Bacterial growth dynamics
- Plotting a growth curve and determining the generation time

Learning Objectives

When you have completed the exercises in Part VII, you will be able to:

- Estimate virus and bacterial cell numbers and determine the generation time for a bacterial cell population.

Watch These Videos

- How to Prepare a Standard Plate Count
- How to Use a Petroff-Hausser Chamber
- How to Use a Spectrophotometer

It is important in microbiology to be able to count—or at least estimate—the number of viruses or bacterial cells in a population. That population could be drinking water, milk, a piece of food, or sewage.

In this section, you are going to use different methods to estimate the numbers of viruses (bacteriophages) and bacterial cells in a liquid sample. Often, the sample taken from the population and suspended in broth must be diluted. Doing so brings the number of viruses or bacterial cells down to a manageable number for estimations. Once you have diluted the sample, you can use different methods to estimate the number of viruses or bacteria. The methods you will focus on in this section are preparing a plaque assay (Exercise 36), using a counting chamber (Exercise 36), and using a spectrophotometer (Exercise 37). Watch the videos in this section to see how each procedure is performed before performing the exercises.

Estimating the Number of Viruses and Bacterial Cells in a Population

EXERCISE **36**

Materials and Equipment

- Young broth cultures of *E. coli* strain B
- Suspensions of bacteriophage T4
- Plates of nutrient hard agar
- Tubes containing 1.8 ml sterile, nutrient soft agar (kept at 45°C)
- Bottles (or tubes) containing 9.9 ml sterile, nutrient broth
- Mechanical pipetters
- 1 ml sterile pipettes
- Petroff-Hausser slide
- Coverslip
- Microscope

- How to Prepare a Standard Plate Count
- How to Use a Petroff-Hausser Chamber

The ability of bacteriophages to replicate in and lyse susceptible host bacteria to form plaques can be used to enumerate (count) the number of phage particles in a solution or sample. Such a method is referred to as a **plaque assay**.

ESTIMATING THE NUMBER OF BACTERIOPHAGES

PURPOSE: to estimate the number of phage particles in a sample.

The plaque assay uses a double layer of agar. A "soft agar," consisting of a mixture of phage and susceptible bacterial cells in molten agar, is poured over a "hard agar" base. This forms an upper layer of softer agar on solidification. During the incubation period at 37°C, susceptible *Escherichia coli* cells will divide rapidly and form an even lawn of bacterial cells. After a phage particle attaches to, penetrates, and replicates in the host cell, phage release lyses the host cell. The released phages then infect surrounding cells and quickly produce a single plaque in the bacterial lawn.

Each plaque formed is referred to as a **plaque-forming unit** (**PFU**) and is assumed to be the result of one initial phage particle. Each PFU then represents a phage particle in the original mixture. Therefore, plaque assay uses the number of PFUs formed to estimate the number of infective phages in the diluted mixture and extrapolated back to estimate the numbers in original sample. Often, the number in the original sample will be too large to write out, so scientists use a method called **scientific notation** to quantify large numbers. Watch the **Microbiology Video: How to Prepare a Standard Plate Count** to see a variation of this procedure in action.

How to Prepare a Standard Plate Count

PROCEDURE

Enumeration requires that there be a sufficient number of phages present to get a statistically accurate count, yet not so many phage particles that counting plaques becomes impossible. Therefore, a series of dilutions of the phage should be made.

1. To prepare dilutions, first label all dilution tubes and media with your name and the dilution designations as described below (FIGURE 36.1).
   a. Four nutrient broth bottles: labeled 10^{-2}, 10^{-4}, 10^{-6}, 10^{-8}.
   b. Four nutrient agar "soft agar" tubes: labeled 10^{-3}, 10^{-5}, 10^{-7}, 10^{-9}.
   c. Four nutrient agar "hard agar" plates: labeled 10^{-3}, 10^{-5}, 10^{-7}, 10^{-9}.

   These plates contain 2 ml of molten agar that was allowed to solidify.
2. Maintain the four nutrient agar "soft agar" tubes in a 45°C water bath.
3. Use a mechanical pipetter and sterile pipettes to make 100-fold serial dilutions of the phage suspension as follows (FIGURE 36.2):
   a. Pipette 0.1 ml of the phage suspension into the nutrient broth bottle labeled 10^{-2}. This is a 1/100 (10^{-2}) dilution. Mix by gently swirling the bottle.

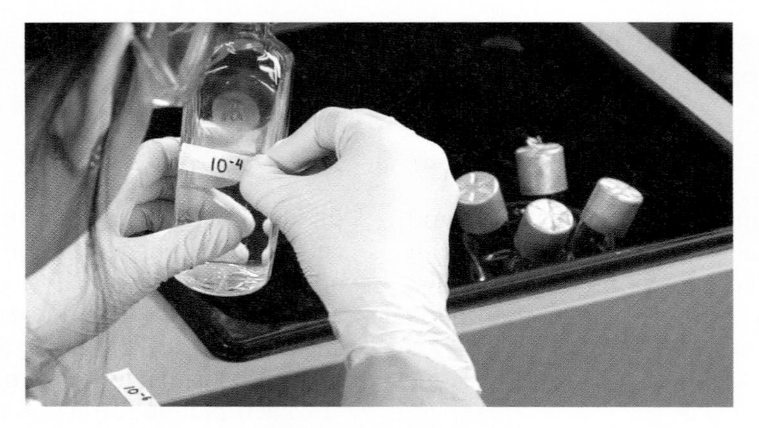

FIGURE 36.1 Label all dilution tubes and media.

FIGURE 36.2 You will need a mechanical pipetter for the dilution process.

b. Using another sterile pipette, transfer 0.1 ml of the phage suspension from the 10^{-2} dilution into the nutrient broth bottle labeled 10^{-4}. This is a 1/100 (10^{-2}) dilution. Mix.

c. With another sterile pipette, transfer 0.1 ml of the phage suspension from the 10^{-4} dilution into the nutrient broth bottle labeled 10^{-6}. This is a 1/100 (10^{-2}) dilution. Mix.

d. With one more sterile pipette, transfer 0.1 ml of the phage suspension from the 10^{-6} dilution into the nutrient broth bottle labeled 10^{-8}. This is a 1/100 (10^{-2}) dilution. Mix.

Don't let the soft agar tubes solidify.

4. To each nutrient "soft agar" tube, add the following:

   a. To the soft agar tube labeled 10^{-3}, aseptically add two drops of the *E. coli* B culture with a sterile pipette and 0.2 ml of the 10^{-2} nutrient broth phage dilution. This is a 1/10 (10^{-1}) dilution. Rapidly mix by rotating the tube between the palms of your hands and pour the contents over the hard nutrient agar plate labeled 10^{-3}. Swirl the plate gently to spread the agar. Allow the soft agar to harden.

   b. To the soft agar tube labeled 10^{-5}, aseptically add two drops of the *E. coli* B culture with a sterile pipette and 0.2 ml of the 10^{-4} nutrient broth phage dilution. This is a 1/10 (10^{-1}) dilution. Rapidly mix by rotating the tube between the palms of your hands and pour the contents over the hard nutrient agar plate labeled 10^{-5}. Swirl the plate gently to spread the agar. Allow the soft agar to harden.

   c. To the soft agar tube labeled 10^{-7}, aseptically add two drops of the *E. coli* B culture with a sterile pipette and 0.2 ml of the 10^{-6} nutrient broth phage dilution. This is a 1/10 (10^{-1}) dilution. Rapidly mix by rotating the tube between the palms of your hands and pour the contents over the hard nutrient agar plate labeled 10^{-7}. Swirl the plate gently to spread the agar. Allow the soft agar to harden.

   d. To the soft agar tube labeled 10^{-9}, aseptically add two drops of the *E. coli* B culture with a sterile pipette and 0.2 ml of the 10^{-8} nutrient broth phage dilution. This is a 1/10 (10^{-1}) dilution. Rapidly mix by rotating the tube between the palms of your hands and pour the contents over the hard nutrient agar plate labeled 10^{-9}. Swirl the plate gently to spread the agar. Allow the soft agar to harden.

5. Once the soft agar over-lay has solidified, incubate the four agar plates in an inverted position for 24 hours at 37°C. If the plates cannot be examined after 24 hours, refrigerate them until the next laboratory session.

6. The number of phage particles contained in the original phage suspension can be estimated by counting the number of plaques formed in the soft agar. This number then is multiplied by the dilution factor. For a valid phage count, the number of plaques per plate should not exceed 300 or be fewer than 30.

   - For example, suppose you count 100 PFUs in the 10^{-7} dilution plate. $(100) \times (10^{7}) = 100 \times 10^{7}$ or 1×10^{9} PFUs per ml of phage suspension.

7. Record your calculations in the Results section.

8. Other samples also could be examined for phage concentrations. The sewage sample used in Exercise 20 might be a very interesting sample to quantify.

One method for counting the number of bacterial cells in a population is to use a counting chamber. In this exercise, you will be using a Petroff-Hausser chamber. Watch **Microbiology Video: How to Use a Petroff-Hausser Chamber** to see how the procedure is done.

How to Use a Petroff-Hausser Chamber

USING A PETROFF-HAUSSER CHAMBER

PURPOSE: to estimate the number of bacterial cells in a broth sample by direct counting.

This method is a process of directly counting cells. You don't know if the bacterial cells are alive or dead because dead cells won't disintegrate; they will just look like regular cells. Thus, in this exercise, you are counting the total cell population, both living and dead.

PROCEDURE

1. Place a coverslip over a Petroff-Hausser slide.
2. Using capillary action from a pipette, add a small volume of bacterial suspension under the coverslip on the Petroff-Hausser slide (FIGURE 36.3). Discard the pipette.
3. Allow the counting chamber to stand for a few minutes to allow the cells to settle.
4. Gently place the chamber on the microscope stage (FIGURE 36.4).
5. Using the 4× lens, center the grid over the light.
6. Switch to the 40× lens and center one of the double-line squares.

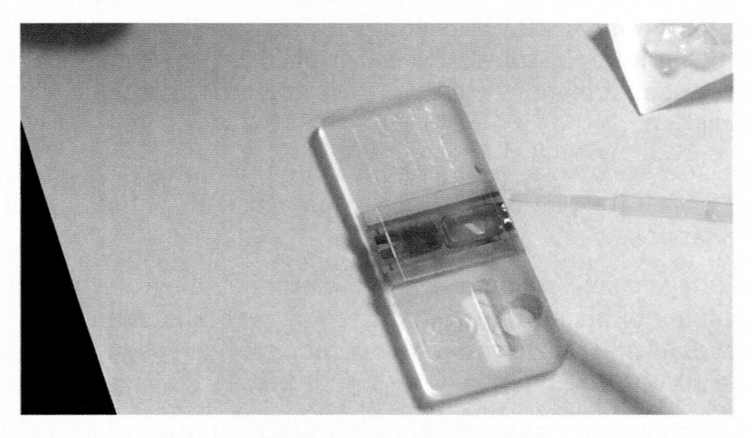

FIGURE 36.3 Use capillary action to add bacterial suspension to the chamber.

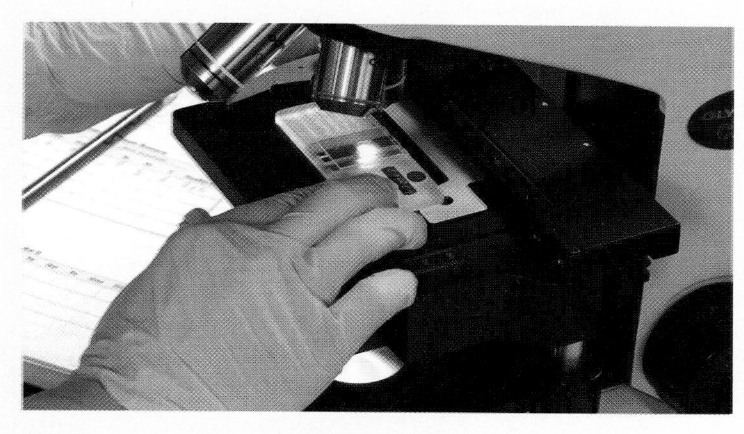

FIGURE 36.4 Place the Petroff-Hausser chamber and slide onto the microscope stage.

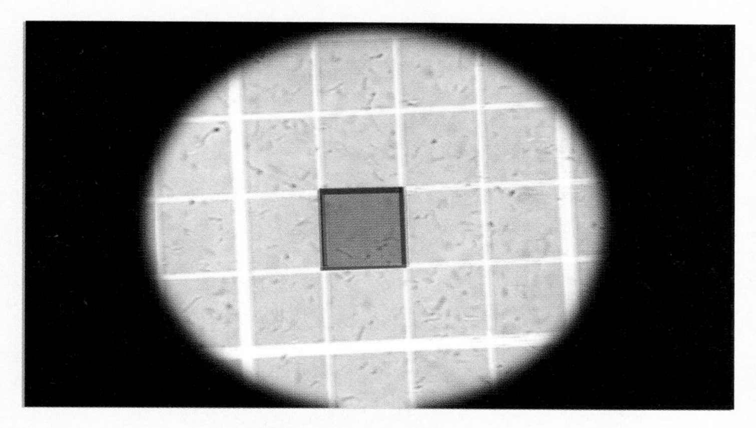

FIGURE 36.5 Count only the cells in the square that are not touching the bottom or right sides.

7. Count the cells in one single-line square, including cells touching the top and left sides. Do not count cells touching the bottom or right sides (FIGURE 36.5).

Do not count cells touching the bottom or right sides of the square, or you will end up overcounting.

8. Continue counting cells in 10 squares. Record the number of cells in the Results section. From that number and the volume of liquid that's in the chamber, you can calculate the actual number of bacterial per milliliter.
9. Record your observations.
10. Disinfect the slide and coverslip.

Name: ______________________ Date: __________ Section: __________

Estimating the Number of Viruses and Bacterial Cells in a Population

EXERCISE RESULTS **36**

RESULTS

ESTIMATING THE NUMBER OF BACTERIOPHAGES

Number of plaques counted: ______________________

Number of estimated bacteriophages in the population: ______________________

Observations and Conclusions

____

____

____

____

____

USING A PETROFF-HAUSSER CHAMBER

Number of cells counted: ______________________

Number of estimated cells in the population: ______________________

Observations and Conclusions

____

____

____

____

____

Questions

1. In the phage enumeration exercise, what is the significance of having the "hard agar" base?

2. Which method is more effective for estimating bacteriophages or bacterial cells? Why?

Microbial Growth: Analysis of a Bacterial Growth Curve

EXERCISE 37

Materials and Equipment

- 5-hour brain-heart infusion (BHI) broths of *Escherichia coli*
- 1 tube with 5.3 ml sterile BHI broth
- 1 tube with 5 ml sterile BHI broth
- 2 sterile spec tubes
- 20° and 37°C incubators
- Spectrophotometer
- 1 ml sterile pipettes

Watch This Video

- How to Use a Spectrophotometer

Bacterial growth is usually defined as the increase in numbers of viable bacterial cells in a population. For optimal growth, several physical factors must be met. These include an optimal pH for the growth medium, an optimal temperature, and adequate oxygen gas (for aerobic microbes). Growth of heterotrophic microorganisms also requires specific nutritional factors, including carbon, nitrogen, sulfur, phosphorus, cofactors, vitamins, and of course, water. These nutritional requirements are supplied in the environment or by the culture medium in which the bacteria are inoculated.

Under these conditions, bacterial cells will undergo rapid **binary fission**. The dynamics of bacterial growth follow a predictable pattern visualized as a **population growth curve**. A growth curve is generated by plotting the increase in cell number versus time of exposure or length of incubation in the growth medium. The curve then can be used to determine the **generation time**, the time required for a microbial population to double in cell number. In the time allotted for this lab, you will not have time to follow a complete growth curve. However, you can study that portion of the curve (log phase) needed to estimate the generation time.

You can appreciate how hard it would be to count individual bacterial cells! Therefore, a simpler but indirect method to estimate bacterial cell number is to use a **spectrophotometer** (FIGURE 37.1). This instrument uses a beam of light that is projected at a tube of liquid (growth medium in this case) placed in the instrument. The medium and any bacteria present will absorb some of the light. The light not absorbed will pass through the tube and be detected by a photoelectric device that converts the light intensity into a measurable number. For the light

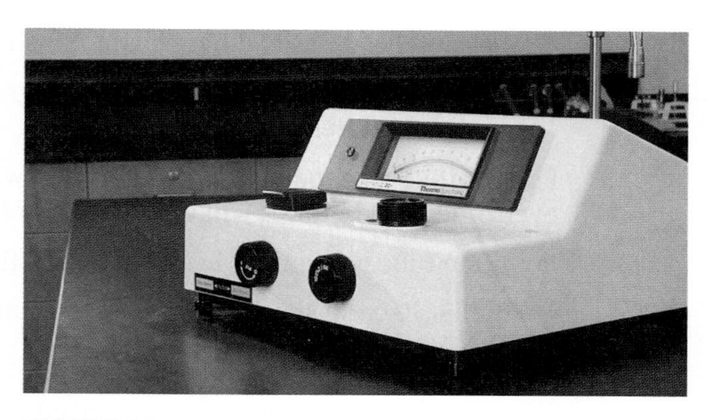

FIGURE 37.1 A spectrophotometer.

absorbed, this is called the **absorbance**. As growth progresses and the number of cells increase, the absorbance will increase also.

BACTERIAL GROWTH DYNAMICS

> **PURPOSE:** to examine the population growth dynamics of bacterial cultures growing under specified growth conditions.

Under a set of permissive growth conditions as described in the earlier lab exercises in this section, bacterial cells in a broth culture will reproduce at a steady rate. Such growth follows a specific set of phases or stages that are described and drawn as a growth curve (FIGURE 37.2). Before starting the procedure, watch **Microbiology Video: How to Use a Spectrophotometer** to see how this exercise is done.

How to Use a Spectrophotometer

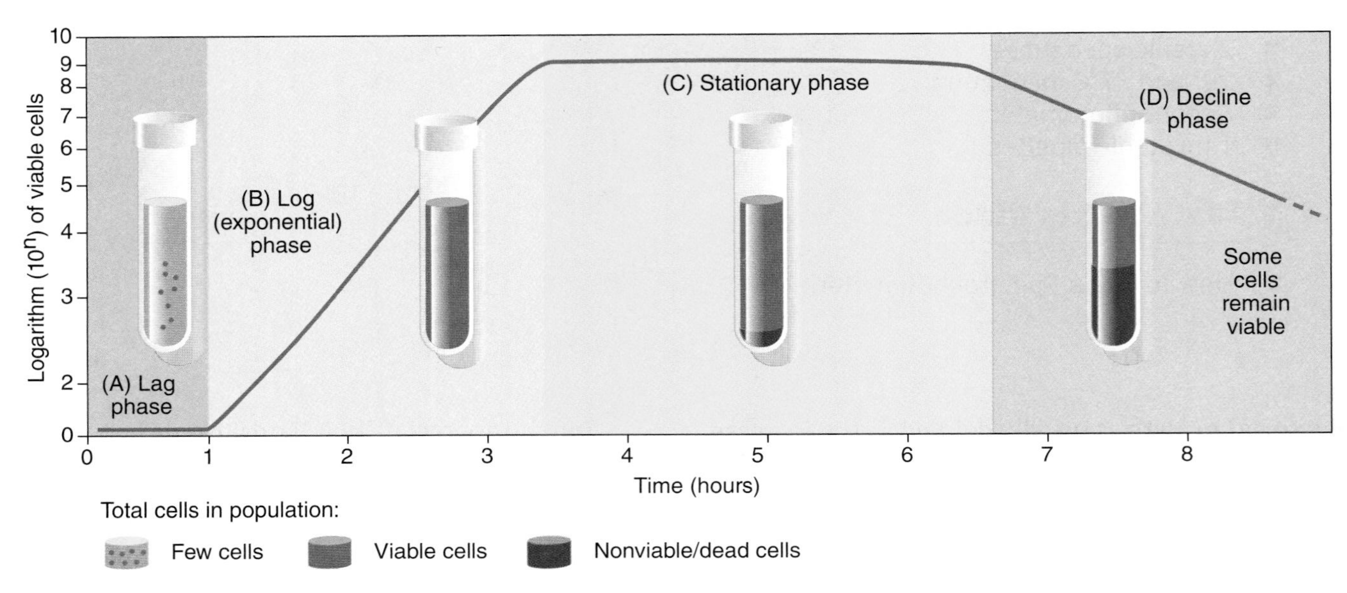

FIGURE 37.2 Four phases of a bacterial growth curve.

PROCEDURE

1. This experiment is best done in lab groups of 2–3 individuals.
   - The exercise described here examines the effect of temperature on growth.
   - Your instructor may have different groups examine other physical or chemical factors governing bacterial growth.
2. For this exercise, odd-numbered groups will incubate their broth culture tubes at 20°C while even-numbered groups will incubate their tubes at 37°C.
3. Pour one tube containing 5.3 ml sterile BHI broth into a sterile spectrophotometer (spec) tube.
   - Standardize the spectrophotometer as directed by your instructor.
4. Label the tube as **Stnd**.
   - Keep this tube because you will need to restandardize the spectrophotometer before every new reading.

5. Label another sterile spec tube with your group's identification.
   - Add 0.3 ml of 5-hour *E. coli* into the tube containing 5 ml of BHI broth.
   - Carefully, but quickly, pour the suspension into the labeled Spec tube. Label the tube as **Expt.**
6. Immediately determine the absorbance of this suspension at 590 nm as demonstrated by the instructor. This will represent the zero time reading for the growth curve.
   - The initial (0 time) absorbance reading should be between 0.05 and 0.15.
   - Record your absorbance in the 0 min time box in TABLE 37.1 in the Results section.
7. After recording the absorbance, incubate the tube at the designated temperature for 20 minutes.
8. Now, every 20 minutes, remove your Spec tube from the incubator and determine the absorbance.
   - **Remember to first standardize the spectrophotometer with your Stnd tube!**
   - Record the absorbance in Table 37.1 and then return the tube to the incubator for another 20 minutes.
9. You will need absorbance readings every 20 minutes through 140 minutes.

PLOTTING A GROWTH CURVE AND DETERMINING THE GENERATION TIME

PURPOSE: to plot a growth curve for such a bacterial culture and estimate the generation time.

The dynamics and phases of microbial growth can be followed and plotted as a population growth curve by graphing the absorbance readings versus time. In other words, the growth curve represents the change in cell numbers with time of incubation.

The curve also can be used to determine the population's generation time. This indirect determination of doubling is made as follows: select two points on the absorbance scale, such as 0.3 and 0.6, that represent a doubling in absorbance (FIGURE 37.3). As shown with the dashed lines in the graph, extrapolate to their respective time intervals. The difference is the generation time; in this example, it would be approximately 20 minutes.

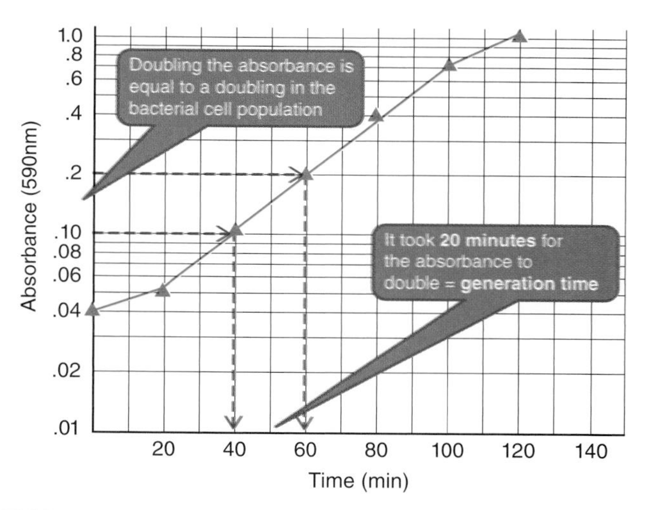

FIGURE 37.3 The estimation of the generation time from the log phase of a bacterial growth curve.

PROCEDURE

1. As you get your absorbance readings, take the data from your table and plot the data on the graph paper provided in the Results section.
    - **You need to plot absorbance (590 nm) versus time (min) of incubation.**
2. After you have determined that the bacterial growth curve is in log phase, calculate the generation time for your culture conditions.
    - Extrapolate from the absorbance scale on the plotted growth curves as explained above.
    - Record your results in TABLE 37.2 in the Results section.

Name: ______________________ Date: __________ Section: __________

Microbial Growth: Analysis of a Bacterial Growth Curve

EXERCISE RESULTS 37

RESULTS

BACTERIAL GROWTH DYNAMICS

TABLE 37.1 Absorbance

Time (min)	Absorbance (590 nm)							
	BHI (20°C)				BHI (37°C)			
	Group 1	Group 3	Group 5	Group 7	Group 2	Group 4	Group 6	Group 8
0								
20								
40								
60								
80								
100								
120								
140								

PLOTTING A GROWTH CURVE AND DETERMINING THE GENERATION TIME

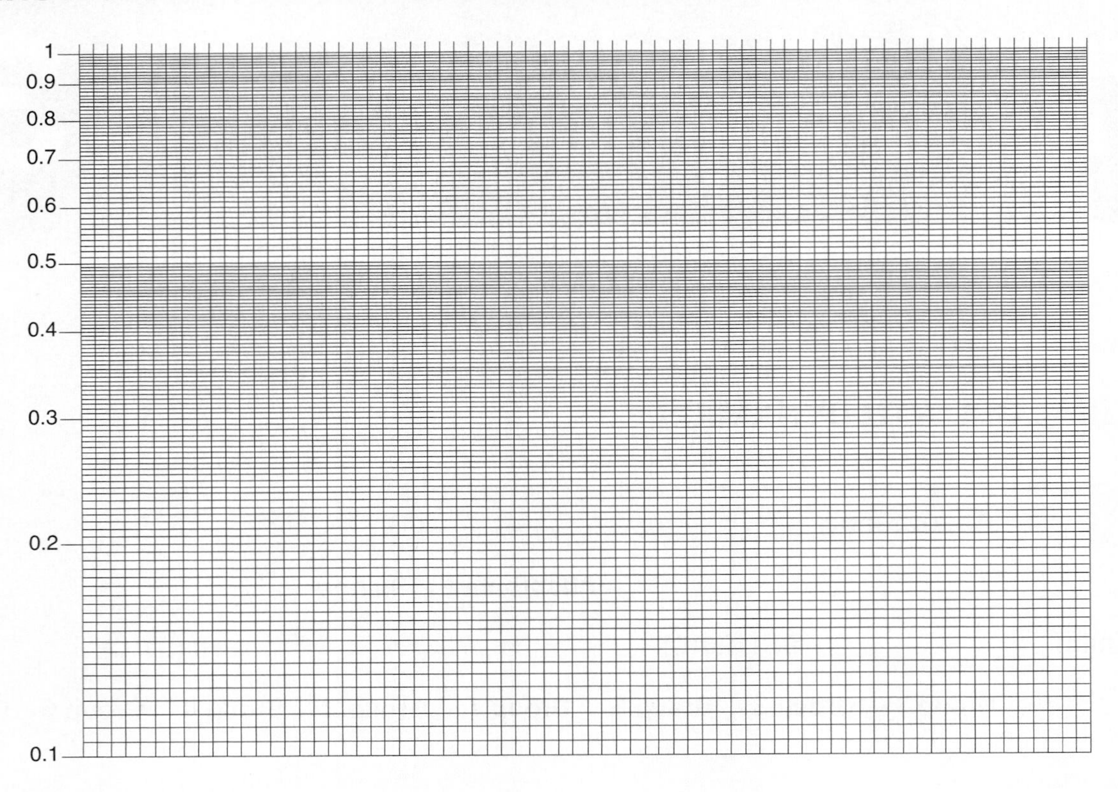

TABLE 37.2 Generation Time

Sample	Calculated Generation Time (min)
BHI (20°C)	
BHI (37°C)	

Observations and Conclusions

___

___

___

___

___

Questions

1. Why do variations in generation time exist in the plotted curves?

2. Does the phrase "microbial growth," such as in bacteria, mean the same thing as when describing human growth? Explain.

3. Propose an explanation for the growth differences at 20°C and 37°C.

4. Answer each of the following questions based on the growth curves (A, B, and C) illustrated in the graph.

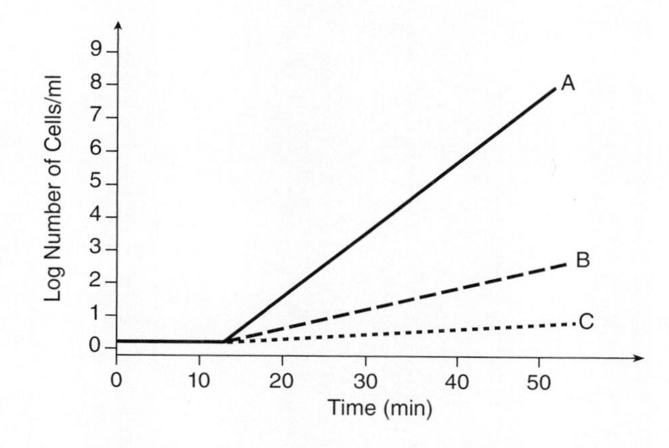

a. Which growth curve (A, B, or C) most closely matches that of a normal bacterium like *Escherichia coli* when growing in the colon? From the growth curve, figure out this bacterium's generation time.

____

____

b. Which growth curve (A, B, or C) would most likely match that of a psychrotrophic bacterium growing at 20°C? From the growth curve, figure out this bacterium's generation time.

____

____

____

Medical Microbiology

PART VIII

Exercises

38. The genus *Mycobacterium*: Colony and cellular morphology

39. The genus *Mycobacterium*: Acid-fast stain technique

40. The genus *Mycobacterium*: Cold acid-fast stain technique

41. The genus *Streptococcus*: Streptococci from the upper respiratory tract

42. The genus *Streptococcus*: Streptococci from the oral cavity

43. The genus *Neisseria*

44. The genus *Staphylococcus*: Isolation of staphylococci

45. The genus *Staphylococcus*: Differentiation between staphylococcal species

46. The enteric bacteria: Isolation of enteric bacteria

47. The enteric bacteria: Differentiation of enteric bacteria on TSI agar

48. The enteric bacteria: The IMViC series

- Indole test
- Methyl red test
- Voges-Proskauer test
- Citrate test

49. The enteric bacteria: Rapid identification of enteric bacteria

50. The genus *Bacillus*

51. The genus *Clostridium*

52. The genus *Lactobacillus*: Isolation of lactobacilli

53. The genus *Lactobacillus*: Caries susceptibility test

Learning Objectives

When you have completed the exercises in Part VIII, you will be able to:

- Perform an acid-fast stain to identify *Mycobacterium* species.
- Isolate and culture streptococci from the upper respiratory tract and oral cavity.
- Identify *Neisseria* species isolated from the throat.
- Use selective and differential media, and employ biochemical tests, to isolate and differentiate staphylococci.
- Use selective and differential media to isolate enteric bacteria, and employ TSI agar and the IMViC series to identify enteric species.
- Identify enteric bacteria using the Enterotube II system.
- Perform the appropriate biochemical tests and staining procedures to isolate and identify members of the genus *Bacillus*.
- Employ anaerobic techniques to isolate and identify *Clostridium* species.
- Isolate lactobacilli from a mixed population and perform a caries susceptibility test.
- Distinguish between selective and differential media.
- Complete the aseptic transfer of microorganisms.

Watch These Videos

- Preparing the Acid-Fast Stain
- Oxidase Test
- IMViC: Indole Test
- IMViC: Methyl Red Test
- IMViC: Voges-Proskauer Test
- IMViC: Citrate Test

Revolutions, be they political, social, or scientific, are events that transform the world in which we live. Certainly, this applies to the Classical Golden Age of microbiology (1854–1914) and the breakthrough in identifying bacteria as a cause of infectious disease. Pasteur, Koch, and their contemporaries were among the first to identify a specific bacterium with a specific disease. Since that time, medical microbiology has made great advances in laboratory diagnosis of disease, and in understanding pathogenesis and the role of immunity in the infectious cycle. Yet, many of the diseases common in the late 1800s remain with us today.

Tuberculosis (TB) was called the white plague when Koch was studying the organism because of the white granules present in the lungs of its victims. TB is caused by just one of more than 80 species in the genus ***Mycobacterium***. Most species, though, are not pathogenic. Identification includes its mold-like morphology and thick, waxy cell walls that can be identified with an acid-fast staining technique (Exercises 39 and 40). The genus ***Streptococcus*** consists of nearly 40 species of gram-positive cocci that typically grow in chains or pairs. Certain groups can be identified by their hemolytic activity in culture and others by their growth on selective and differential media (Exercises 41, 42).

The genus ***Neisseria*** consists of gram-negative diplococci. The pathogenic members are difficult to distinguish using morphological and cultural characteristics, although they are recognizable by the diseases they cause—a form of meningitis and gonorrhea. Many species are commensals of the upper respiratory tract (Exercise 43).

The genus ***Staphylococcus*** consists of three major species that can be identified from other bacteria and from one another by using selective and differential growth media (Exercises 44, 45).

The **enteric bacteria** consist of a large number of genera that live within or are responsible for infections in the intestinal tract. Species can be separated and identified on differential and selective growth media (including a rapid Enterotube II system), by the presence or absence of gas and acid production, and with other special tests (Exercises 46, 47, 48, 49).

The anthrax bacillus, the subject of some of Koch's most elegant work, represents the most pathogenic species of the genus ***Bacillus***. However, many species are common soil inhabitants and can be identified by several typical characteristics (Exercise 50). The genus ***Clostridium*** also consists of sporeforming species but it can be separated from *Bacillus* by its necessity for growth under anaerobic conditions. Most of the clostridia are saprobic species that naturally grow in the soil and can be isolated and studied using anaerobic techniques such as the GasPak system (Exercise 51).

Unlike some of the other species in this discussion, the genus ***Lactobacillus*** is not a significant pathogen, although it may be associated with dental caries (Exercise 53). Still, it is of medical significance because it is part of the human microbiome. *Lactobacillus* species normally residing in the vagina can protect against fomite-transmitted gonorrhea and produce a low pH that is inhibitory to other potential disease-causing microorganisms. The ability of *Lactobacillus* species to produce acid is of great significance in dairy microbiology and the genus can be easily isolated from dairy products like yogurt (Exercise 52).

Many of the exercises in Part VIII involve potential human pathogens. Therefore, it is imperative that these exercises be done in a lab equipped to handle biosafety level 2 (BSL-2) organisms (see Exercises 1 and 2). A biosafety level 1 (BSL-1) lab is not sufficient for working with BSL-2 organisms.

Many of the exercises in Part VIII also can be used to demonstrate the usefulness of special media formulations to select for or differentiate between bacterial species. **Selective media** are used to isolate (select) specific groups of bacteria growing on agar (Exercises 41, 42, 52), while **differential media** facilitate a visual differentiation of different species of bacterial colonies growing on an agar plate (Exercises 44, 46). **Enriched media**, where additional growth factors are added to a medium, also can act as a selective or differential medium (Exercises 41, 42, 43).

The Genus *Mycobacterium*: Colony and Cellular Morphology

EXERCISE 38

Materials and Equipment

- Agar cultures of *Mycobacterium smegmatis*
- Plates of Lowenstein-Jensen medium
- Sterile inoculating needle

Members of the genus ***Mycobacterium*** include several saprobic commensals of the body as well as *Mycobacterium tuberculosis*, the agent of tuberculosis, and *Mycobacterium leprae*, the cause of leprosy. The cells are straight or slightly curved rods without flagella, spores, or capsules. They often display a filamentous, funguslike growth on agar media but quickly fragment to rods when disturbed. The prefix *myco-* refers to the funguslike property. *Mycobacterium* species grow slowly in agar media and may require special supplements such as egg, potato, or serum. **Lowenstein-Jensen medium** commonly is used to cultivate them.

In this exercise, laboratory cultures of *Mycobacterium* will be examined for typical colony and cellular appearance.

COLONY AND CELLULAR MORPHOLOGY

PURPOSE: to recognize the colony and cellular characteristics of the mycobacteria.

The colony and cellular morphology of an organism are inherited traits derived from biochemical information stored in the chromosomal DNA. They are distinctive for each species and are passed from generation to generation in the genes. In this exercise, the colony and cellular morphology of *Mycobacterium* species will be studied by noting the characteristics of colonies isolated on agar media and by preparing simple stains.

PROCEDURE

1. Examine the cultures of *Mycobacterium* for unique features of this organism, including pigmentation and consistency of the growth. Use information in Exercise 5 (Figure 5.2) as a guide.
   - With a sterile inoculating needle, touch the growth to determine its texture.
   - Note your observations in the Results section.
2. Isolate colonies of *Mycobacterium* by streaking a plate of Lowenstein-Jensen medium using the streak plate technique described in Exercise 5.
   - Incubation should be at 37°C for at least 7 days because *Mycobacterium* grows slowly.
   - Examine the isolated colonies with reference to Exercise 5, and enter your observations in the Results section together with a diagram of individual colonies.

3. Examine cellular morphology by preparing an air-dried, heat-fixed smear of the organism and staining by the simple stain procedure outlined in Exercise 13.
   - Heat is necessary to assist stain penetration. Therefore, thoroughly heat-fix the slide and apply the stain immediately.
   - In the Results section, enter a representation of typical *Mycobacterium* cells.

Name: ______________________ Date: __________ Section: __________

The Genus *Mycobacterium*: Colony and Cellular Morphology

EXERCISE RESULTS **38**

RESULTS

COLONY AND CELLULAR MORPHOLOGY

Cultural characteristics:

____

____

____

Colony characteristics:

____

____

____

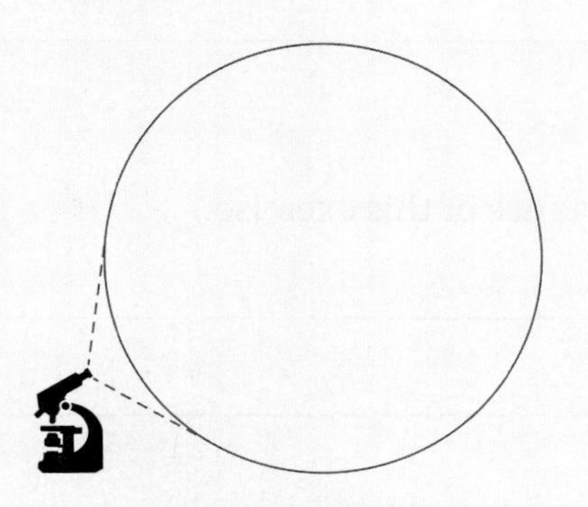

Simple Stained Smear

Organism: ______________________

Magnif.: ______________________

Colonial Morphology

Organism: ______________________

Form: ______________________

Elevation: ______________________

Margin: ______________________

Observations and Conclusions

____

____

____

____

____

Questions

1. Explain the diagnostic importance of this exercise.

____

____

____

____

____

The Genus *Mycobacterium*: Acid-Fast Stain Technique

EXERCISE **39**

Materials and Equipment

- Agar cultures of *Mycobacterium smegmatis*
- Selected nonacid-fast organisms as controls
- Ziehl-Neelsen carbolfuchsin
- Inoculating loop
- Acid alcohol
- Distilled water
- Methylene blue
- Steaming apparatus
- Microscope

Watch This Video

- Preparing the Acid-fast Stain

Members of the genus *Mycobacterium* contain an abundance of fats and waxes in their cell walls, including mycolic acid. Excessive physical treatment such as heat is necessary to penetrate this layer. However, once stain has penetrated and combined with the mycolic acid, the cells resist decolorization even when a dilute acid-alcohol solution is applied. The organisms therefore are said to be **acid-resistant** or **acid-fast**.

ACID-FAST STAIN TECHNIQUE

PURPOSE: to identify *Mycobacterium* species by the acid-fast procedure.

The **acid-fast technique** is important in the diagnosis of such diseases as tuberculosis and leprosy. The procedure is essentially similar to that devised by Ziehl and Neelsen in the 1880s. In a mixed culture, the primary stain colors all cells, but only the acid-fast organisms resist the subsequent treatment with acid-alcohol. A counterstain is then applied for bacteria that have been decolorized. These bacteria are said to be nonacid-fast. **Watch Microbiology Video: Preparing the Acid-fast Stain** for a demonstration of how to perform this exercise.

Commercially-available prepared slides can be used to replace the staining procedure.

Preparing the Acid-fast Stain

PROCEDURE

1. Set up a steaming apparatus as shown in FIGURE 39.1A and explained in Exercise 16 (for spore staining).
   - Bring the water to a rolling boil.
2. On a clean slide, prepare three air-dried, heat-fixed smears: a *Mycobacterium* species, a nonacid-fast bacterium, and a mixture of the two. The mixture is prepared by adding a loopful of one bacterium to a loopful of water, and then adding a loopful of the second bacterium before the water dries.
3. Cut a piece of blotting paper just large enough to cover the smears.
   - Place the paper over the smears, and saturate it with **carbolfuchsin**, the primary stain (FIGURE 39.2).
   - Place the slide on the rack over the boiling water.

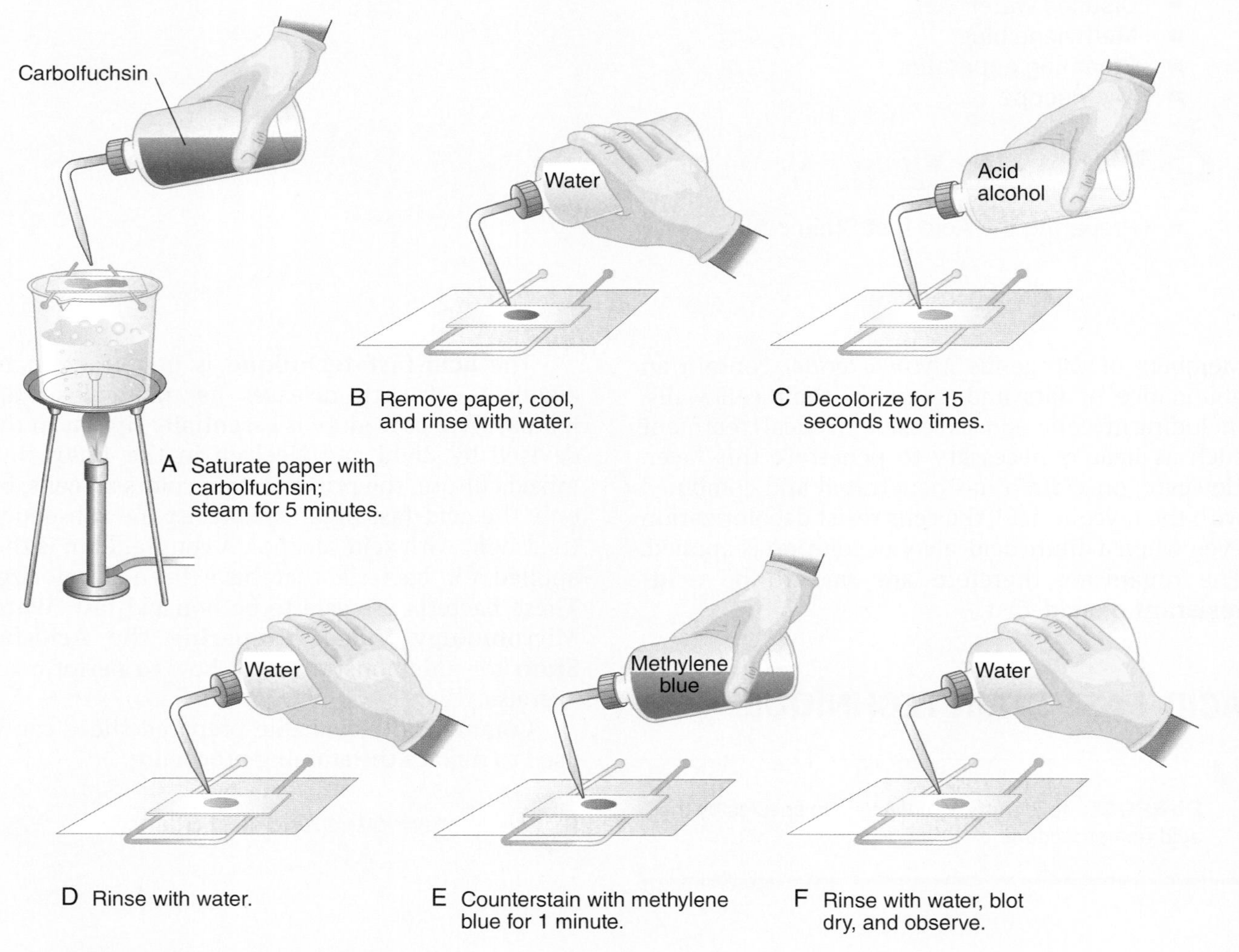

FIGURE 39.1 The acid-fast stain technique.

FIGURE 39.2 Carbolfuchsin is the primary stain for this exercise.

4. Allow the slide to remain over the boiling water for 5 minutes while adding sufficient carbolfuchsin to keep the smears wet.

The acid-fast staining should only be performed in a well-ventilated area, preferably a fume hood.

   - At the conclusion of the staining period, remove the paper and wash the slide with a gentle stream of distilled water (**FIGURE 39.1B**).

5. Decolorize the slide with **acid-alcohol** (**FIGURE 39.3**) by placing several drops on the smears and rocking them back and forth for 15 seconds as performed in the Gram stain technique (Exercise 15).
   - Allow the excess to drip off, then repeat the decolorization for another 15-second interval or until the drippings are clear (**FIGURE 39.1C**).
   - Gently wash the slide with distilled water (**FIGURE 39.1D**).

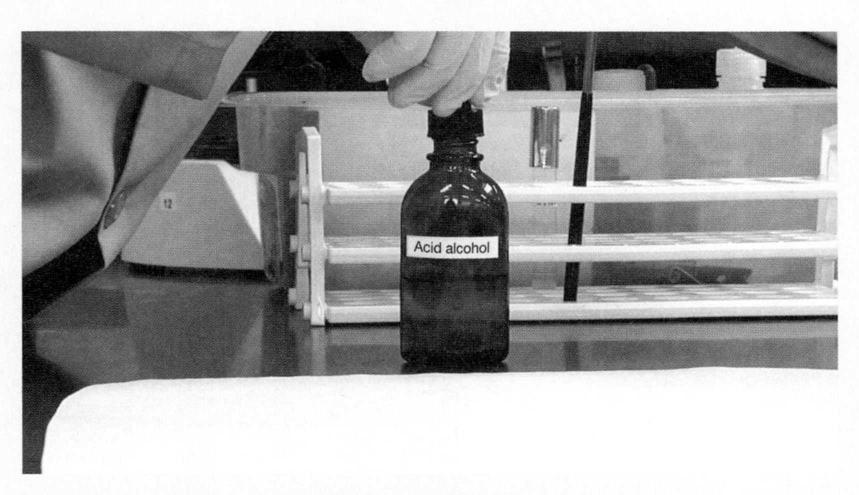

FIGURE 39.3 Use acid-alcohol to decolorize the slide.

6. Stain the slide with **methylene blue**, the counterstain, for 1 minute (FIGURE 39.1E). Rinse the slide and blot it dry (FIGURE 39.1F).
7. Locate the stained cells using the low power (10×) lens.
   - Examine the smears under 40× and oil immersion, and note the red acid-fast rods and the blue nonacid-fast organisms. The slide containing the bacterial mixture should contain both red and blue forms.
   - Enter representations of the organisms in the appropriate space in the Results section indicating whether they are acid-fast or nonacid-fast.

Quick Procedure

Acid-Fast Stain

1. Place the heat-fixed slide over boiling water.
2. Stain with carbolfuchsin for 5 min; wash.
3. Decolorize for two 15-sec periods with acid-alcohol; wash.
4. Stain with methylene blue for 1 min; wash; dry; observe.

Name: ______________________ Date: __________ Section: __________

The Genus *Mycobacterium*: Acid-Fast Stain Technique

EXERCISE RESULTS **39**

RESULTS

ACID-FAST STAIN TECHNIQUE

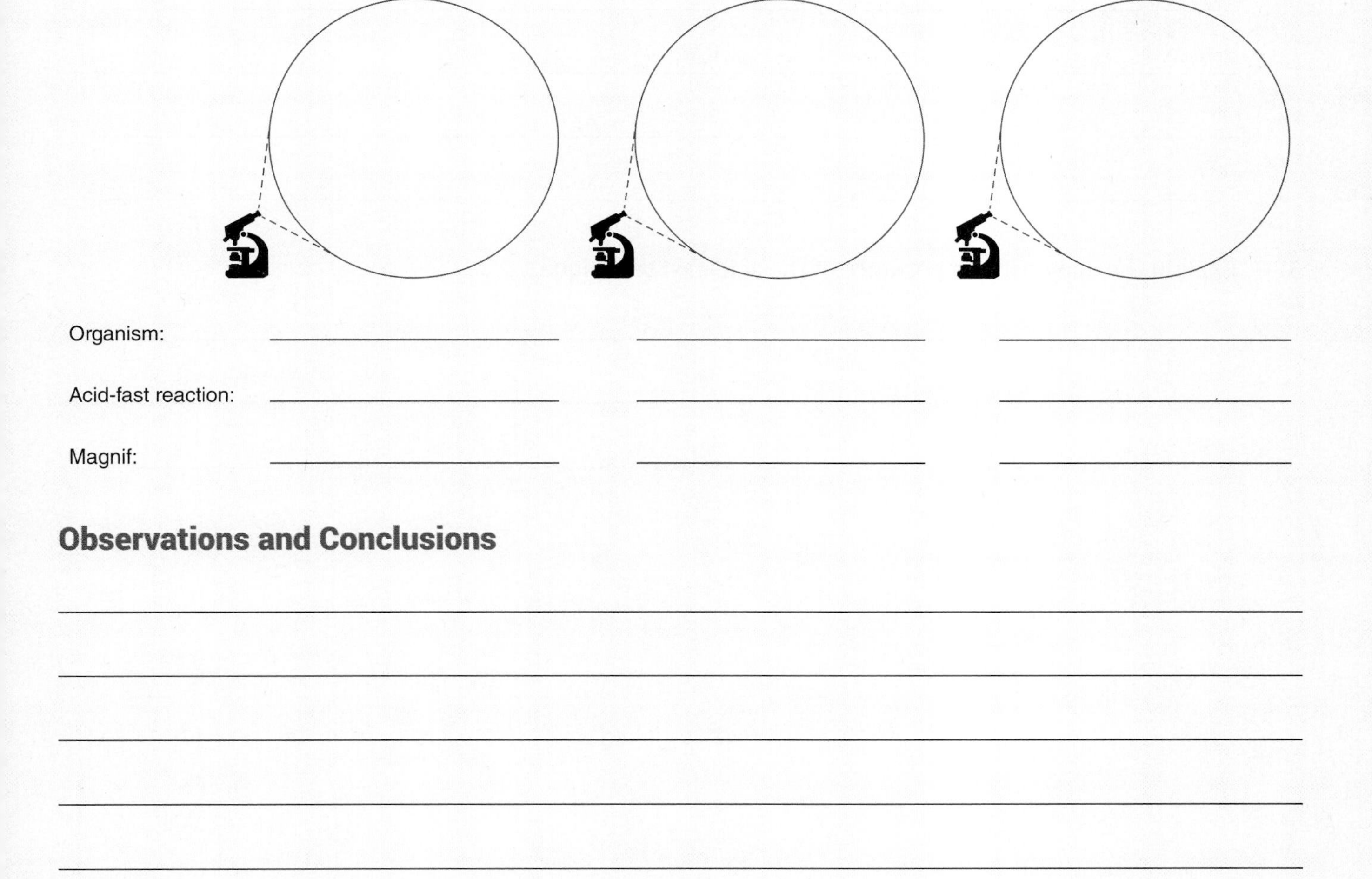

Observations and Conclusions

Questions

1. In what ways is the acid-fast technique similar to the Gram stain technique? How does it differ?

2. Suppose alcohol instead of acid-alcohol were used in the acid-fast technique. Would decolorization take place? How would this affect the results?

3. Explain the diagnostic importance of the acid-fast technique.

The Genus *Mycobacterium*: Cold Acid-Fast Stain Technique

EXERCISE 40

Materials and Equipment

- Agar cultures of *Mycobacterium smegmatis*
- Selected nonacid-fast organisms as controls
- Acid alcohol
- Distilled water
- Methylene blue
- Microscope
- Kinyoun carbolfuchsin

In this exercise, laboratory cultures of *Mycobacterium* will be examined for typical colony and cellular appearance, and the cold acid-fast technique will be performed to demonstrate an important property used for detecting *Mycobacterium* species.

COLD ACID-FAST STAIN TECHNIQUE

PURPOSE: to identify *Mycobacterium* species without using steam.

The acid-fast technique may also be performed by a cold method that eliminates the steaming step. The carbolfuchsin used in this method contains the detergent tergitol 7. This substance dissolves the waxes in the cell walls of *Mycobacterium* species and thereby facilitates penetration of the dye. The staining reagent is called Kinyoun carbolfuchsin, after its developer.

PROCEDURE

1. Perform the acid-fast technique as described in Exercise 39, with the following exception: Instead of saturating a piece of paper with stain and steaming it for 5 minutes, flood the smears with Kinyoun carbolfuchsin and permit the stain to remain for 5 minutes. Wash the stain free, and continue with the acid-alcohol decolorization step.
2. Conclude the procedure by staining with Brilliant Green for 1 minute.
   - The colors of acid-fast bacteria will be similar to those in the hot staining method; nonacid-fast bacteria will be green.
   - Enter your drawings in the Results section.

Name: ______________________ Date: __________ Section: __________

The Genus *Mycobacterium*: Cold Acid-Fast Stain Technique

EXERCISE RESULTS **40**

RESULTS

COLD ACID-FAST STAIN TECHNIQUE

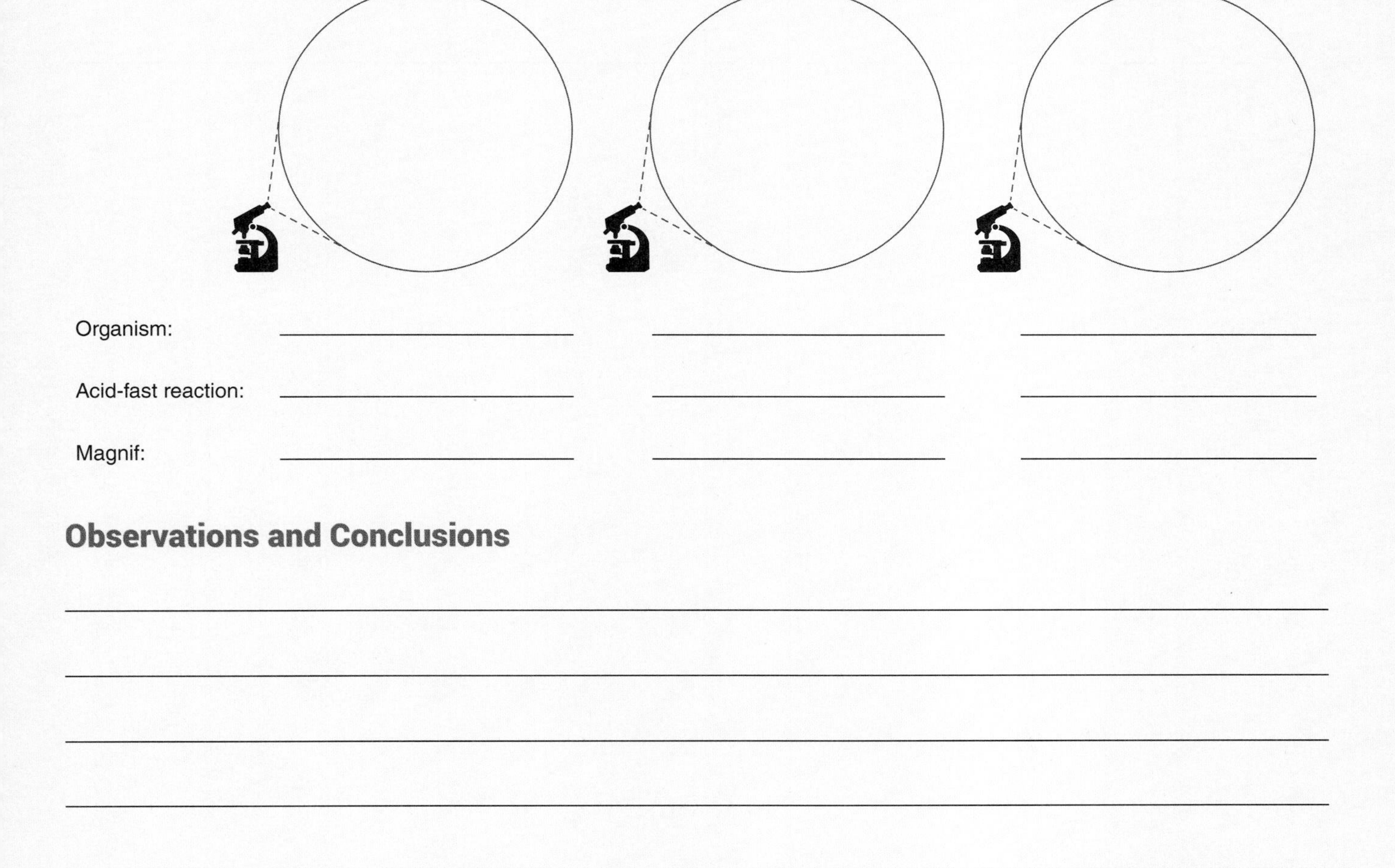

Observations and Conclusions

Questions

1. If the cells from the mixed smear were observed after each step of the acid-fast or cold acid-fast procedure, predict the colors of the cells at each step.

2. Why are the acid-fast procedures considered differential staining techniques?

The Genus *Streptococcus*: Streptococci from the Upper Respiratory Tract

EXERCISE 41

Materials and Equipment

- Blood agar plates
- Sterile cotton swabs and tongue depressors
- Candle jar (optional)
- Bacitracin disks (0.04 units per disk)
- Gram stain reagents

The genus ***Streptococcus*** includes a broad group of gram-positive, facultatively anaerobic cocci that form chains of varying length. Species of streptococci may be located on the human skin as well as in the mouth, upper respiratory tract, and intestine. Cultivation in a rich medium such as **blood agar** is often necessary because the organisms lack certain enzyme systems for nutrition (TABLE 41.1). This exercise is concerned with the isolation of streptococci from the upper respiratory tract, and with the study of some of their properties. Special precautions should be taken because the bacteria isolated may be pathogenic.

STREPTOCOCCI FROM THE UPPER RESPIRATORY TRACT

PURPOSE: to distinguish hemolytic streptococcus isolated from the throat.

Streptococci are an important component of the microbiome of the upper respiratory tract. The organisms may be cultivated on an enriched medium that contains blood and is incubated in an environment rich in carbon dioxide and low in oxygen. Certain streptococci will destroy the red blood cells in the medium completely and form a clear zone around the colonies. These are **beta-hemolytic streptococci**, as typified by pathogenic *Streptococcus pyogenes*. Other streptococci cause incomplete destruction of the blood cells, and their colonies are surrounded by an olive green or brown discoloration of the medium. These are **alpha-hemolytic streptococci**. They include *S. mitis* and *S. pneumoniae*, the agent of bacterial (lobar) pneumonia. A final group, the **nonhemolytic streptococci**, cause no hemolysis of red blood cells. *S. lactis*, the organism in yogurt, is a member of the group. This method of classification (hemolysis) was first developed by J. H. Brown in 1919 (see TABLE 41.2).

In this section, streptococci will be isolated from the upper respiratory tract, and the various types of hemolysis will be examined.

TABLE 41.1 Characteristics of Streptococcus
Gram-positive cocci 0.5 to 2.0 µm in diameter that occur in pairs or chains
Nonmotile
Nonspore forming
Facultative anaerobes
Some species show enhanced growth under increased 1%-10% CO_2
Require supportive or enriched media, such as blood agar, for growth
Catalase negative

TABLE 41.2	Brown's Classification of Hemolysis
Alpha (α)	Incomplete or partial hemolysis; green or brown color surrounding the colony. This pattern is seen with the viridans streptococci, which are the major normal flora of the oropharynx, and *Streptococcus pneumoniae*, an important cause of bacterial pneumonia; alpha-hemolysis occurs as a result of hydrogen peroxide production by the organism, which destroys the red cells and releases hemoglobin. Example: *S. mitis* and *S. pneumoniae*.
Beta (β)	Complete hemolysis of red blood cells; clearing or colorless zone around colony. An example of this type of hemolysis is exhibited by *Streptococcus pyogenes*, the agent of streptococcal pharyngitis.
Nonhemolytic	Lack of hemolysis; no apparent change in color of area surrounding colony; has been referred to as gamma (γ) hemolytic. Example: *S. lactis*.

PROCEDURE

1. If blood agar has not been prepared before the laboratory session, the instructor will demonstrate its preparation. A blood agar base such as trypticase soy agar and defibrinated whole sheep blood will be used. A 5% blood agar medium will be prepared.
2. Select or prepare a plate of blood agar and label the bottom side with your name, the date, the name of the medium, and the designation "throat swab."
   - Obtain a sterile cotton swab and sterile tongue depressor.

When taking a throat culture, be certain the cotton swab and tongue depressor are sterile. They should come from an unopened package or from a container designated "sterile."

3. Have a fellow student swab your pharynx (throat) using the sterile cotton swab and tongue depressor. The light should permit good visibility of the pharynx, and the swab should be rolled across the posterior membranes in the region of the tonsils.
   - The student should take care to avoid touching the tongue or other tissues.
   - Refer to FIGURE 41.1 for the proper technique.

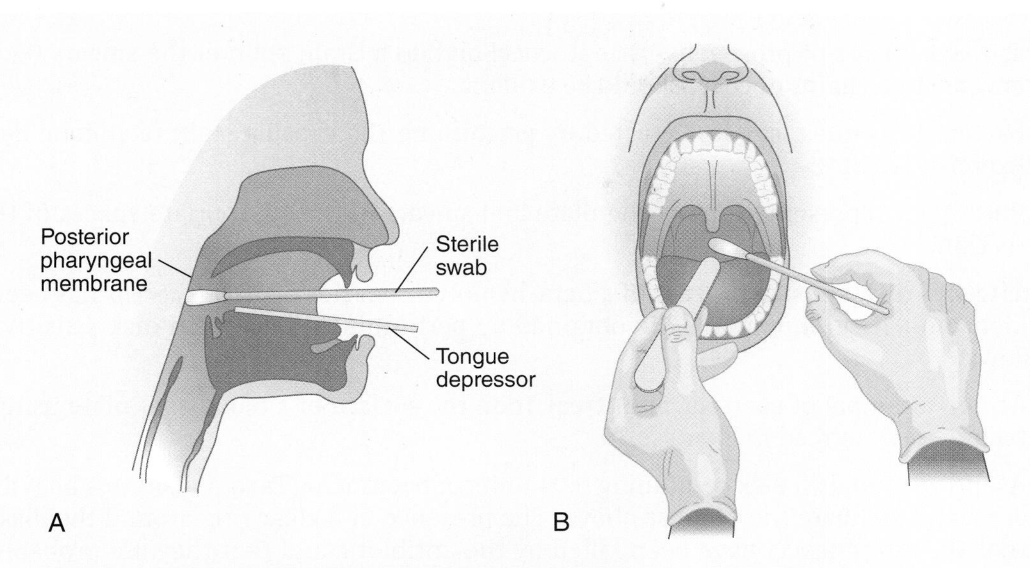

FIGURE 41.1 Two views of the procedure for obtaining a throat swab.

4. Apply the bacteria from the swab to one area of the blood agar plate by rubbing the swab on the agar gently as you turn it.
   - Discard the swab in the disinfectant.
   - Perform the **streak plate technique** as follows:
     - Take a sterile loop and pass it a couple of times across the swabbed area, then continue streaking back and forth into a second area of the plate, as described in Exercise 5.
     - Sterilize the loop and continue the streaking in order to obtain isolated colonies, as explained in Exercise 5.
   - As a final step, stab the loop into the agar in a place of high bacterial concentration. This will force bacteria into the medium, and if hemolysis occurs during incubation, it will be particularly visible in this area.
5. Invert the plate and place it in the incubator at 37°C for 24 to 48 hours.
   - Alternately, to obtain an environment rich in CO_2, a wide-mouthed **candle jar** may be used.
   - Place the plate in the jar and stand a candle on the plate.
   - Light the candle and replace the lid tightly (FIGURE 41.2). When the candle burns out, the oxygen level will have been reduced, and the carbon dioxide level will be increased.
   - Place the in the incubator at 37°C.
6. Following the incubation, examine the plate for evidence of alpha-hemolytic and beta-hemolytic colonies as described in the introductory paragraph to this exercise. To verify the presence and purity of streptococci, prepare air-dried,

FIGURE 41.2 A candle-jar apparatus made from a wide-mouthed jar.

heat-fixed smears of prospective streptococci and do a Gram stain of the smears (Exercise 14). Gram-positive chains of cocci should be evident.

- Bacterial capsules may be detected by performing the capsule stain technique described in Exercise 17.
- Enter your representations of the plate and smears in the appropriate spaces of the Results section.

7. **Bacitracin disk sensitivity test.** If a beta-hemolytic species of *Streptococcus* has been isolated, evidence for its grouping may be obtained by performing a bacitracin disk sensitivity test as follows:

- Obtain a sample of bacteria, and streak it on the surface of a blood agar plate using the lawn technique described in Exercise 19.
- Aseptically apply a disk containing 0.04 units of bacitracin (Taxo A disk) to a heavily streaked area, and incubate the plate as above. The presence of a clear ring around the disk indicates that the streptococci have been killed by the antibiotic and that they are probably Group A beta-hemolytic streptococci known as *Streptococcus pyogenes*. The absence of a ring provides evidence that they belong to another group.
- Enter your results in the appropriate space in the Results section.

Name: ______________________ Date: __________ Section: __________

The Genus *Streptococcus*: Streptococci from the Upper Respiratory Tract

EXERCISE RESULTS **41**

RESULTS

STREPTOCOCCI FROM THE UPPER RESPIRATORY TRACT

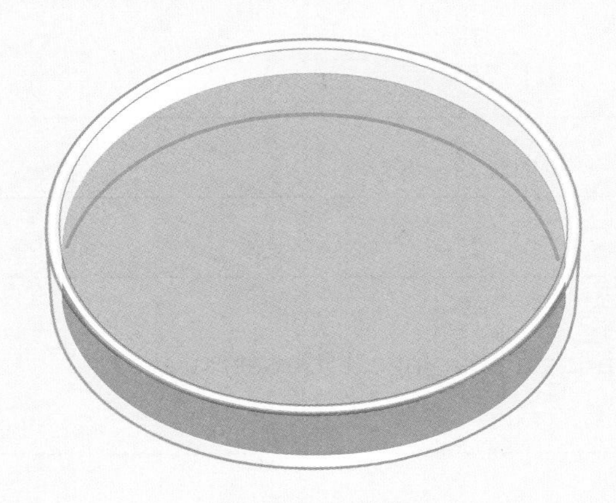

Blood Agar Plate

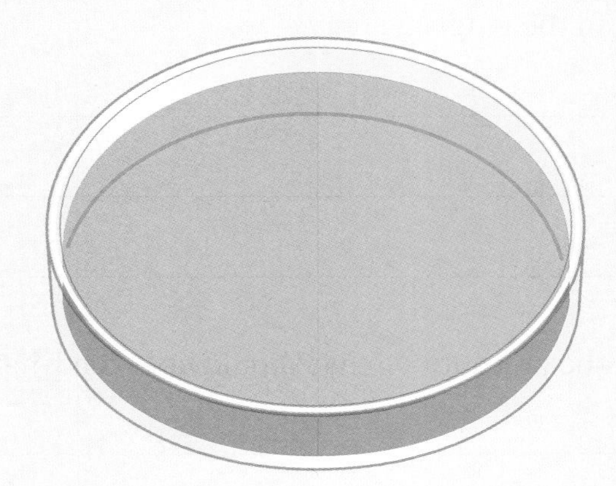

Bacitracin Disk Plate

Stained Smears of Streptococci

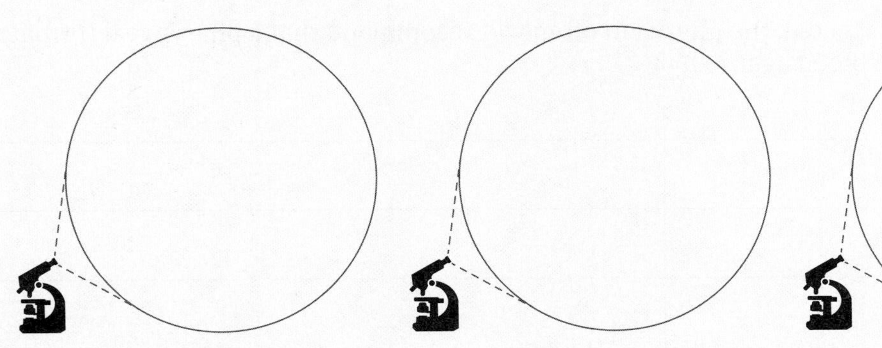

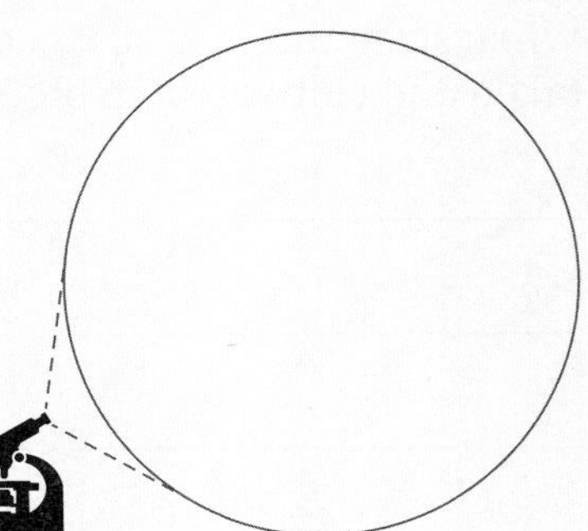

Source: ______________ ______________ ______________

Magnif.: ______________ ______________ ______________

Observations and Conclusions

Questions

1. What positive role may be played by the nonpathogenic streptococci normally found between the teeth and in the gingival crevice?

2. List the precautions that should be taken when obtaining a pharyngeal (throat) swab.

3. When strep throat is suspected, the physician often will recommend that a pharyngeal (throat) swab be taken and cultivated on blood agar. Why?

4. Explain how beta-hemolysis may be distinguished from alpha-hemolysis.

The Genus *Streptococcus*: Streptococci from the Oral Cavity

EXERCISE 42

Materials and Equipment

- Plates of mitis salivarius agar
- Tubes of Todd-Hewitt broth
- Sterile cotton swabs
- Gram stain reagents

Many species of streptococci, especially *Streptococcus mitis* (alpha-hemolytic) and *S. salivarius* (nonhemolytic), are dominant species in the microbiome of the oral cavity where they are found in the spaces between the teeth and underthe gums (the gingival crevices). Another nonhemolytic species, *S. mutans*, ferments sucrose to acid and is an important cause of tooth decay. Streptococci such as these will be isolated using **mitis salivarius agar**. This is a selective medium because it contains crystal violet, potassium tellurite, and trypan blue, which inhibit most gram-negative bacilli and gram-positive bacteria other than the streptococci. **Todd-Hewitt broth** is an enrichment medium containing glucose, as well as sodium carbonate, and disodium phosphate buffers that favor the growth of hemolytic streptococci.

STREPTOCOCCI FROM THE ORAL CAVITY

PURPOSE: to identify streptococci using selective and enrichment media.

PROCEDURE

1. Select one plate of mitis salivarius agar (or pour one plate) and one tube of Todd-Hewitt broth, and label them as described in Exercise 41.
2. Use a sterile cotton swab to rub vigorously in the gingival crevice and space between the teeth.
   - Apply the swab to one small area of the mitis salivarius agar plate.
   - Use a sterile loop to streak for isolated colonies, as outlined in Exercise 5.
   - Incubate the plate at 37°C for 24 to 48 hours.
   - A candle jar may be used to decrease the oxygen concentration (as in Exercise 41).
3. Take a fresh sterile cotton swab and rub an area of the gingival crevice and space between the teeth.
   - Place the swab in the tube of Todd-Hewitt broth and break off the end, thereby allowing the swab to fall into the medium. Replace the cap on the tube.
   - Incubate the tube at 37°C for 24 to 48 hours.

4. Observe the mitis salivarius plate for evidence of streptococcal colonies. *S. mitis* will appear as flat, blue colonies with dark centers. *S. salivarius* produces large, blue domed-shaped colonies with a gumdrop appearance. *S. mutans* can be recognized by blue colonies having a domed brown center. Gram-negative and non-streptococcal gram-positive bacterial growth is inhibited in this agar medium. Gram stains should be prepared from the colonies to verify the presence and purity of streptococci.
   - Place representations in the Results section together with your detailed observations.
5. Observe the Todd-Hewitt broth for evidence of growth as a sediment at the bottom of the tube. The broth is used primarily to enrich the growth of pathogenic (beta-hemolytic) streptococci. Because streptococci form long chains in broth, Gram-stained smears should reveal characteristic streptococcal arrangements.
   - Be careful to avoid cotton fibers from the swab when taking samples.
   - Enter drawings in the Results section.

Name: ____________________ Date: __________ Section: __________

The Genus *Streptococcus*: Streptococci from the Oral Cavity

EXERCISE RESULTS **42**

RESULTS

STREPTOCOCCI FROM THE ORAL CAVITY

Mitis Salivarius Agar Plate

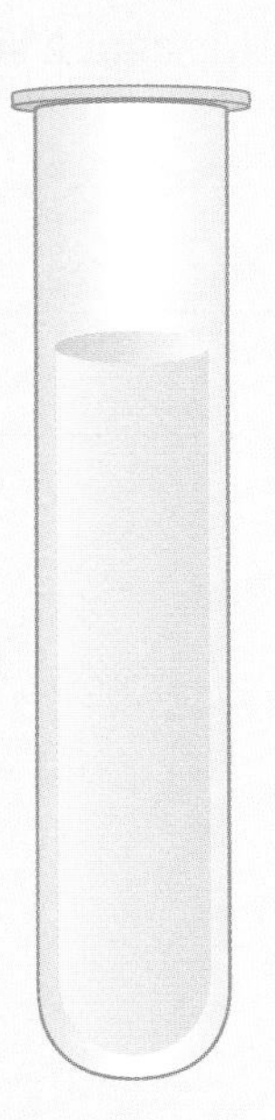

Todd-Hewitt Broth

Stained Smears of Streptococci

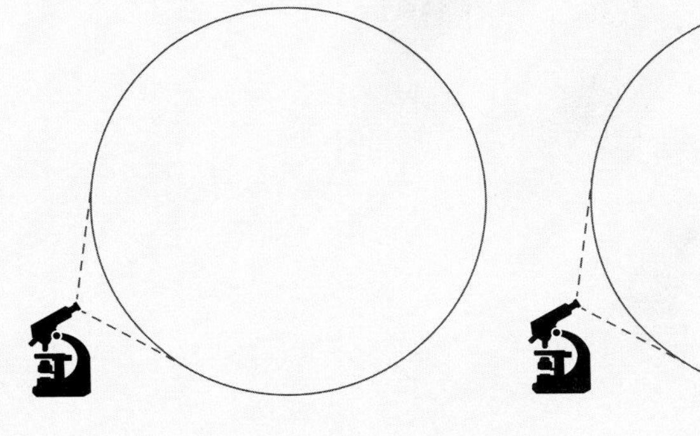

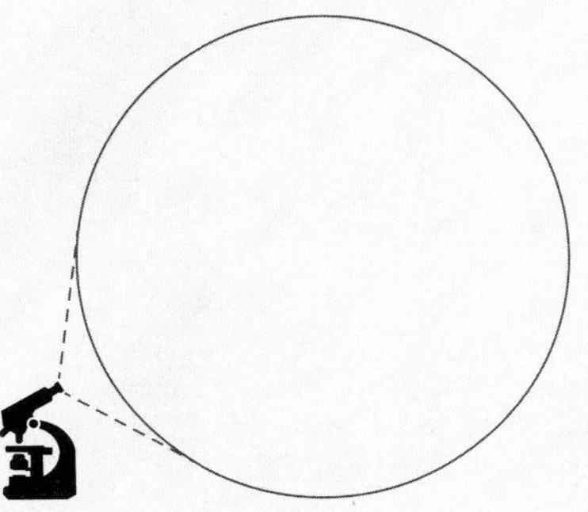

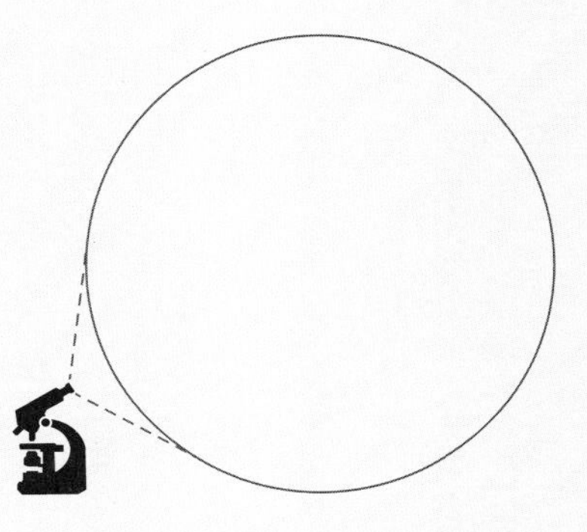

Source: __________ __________ __________

Magnif.: __________ __________ __________

Observations and Conclusions

Questions

1. Is the mitis salivarius agar a selective or differential medium for streptococci? Explain.

The Genus *Neisseria*

EXERCISE

43

Materials and Equipment

- Chocolate agar plates
- Blood agar plates
- Candle jar (optional)
- Sterile cotton swabs and tongue depressors
- Oxidase reagent in dropper bottles
- Gram stain reagents

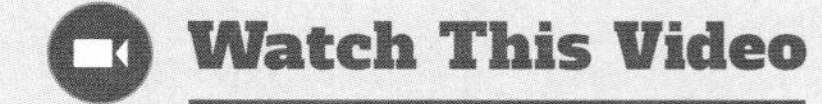

Watch This Video

- Oxidase Test

The flora of the upper respiratory tract often includes a number of species of the genus ***Neisseria*** (see FIGURE 43.1). These include *N. sicca, N. mucosa*, and *N. subflava*. These organisms are structurally similar to *N. meningitidis*, the agent of meningococcal meningitis, and *N. gonorrhoeae*, the cause of gonorrhea. Another common inhabitant of the upper respiratory tract is *Moraxella (Branhamella) catarrhalis*, which also is a member of the family Neisseriaceae.

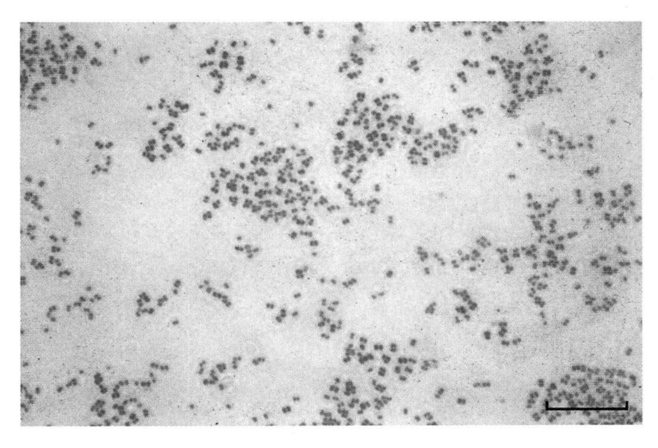

FIGURE 43.1 Gram-negative diplococci typical of **Neisseria** species. (Bar = 10 µm.).

Members of the genus *Neisseria* are small gram-negative diplococci with flattened adjacent sides. They sometimes give the appearance of two tiny beans lying face to face. The organisms grow on enriched media such as **blood agar** and **chocolate agar**, and produce oxidase, an enzyme that changes the color of a special reagent. There are several *Neisseria* species found in the human body. These are listed in TABLE 43.1. In this exercise, *Neisseria* species will be isolated from the throat and their properties observed.

THE GENUS *NEISSERIA*

PURPOSE: to identify *Neisseria* species isolated from the throat.

TABLE 43.1 *Neisseria* Species Found in Humans
N. gonorrhoeae
N. meningitidis
N. lactamica
N. sicca
N. subflava
N. mucosa
N. flavescens
N. cinerea
N. polysaccharea
N. elongata
N. kochii

PROCEDURE

Wear gloves when adding the blood to the agar base to prepare blood or chocolate agar.

1. Blood agar plates are prepared according to the method explained in Exercise 41. Chocolate agar is prepared by heating a rich medium such as trypticase soy agar to 80°C for 10 minutes, and then adding defibrinated sheep blood to a 5% concentration. The heat lyses the red blood cells and releases the hemoglobin. The hemoglobin chars and causes the medium to become brown; hence, the name chocolate agar.
2. Select or prepare a plate of blood agar and/or one of chocolate agar.
   - Label the bottom side of each plate with your name, the date, the name of the medium, and the designation "throat swab."
   - Obtain a sterile cotton swab and a sterile tongue depressor.
3. Have a fellow student swab your pharynx (throat) according to the method outlined in Exercise 41.
   - Apply the bacteria on the swab to one small area of the agar plate by rubbing it gently.
   - Perform the **streak plate technique**:
     - Using a sterile loop, streak for isolated colonies as described in Exercises 5 and 41 (FIGURE 43.2).
   - Incubate the plate(s) at 37°C for 24 to 48 hours in the inverted position.
   - If a candle jar is available, it may be used as described in Exercise 41 to increase the CO_2 tension and encourage growth of the *Neisseria* species.

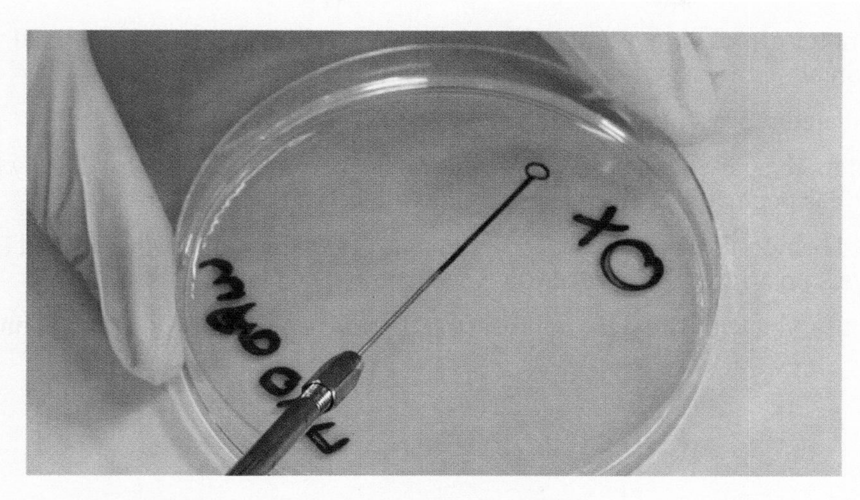

FIGURE 43.2 Streak the plate for isolated colonies.

4. Observe the plate(s) for grayish white and light yellow colonies, which may be *Neisseria* species. Some colonies may be wrinkled, others mucoid in texture.
   - To verify the presence of *Neisseria*, perform the **oxidase test** as follows (**Watch Microbiology Video: Oxidase Test** to see this test performed):
     - Place several drops of freshly prepared oxidase reagent (tetramethyl-*p*-phenylenediamine dihydrochloride) onto the colonies (FIGURE 43.3).
     - The oxidase present in *Neisseria* colonies will cause the colonies to become pink, then maroon, and finally blue-black. These changes should occur rapidly, and they will be complete within several minutes.
   - Identify colonies of *Neisseria* and prepare representations of the plates in the appropriate space in the Results section.

Oxidase Test

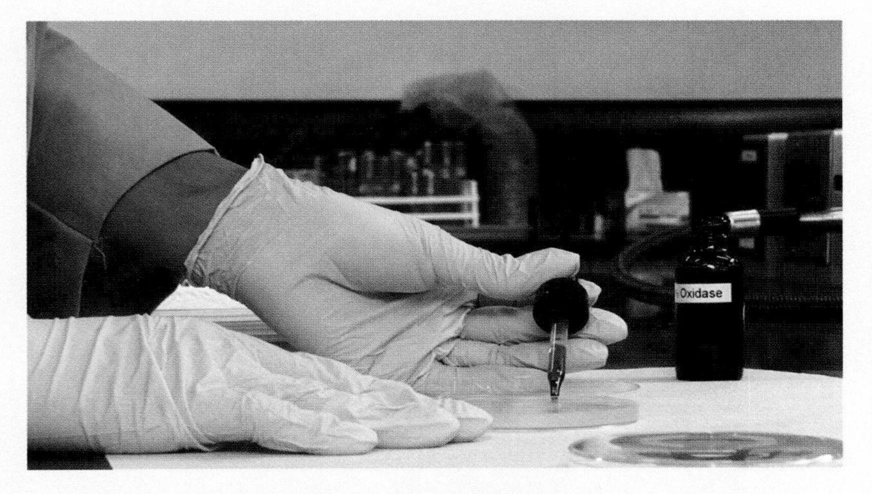

FIGURE 43.3 Place several drops of oxidase reagent onto the colonies.

5. Select samples of possible *Neisseria* species, and prepare air-dried, heat-fixed smears for Gram staining. Small gram-negative diplococci should be observed.
   - Draw labeled representations in the Results section.
   - If transfers to agar slants are to be made, these should be done immediately after the oxidase reagent has been added, since the reagent will kill the cells in the colonies.
   - It should be noted that rod-shaped organisms of the genera *Alcaligenes* and *Pseudomonas* will also give a positive oxidase reaction if present on the plates.
   - When observed with the light microscope, these bacteria will appear as gram-negative rods.

Name: ______________________ Date: __________ Section: __________

The Genus *Neisseria*

EXERCISE RESULTS 43

RESULTS

NEISSERIA SPECIES FROM THE UPPER RESPIRATORY TRACT

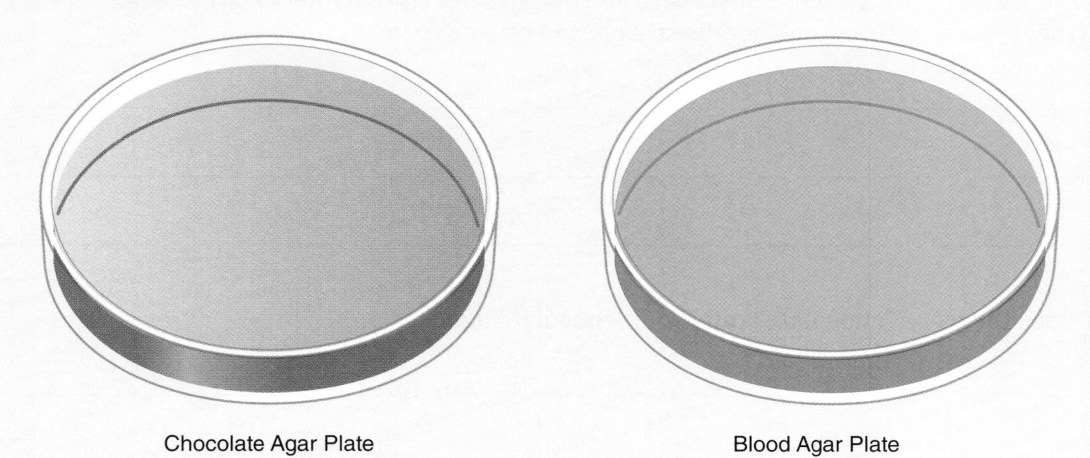

Chocolate Agar Plate

Blood Agar Plate

Stained Smears of *Neisseria* Species

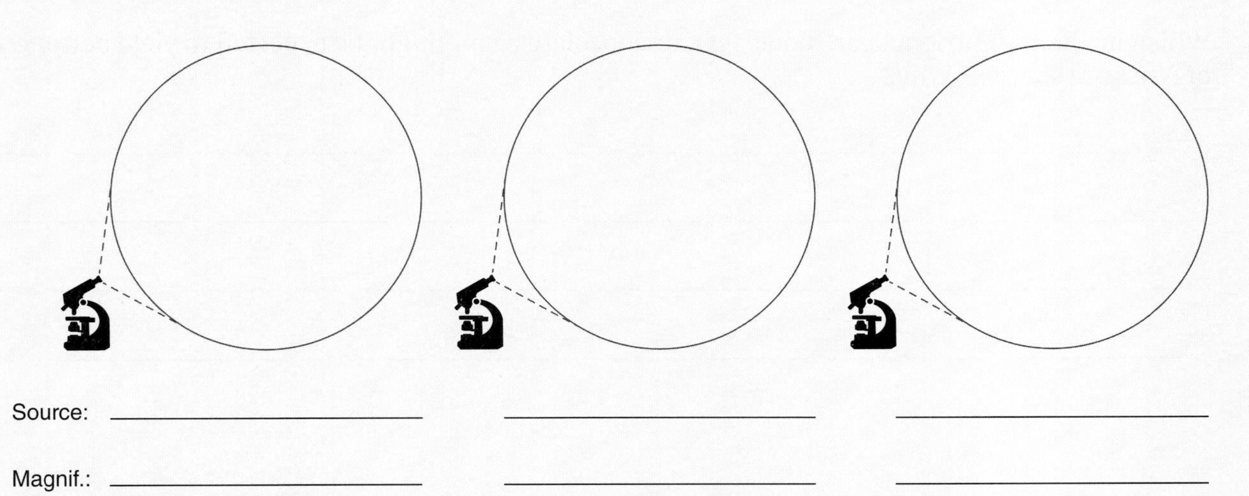

Source: __________ __________ __________

Magnif.: __________ __________ __________

Observations and Conclusions

Questions

1. Does the isolation of *Neisseria* species from the pharynx (throat) necessarily mean that a person has gonorrheal pharyngitis or another disease caused by *Neisseria*?

2. What does the word "chocolate" refer to in chocolate agar?

3. Which medium—nutrient agar, blood agar, or chocolate agar—might be expected to yield better growth of *Neisseria* species? Why?

4. Would the observation of oxidase-positive colonies on chocolate agar necessarily represent final proof that *Neisseria* colonies were present?

5. What is the microscopic appearance of Gram-stained cells of *Neisseria* species?

The Genus *Staphylococcus*: Isolation of Staphylococci

EXERCISE 44

Materials and Equipment

- Plates of mannitol salt agar
- Sterile cotton swabs
- Tubes of sterile saline
- Gram stain reagents
- Nutrient agar slants

Members of the genus ***Staphylococcus*** are gram-positive spherical organisms about 1 micrometer in diameter. They occur singly, in pairs, and in irregular clusters, and form yellow, orange, or white colonies on agar media. They are salt tolerant and grow on ordinary bacteriological media as well as on the selective media used in this exercise (see TABLE 44.1).

Up to three species of *Staphylococcus* are studied in this exercise. Certain strains of *Staphylococcus aureus* are the cause of food poisoning and toxic shock syndrome. They also are the cause of boils and carbuncles. A second species, *S. epidermidis*, usually is a saprobe of the skin that is rarely involved in human infection. The third species, *S. saprophyticus*, is an opportunistic species that may cause urinary tract infections in women of childbearing years. In this exercise, staphylococcal species will be isolated from the body's environment and their properties examined.

TABLE 44.1	Important Characteristics of *S. aureus*
Description	Medium to large, raised colonies on blood agar; cream to golden yellow pigmentation on colistin naladixic acid agar
Gram stain	Gram-positive cocci arranged in clusters
Hemolysis	Beta-hemolytic on sheep blood agar
Bound coagulase (clumping factor)	Positive
Free coagulase	Positive
MSA	Growth and fermentation
DNase	Positive

ISOLATION OF STAPHYLOCOCCI

PURPOSE: to isolate and identify staphylococcal species from the nasal cavity and other environments.

Species of *Staphylococcus* are tolerant to salt and, therefore, can be selected out from a mixture of bacteria in a high-salt medium. In addition, *S. aureus* ferments mannitol, an alcoholic derivative of the hexose mannose, while *S. epidermidis* and *S. saprophyticus* do not. Therefore, if the differential medium contains mannitol, the two species may be differentiated from one another. In this section, we will use **mannitol salt agar**, a medium that is both selective and differential. The salt concentration is a high 7% to inhibit other organisms, and mannitol is available for fermentation to acid. The indicator **phenol red** is included in the medium to detect acid production. It will change from red to yellow in the presence of acid-producing *S. aureus*. No color change will occur with the other two species.

PROCEDURE

1. Select or prepare two plates of mannitol salt agar and, on the bottom sides, label them in the prescribed manner with your name, the date, and the name of the medium.
   - Label one plate for a "nose swab" and the second for an environment of your choice.

Do not swab the nose too vigorously, to avoid breaking one or more blood vessels.

2. Moisten a sterile cotton swab in sterile saline solution, and swab the internal perimeter of the nose cavity.
   - Apply the swab to one area of the appropriate mannitol salt agar plate, and place the swab in disinfectant.
   - Perform the **streak plate technique**:
     - Using a sterile loop, pass it a couple of times across the swabbed area and then streak into a second area of the plate to obtain isolated colonies, as described in Exercise 5.
     - Continue streaking into the third and fourth areas of the plate.
   - Incubate the plate in the inverted position at 37°C for 24 to 48 hours, then refrigerate it.
3. Moisten a second swab, and attempt to isolate staphylococci from another environment.
   - Consider a telephone receiver, a doorknob, the surface of a dollar bill, the area under your fingernail or behind your ear, or another source.
   - If you wish to test a food sample, use the custard filling from a donut or a sample of hamburger meat or something similar. (If food is left out to "ripen" for a day or two, the chances of obtaining staphylococci will be greater.)
   - Apply the swab to the second plate as in step 2, and continue streaking with the loop to maximize the chances that individual colonies will develop.
   - Incubate the plate as above.
4. Observe the plates for the presence of round opaque colonies that are white or pigmented. These are probably colonies of *Staphylococcus* species. *S. aureus* produces golden-yellow pigmented colonies, while the other species produce white colonies. Also, colonies surrounded by a yellow halo contain bacteria that fermented the mannitol (the pH indicator turned from red to yellow) and probably contain *S. aureus*. Where no color change to yellow has taken place, the colonies are possibly *S. epidermidis* or *S. saprophyticus*, because these species do not ferment mannitol. Another possibility is *Micrococcus luteus*, as noted below.

   Enter a labeled representation of the plate indicating colony color and any color changes from mannitol fermentation in the Results section and TABLE 44.2.
5. The presence of staphylococci may be verified by performing the Gram stain technique on samples from isolated colonies. Typical clusters of gram-positive cocci should appear.
   - Place illustrations and your notes in the Results section.
   - On occasion, *Micrococcus luteus* will be isolated on the medium. These bacteria also are gram-positive cocci, but they occur in characteristic tetrads (clusters of four cells) or in cubical packets of eight cells (the sarcina configuration). The colonies on agar are bright yellow.

6. Other characteristics can be used to identify or differentiate between specific species of *Staphylococcus*. However, to do this requires pure cultures of the isolated species.

Extra caution should be observed when working with potential pathogens.

7. Obtain three nutrient agar slant tubes and label each tube with your name.
    - Label one tube A, another B, and the third C. These will be used to produce three pure cultures of *Staphylococcus*.
8. Based on colony color (pigmentation), mannitol fermentation results, and the purity of the colony based on Gram staining, select three separate colonies to pure culture.
    - One colony should be positive for mannitol fermentation (yellow halo). Identify this colony as isolate A.
    - The other two should be white colony isolates that cannot ferment mannitol (no yellow halo). Identify these colonies as isolates B and C.
9. Sterilize your inoculation loop and transfer a loopful of colony isolate A to the agar slant labeled A.
    - Streak the loop up the length of the agar, being careful not to gouge the agar (Exercise 4).
    - Repeat the process with a sterile loop from colonies B and C to slants B and C.
    - Incubate the three slants at 37°C for 24 to 48 hours. If they will not be used until the following week, they can be refrigerated after 48 hours.

Name: ______________________ Date: __________ Section: __________

The Genus *Staphylococcus*: Isolation of Staphylococci

EXERCISE RESULTS 44

RESULTS

ISOLATION OF STAPHYLOCOCCI

Mannitol Salt Agar Plates

Source: ______________________ ______________________

Stained Smears of Staphylococci

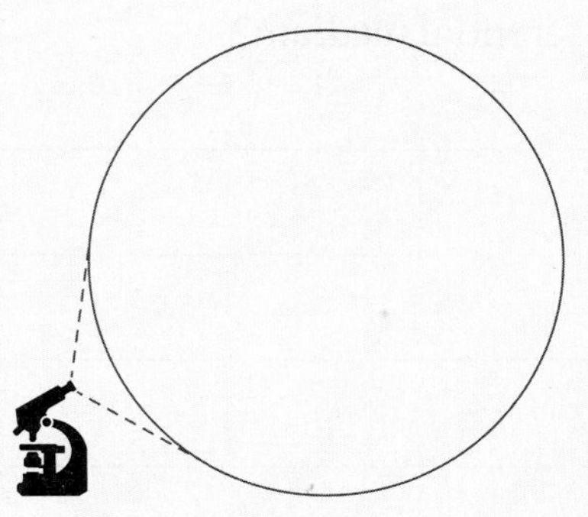

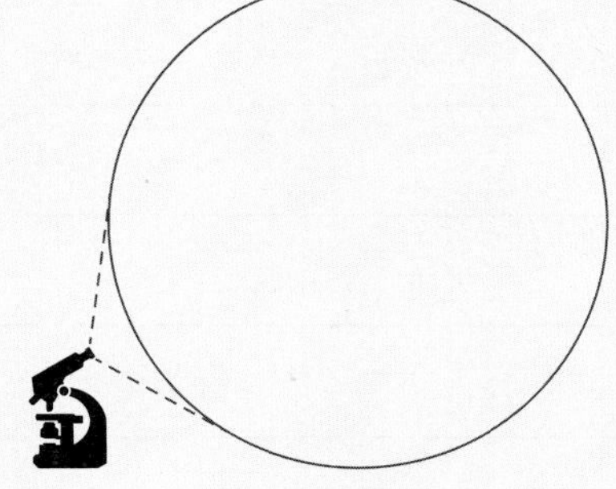

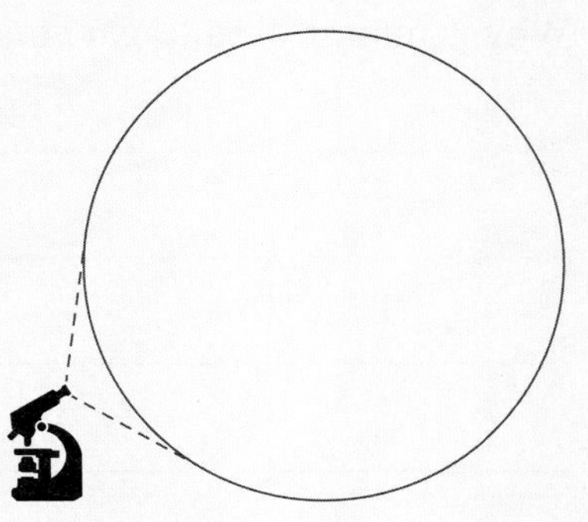

Source: ______________ ______________ ______________

Magnif.: ______________ ______________ ______________

TABLE 44.2 Characteristics of Staphylococcal Isolates A, B, and C

Characteristics	A	B	C
Colony color			
Acid production from mannitol			

Observations and Conclusions

Questions

1. Why are species of *Staphylococcus* often found as contaminants of ham and ham products?

2. Why is mannitol salt agar considered a selective as well as a differential medium?

The Genus *Staphylococcus*: Differentiation between Staphylococcal Species

EXERCISE 45

Materials and Equipment

- Isolated staphylococci from Exercise 44
- Tubes containing 0.5 ml of coagulase plasma
- DNase test agar
- 1N HCl
- Mueller-Hinton agar
- 30 μg Novobiocin discs
- Sterile forceps

Members of the genus ***Staphylococcus*** are found in humans, other animals, and the environment. On humans, these organisms may be found on the skin or in the nasal cavity, and may lead to infection.

DIFFERENTIATION BETWEEN STAPHYLOCOCCAL SPECIES

PURPOSE: to distinguish between species of *Staphylococcus* based on biochemical characteristics.

Each of the three species of *Staphylococcus* has specific biochemical characteristics that can be used to separate it from the other two species. *S. aureus* has the ability to produce **coagulase**, an enzyme that clots blood plasma and enhances its virulence by providing resistance to phagocytosis. The species also produces **DNase** to digest DNA and is sensitive to the antibiotic **novobiocin**. Of the three species, only *S. saprophyticus* is resistant to novobiocin (TABLE 45.1).

In this exercise, the three samples of isolated staphylococci from Exercise 44 will be tested for coagulase and DNase activity as well as sensitivity to novobiocin.

TABLE 45.1 Characteristics of Staphylococcal Species

Characteristic	*S. aureus*	*S. epidermidis*	*S. saprophyticus*
Colony color	Golden-yellow	White	White
Acid production from mannitol	Yes	No	No
Coagulase production	Yes	No	No
DNase activity	Yes	No	No
Novobiocin sensitivity	Yes	Yes	No

PROCEDURE

Obtain the three slant cultures (isolates A, B, and C) that you made in Exercise 44.

I. Coagulase Activity

1. Obtain three tubes of **coagulase plasma** and label them with your name.
   - Label one tube A, one B, and one C.
2. Inoculate coagulase tube A with a heavy loopful of bacteria from slant A. Sterilize the loop.
3. Repeat step 2 to transfer bacteria from slants B and C to coagulase tubes B and C.
4. Observe the tubes at intervals designated by the instructor. Incubation can continue for up to 24 hours. Coagulase activity is determined by tilting each coagulase tube to the side and noting the presence of a solid immovable clot (FIGURE 45.1). If the plasma still flows, clot formation and coagulase activity have not taken place.
5. Enter your results in the Results section, in TABLE 45.2, using (+) for coagulase activity and (−) for lack of activity. Include the time for clotting, if known.

II. DNA Digestion

1. Obtain a **DNase test agar plate** and divide the bottom of the plate into four sections.
   - Label each of three sections with a colony designation (A, B, or C). Label the fourth section control. Label the plate "DNA test" and make sure your name also is on the plate.
2. Using aseptic technique, make a single line of inoculation of each staph isolate (A, B, and C) in its respective sector on the DNase test agar plate. Leave the fourth section blank.
3. Incubate the plate in the inverted position for 24 to 48 hours at 37°C.
4. Test for DNA digestion by flooding the plate with 1N hydrochloric acid (HCl). DNA will normally react with HCl and form a very fine precipitate, which will give a cloudy appearance to the agar after several minutes.
   - If DNA digestion has occurred, the area near the streak will remain clear. The remainder of the plate, however, becomes cloudy.
   - A dark background is recommended, and observations should be made from an angle to see the cloudiness.
   - Compare your results with the control section containing no bacteria.

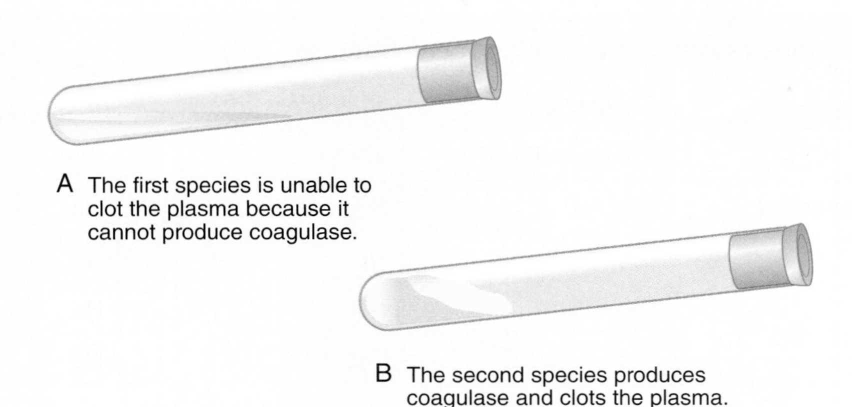

FIGURE 45.1 The coagulase test performed with two species of staphylococci.

5. Determine whether DNA has been digested in each of the sectors of the plate and record each result in the Results section. Enter a (+) for DNase activity and (–) for lack of activity.

III. Novobiocin Sensitivity

1. Obtain three **Mueller-Hinton agar plates** and label the bottom side of each plate with your name and "novo sensitivity test."
   - Label one plate A, one B, and one C for each staph isolate.
2. Inoculate the plates with the respective staph isolate following the "3+1" procedure for making a bacterial lawn as outlined in Exercise 19.
3. Using alcohol-dipped and flamed forceps, aseptically apply a novobiocin antibiotic disc to the center of each plate.
   - Gently tap the forceps on each disc to ensure it is firmly in contact with the agar.
4. Incubate the plates in the inverted position for 24 to 48 hours at 37°C.
5. Observe the novobiocin plates. Novobiocin sensitivity is determined by measuring the diameter of the halo (**zone of inhibition**) around the disc where no bacterial growth has occurred. If the zone diameter is less than 17 mm, the bacterium is considered resistant to novobiocin. If the zone diameter is greater than 17 mm, the bacterium is considered sensitive to the antibiotic.
   - Use a metric ruler to measure each zone diameter.
6. Record the measurements and whether the bacterial isolate is resistant (R) or sensitive (S) to novobiocin in the appropriate Results section.

Name: ______________________ Date: __________ Section: __________

The Genus *Staphylococcus*: Differentiation between Staphylococcal Species

EXERCISE RESULTS **45**

RESULTS

DIFFERENTIATION BETWEEN STAPHYLOCOCCAL SPECIES

TABLE 45.2 Characteristics of Staphylococcal Isolates A, B, and C

Characteristics	A	B	C
Colony color			
Acid production from mannitol			
Coagulase production			
DNase activity			
Novobiocin sensitivity			

Observations and Conclusions

____

____

____

____

____

Questions

1. What is the relationship of coagulase activity to the pathogenicity of *Staphylococcus aureus*?

2. Suppose that a coagulase test were performed without a control tube. How would this omission decrease the reliability of the test?

3. Explain how the ability to produce DNase would benefit the pathogenicity of *S. aureus*.

4. From the series of tests and observations made of the three *Staphylococcus* isolates, determine which is *S. aureus*. From the results, is one of the other isolates either *S. epidermidis* or *S. saprophyticus*—or are both present?

The Enteric Bacteria: Isolation of Enteric Bacteria

EXERCISE 46

Materials and Equipment

- Fresh fecal samples and small spatulas
- Plates of MacConkey agar and EMB agar
- Inoculating loop
- Tubes of sterile saline

Some of the bacteria normally inhabiting the animal or human intestinal tract are gram-negative nonspore–forming rods called **enteric bacteria**. These organisms belong to the family Enterobacteriaceae, described in Section 5 of *Bergey's Manual of Systematic Bacteriology*. Many species are facultatively anaerobic, and certain species cause serious human diseases such as typhoid fever, shigellosis, and infantile and traveler's diarrheas.

In this exercise, enteric bacteria will be isolated from fecal samples.

ISOLATION OF ENTERIC BACTERIA

PURPOSE: to isolate and identify enteric bacteria on selective and differential media.

Enteric bacteria may be isolated by cultivating fecal samples on differential media containing lactose. In this section, MacConkey agar and Levine eosin methylene blue (EMB) agar will be used. **MacConkey agar** (FIGURE 46.1) is a differential medium containing bile salts to inhibit nonenteric bacteria. In addition, it has two dyes, neutral red and crystal violet, which are taken up by lactose-fermenting bacteria. Crystal violet is inhibitory to the growth of gram-positive bacteria. The enteric bacteria form pigmented colonies while the nonlactose fermenters form colorless colonies. In **EMB agar**, pigmented colonies also are formed by lactose fermenters, but the type of pigmentation is distinctive to various genera, as explained presently. EMB agar is another differential medium and does not support the growth of gram-positive bacteria.

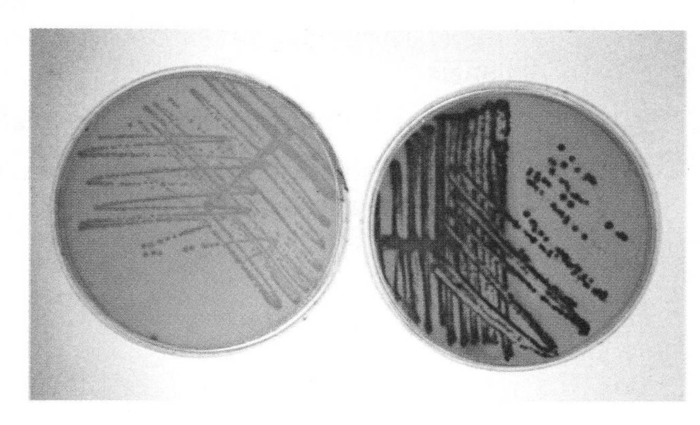

FIGURE 46.1 MacConkey agar plates. The plate on the left shows colorless colonies, which indicate negative lactose fermentation. The plate on the right shows red colonies which indicate positive lactose fermentation.

PROCEDURE

Handle fecal samples with care, as they may contain microbial pathogens.

1. Collect a fecal sample from an animal or human within 1 hour of the laboratory period.
   - A nonporous, waxed, clean vessel should be used for collection.
2. Using a small spatula, transfer a pellet of feces of approximately 1 gram weight to a tube of sterile saline.
   - Shake the tube to break up the fecal material.
   - Allow the solid particles to fall to the bottom of the tube.
3. Obtain or prepare plates of MacConkey agar and EMB agar, and label them on the bottom side with your name, the name of the medium, the date, and the designation "fecal sample."
   - Obtain a loopful of the fecal suspension and streak each plate for isolated colonies according to the streak-plate method outlined in Exercise 5.
   - Invert the plates, and incubate them at 37°C for 24 to 48 hours.
4. Observe the MacConkey agar for evidence of lactose-fermenting bacteria, which will appear as red colonies. *Escherichia coli* will usually be brick red, while *Enterobacter, Klebsiella*, and other lactose fermenters will be pink to red with a mucoid texture.
   - Colonies of nonlactose fermenters will appear colorless.
   - Enter a labeled representation of the plate in the Results section.
   - Gram stains should be prepared to confirm the presence of gram-negative rods.
5. Inspect the EMB plate and take note of pigmented colonies that appear blue-black with a green metallic sheen when viewed in reflected light. These are probably colonies of *Escherichia coli.*
   - Other pigmented colonies may be convex and mucoid with dark centers and creamy pink borders. These may be *Enterobacter aerogenes.* Nonlactose fermenters will yield colorless colonies.
   - Make a labeled representation of the plate in the Results section.
   - Gram stains should be prepared and observed for gram-negative rods, and diagrams should be placed in the appropriate spaces.

Name: ______________________ Date: __________ Section: __________

The Enteric Bacteria: Isolation of Enteric Bacteria

EXERCISE RESULTS 46

RESULTS

ISOLATION OF ENTERIC BACTERIA

MacConkey Agar Plate

EMB Agar Plate

Source: ______________________ ______________________

Stained Smears of Enteric Bacteria

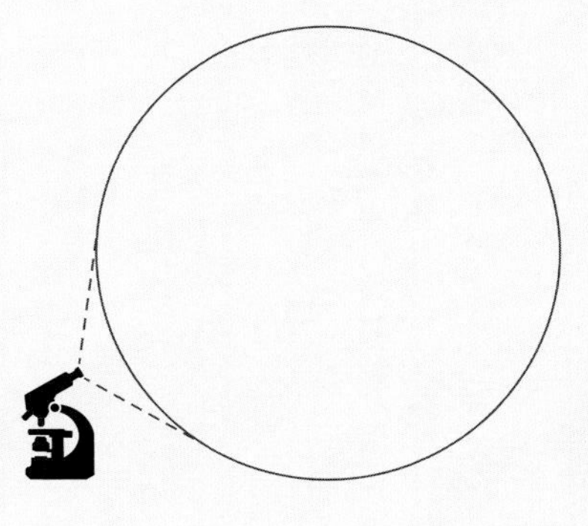

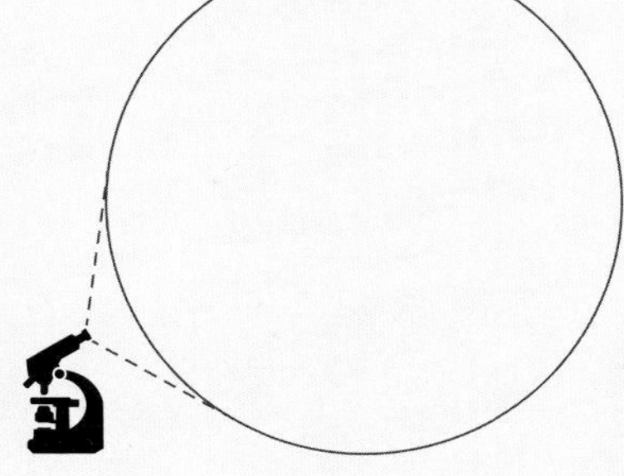

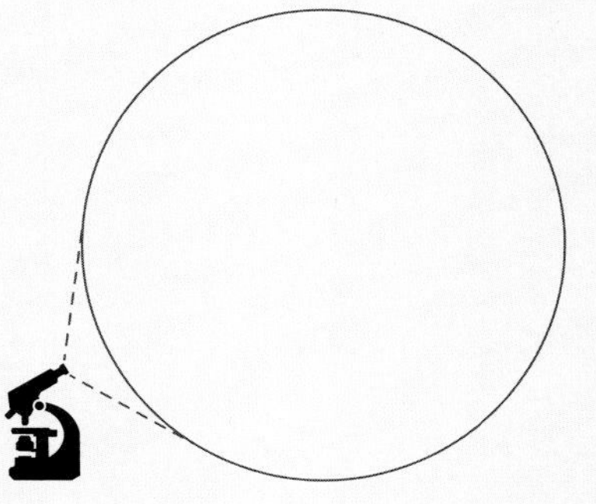

Source: ______________ ______________ ______________

Magnif.: ______________ ______________ ______________

Observations and Conclusions

____

____

____

____

____

Questions

1. What conclusions may be drawn if no growth appeared on MacConkey agar and EMB agar after inoculation of the media and an incubation period?

   ____

   ____

   ____

   ____

The Enteric Bacteria: Differentiation of Enteric Bacteria on TSI Agar

EXERCISE 47

Materials and Equipment

- Slants of triple sugar iron (TSI) agar
- Isolated enteric bacteria from Exercise 46
- Inoculating loop and needle

Triple sugar iron (**TSI**) **agar** is a valuable medium for differentiating enteric bacteria. The medium contains three carbohydrates—glucose, lactose, and sucrose—as well as the pH indicator phenol red. Bacteria are inoculated into the slant of the medium and into the deep portion (the "butt"), where anaerobic conditions prevail.

DIFFERENTIATION OF ENTERIC BACTERIA ON TSI AGAR

PURPOSE: to differentiate enteric bacteria based on acid and gas production.

Three reactions are possible in TSI agar depending on the species of bacterium inoculated. First, species like Salmonella can only ferment glucose. Although, both the slant and the butt will turn yellow after several hours, the butt will remain yellow and the slant will soon revert back to red as alkaline conditions reappear from the digestion of peptones and the production of ammonium compounds (FIGURE 47.1A). Second, if glucose and lactose and/or sucrose are fermented, sufficient acid will be produced to cause both the slant and butt to remain yellow (FIGURE 47.1B, C). Third, if no carbohydrate is fermented, the slant and butt will remain a red color.

Gas (CO_2) production from the carbohydrates also may be determined in TSI agar by noting the presence of cracks and fissures in the medium after the incubation period. If large amounts of gas are produced, portions of the medium may be pushed up in the tube (see Figure 47.1B, C). Also, hydrogen sulfide production may be tested because TSI agar contains iron ions and sodium thiosulfate. Certain bacteria use thiosulfate in their metabolism and release hydrogen sulfide (H_2S). The H_2S then reacts with iron, yielding iron sulfide, which is seen as a black precipitate in the butt (see Figure 47.1A, B). TABLE 47.1 lists the characteristics of different genera of isolated bacteria based on lactose fermentation.

In this exercise, bacteria isolated on the MacConkey and EMB plates will be differentiated by their reactions in TSI medium.

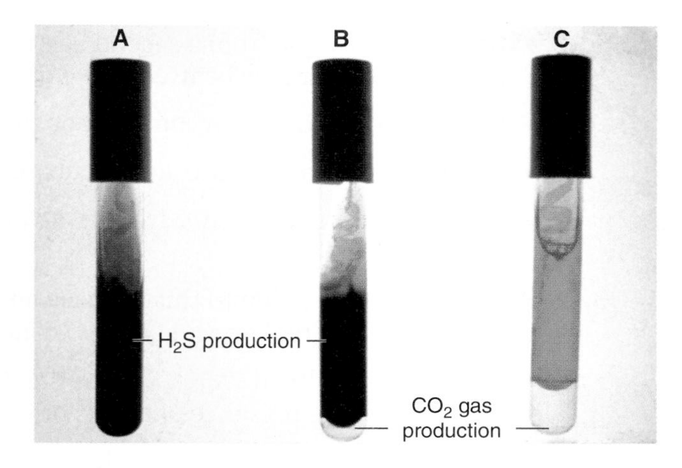

FIGURE 47.1 Typical TSI reactions for selected members of the Enterobacteriaceae. Tube **A** is typical of *Salmonella*. Tube **B** is representative for *Citrobacter fruendii*. Tube **C** is typical of the lactose fermenters such as *E. coli*, *Enterobacter*, and *Klebsiella*.

TABLE 47.1 Characteristics of Genera of Enteric Bacteria

Genus of Bacteria	Lactose Fermentation	TSI Reactions				IMViC Reactions			
		SLANT	BUTT	GAS	H_2S	I	M	VP	C
Escherichia	+	Acid	Acid	+	–	+	+	–	–
Enterobacter	+	Acid	Acid	+	–	–	–	+	+
Klebsiella	+	Acid	Acid	+	–	±	±	±	±
Serratia	–	Acid	Acid	–	–	–	–	+	+
Citrobacter	+	Alk.	Acid	+	+	±	+	–	+
Salmonella	–	Alk.	Acid	–	±	–	+	–	+
Shigella	–	Alk.	Acid	–	–	–	±	+	–
Proteus	–	Alk.	Acid	±	±	±	+	–	±
Pseudomonas	–	Alk.	Alk.	–	–	–	–	–	+

± Certain species positive, others negative

PROCEDURE

1. Select two tubes of TSI agar and label each.
   - Using the plates from Exercise 46, choose a lactose-fermenting colony of bacteria and a nonlactose fermenter.
2. Using a sterile inoculating loop, streak a sample from a bacterial colony onto the surface of the slant.
   - With a sterile inoculating needle, obtain a sample from the same colony and stab the needle into the butt.
   - Inoculate a sample from a second bacterial colony to TSI agar in a similar manner. Additional tubes of TSI agar may be used at the direction of the instructor.
   - Include an uninoculated control tube in the experiment.
   - Be certain the cap is loose, and incubate the tubes at 37°C.
   - Examine the tubes after 24 hours of incubation or, if that is not possible, they should be refrigerated.
3. Reread the introduction to this exercise and observe the slant and butt of the tube to determine whether alkaline (red) or acid (yellow) conditions exist.
   - Enter your results in TABLE 47.2 in the Results section. Note whether CO_2 gas was evolved as indicated by the presence of cracks or fissures in the medium.
   - Also, note whether H_2S was produced by noting a blackening of the medium.
   - Enter these results in the table. At this point it may be possible to narrow down the list of possible genera by referring to Table 47.1.
   - Gram stains may also be prepared to confirm the presence of gram-negative rods.

Name: ______________________ Date: __________ Section: __________

The Enteric Bacteria: Differentiation of Enteric Bacteria on TSI Agar

EXERCISE RESULTS **47**

RESULTS

TSI AGAR TUBES

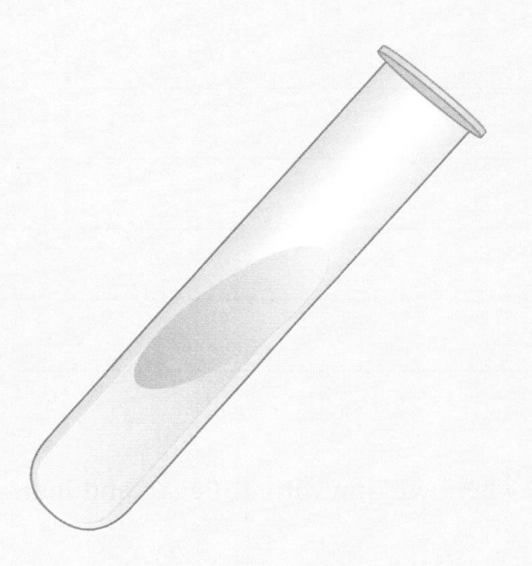

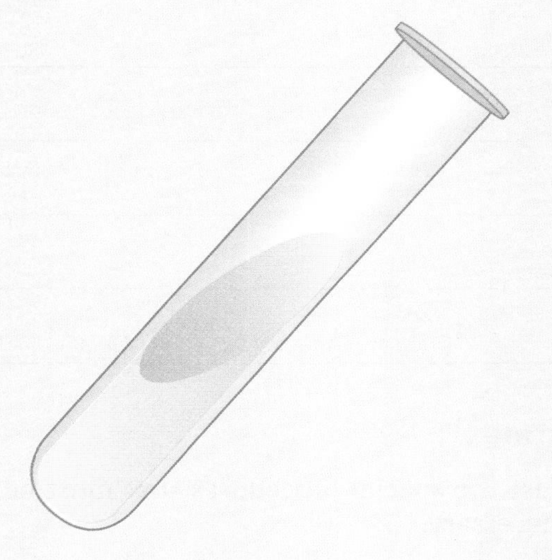

Source: ______________________ ______________________

Source: ______________________

TABLE 47.2 Characteristics of Enteric Bacteria

Organism	Lactose Fermentation	TSI Reactions			
		Slant	Butt	Gas	H_2S
Unknown #1					
Unknown #2					
Unknown #3					

Observations and Conclusions

Questions

1. List the special procedures that must be used when working with TSI agar, and indicate why each is necessary.

2. Could reliable results be obtained if TSI agar were inoculated with a sample directly from a fecal suspension? Explain.

The Enteric Bacteria: The IMViC Series

EXERCISE **48**

Materials and Equipment

- Tubes of MR-VP medium, each with 5.0 ml
- Tubes of trypticase soy or tryptone broth
- Kovac's reagent
- Methyl red solution
- Barritt's reagents A (Barr A) and B (Barr B)
- Slants of Simmons citrate agar
- Broth cultures of *E. coli* and *E. aerogenes*
- TSI agar slants from Exercise 47

Watch These Videos

- IMViC: Indole Test
- IMViC: Methyl Red Test
- IMViC: Voges-Proskauer Test
- IMViC: Citrate Test

Additional biochemical tests provide more information for the identification of an enteric bacterium. Though primarily used to distinguish *Escherichia coli* from *Enterobacter aerogenes*, the results from these tests may also be used for other organisms, as indicated in Exercise 47, Table 47.1.

The **IMViC series** consists of four tests: indole production (I), the methyl red test (M), the Voges-Proskauer test (V), and the citrate test (C). The "i" is included in the acronym for pronunciation purposes. The **indole test** identifies species that can digest tryptophan to indole and other products. The **methyl red test** depends on an organism's ability to ferment glucose and produce large amounts of acid. Methyl red indicator, added at the end of the incubation period, remains red in acid solution but becomes yellow in alkaline or neutral solution.

The **Voges-Proskauer test** uses the digestion of glucose to acetylmethylcarbinol to differentiate between enteric species. If present in the culture tube, acetylmethylcarbinol will react with alpha naphthol and potassium hydroxide to form a red chemical compound.

The **citrate test** is based on the ability of certain bacteria to use citrate, a salt of citric acid, as a sole carbon and energy source in growth. When this occurs, the pH of the medium rises and the indicator, bromthymol blue, becomes deep blue. If citrate use has not occurred, the original green color of the medium remains.

Before starting this exercise, watch the related **Microbiology Videos**: **Indole Test**, **Methyl Red Test**, **Voges-Proskauer Test**, and **Citrate Test** to understand how these tests are performed.

IMViC: Indole Test

IMViC: Methyl Red Test

IMViC: Voges-Proskauer Test

IMViC: Citrate Test

THE IMViC SERIES

PURPOSE: to further differentiate enteric bacteria based on additional biochemical tests.

In this exercise, enteric bacteria from Exercises 46 and 47 thought to be *Escherichia coli* or *Enterobacter aerogenes* may be used as test organisms, or a different isolate may be employed. The instructor also may supply laboratory cultures of *E. coli* or *E. aerogenes*.

PROCEDURE

1. Select one tube of trypticase soy or tryptone broth for the indole test; two tubes of MR-VP broth, one labeled for the methyl red (MR) test and one for the Voges-Proskauer (VP) test; and one slant of Simmons citrate agar for the citrate test.
   - Label each tube with your name, the date, the name of the test, and the name or source of the inoculum.
2. Aseptically inoculate each medium with a loopful of enteric bacteria from a TSI agar slant from Exercise 47 or with a bacterial culture supplied by the instructor.
   - Incubate all tubes at 37°C for 24 to 48 hours.
   - Uninoculated control tubes also should be included.
3. **Indole Test.** Complete the indole test by adding 5 drops of **Kovac's reagent** to the trypticase soy or tryptone broth tube (FIGURE 48.1). The presence of a red surface layer indicates indole production.
   - Enter a (+) for indole production and a (−) for nonproduction in TABLE 48.1 in the Results section.
4. **Methyl Red (MR) Test.** Complete the methyl red test by adding five drops of methyl red solution to one MR-VP (MR) tube (FIGURE 48.2).
   - Compare this to the control tube, and note whether the color remains red, indicating a positive test for acid production, or whether the red color has disappeared, indicating a negative reaction.
   - Enter your results in Table 48.1.

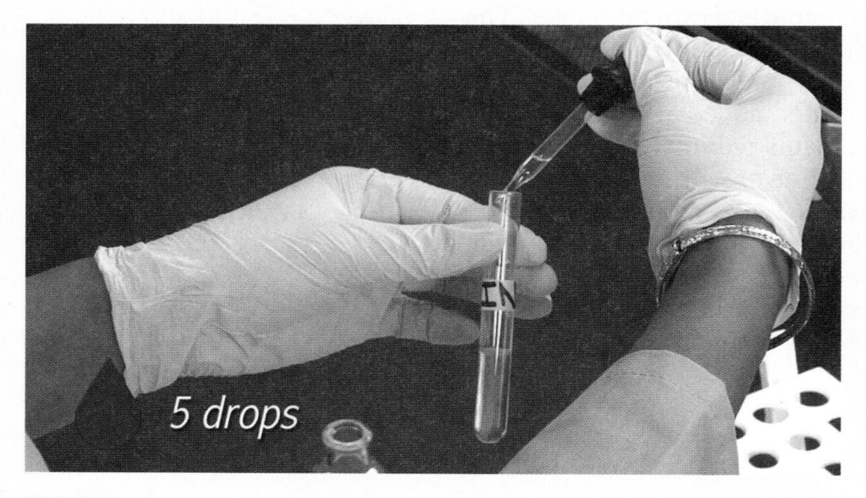

FIGURE 48.1 Add 5 drops of Kovac's reagent to the trypticase soy or tryptone broth tube.

FIGURE 48.2 Add five drops of methyl red indicator to one MR-VP tube.

5. **Voges-Proskauer (VP) Test.** Complete the Voges-Proskauer test by adding 10 drops of Barritt's A (Barr A) reagent to the VP broth tube. Mix and the add 10 drops of Barritt's B reagent (Barr B) to the tube. Mix and look for the development of a red color after several minutes, which indicates a positive test.
   - Use the control tube for comparison. Lack of color change reflects a negative test.
   - Enter the results in Table 48.1
6. **Citrate Test.** Complete the citrate test by noting the presence of growth on Simmons citrate agar and the color of the medium as compared to the control (FIGURE 48.3). A blue color indicates the ability to use citrate and constitutes a positive test. A green color represents a negative result.
   - Complete Table 48.1 in the Results section, carry over data from Exercise 47, and indicate which enteric bacterium was present in the TSI slant or assigned culture.

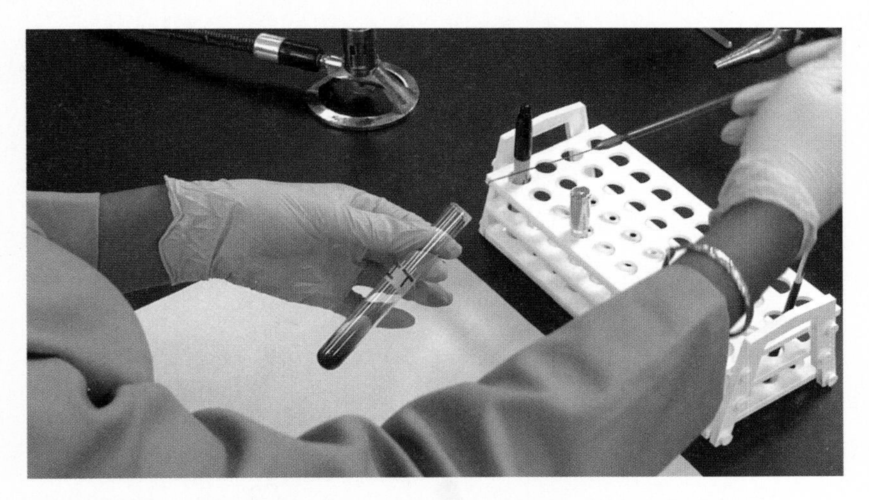

FIGURE 48.3 The citrate test will determine whether the bacteria can use sodium citrate as the only carbon and energy source.

Name: ______________________ Date: __________ Section: __________

The Enteric Bacteria: The IMViC Series

EXERCISE RESULTS 48

RESULTS

THE IMVIC SERIES

TABLE 48.1 Characteristics of Enteric Bacteria

Organism	Lactose Fermentation	TSI Reactions				IMViC Reactions				
		Slant	Butt	Gas	H_2S	I	M	VP	C	Possible Genera
Unknown #1										
Unknown #2										
Unknown #3										

Observations and Conclusions

____

____

____

____

____

Questions

1. Summarize the chemistry of the tests that make up the IMViC series. How might the IMViC series be of value in the bacteriological analysis of water?

2. Which procedures in this exercise may be used to isolate *Escherichia coli* from a fecal suspension and verify its presence?

The Enteric Bacteria: Rapid Identification of Enteric Bacteria

EXERCISE 49

Materials and Equipment

- Enterotube II
- Plate of isolated enteric bacteria from Exercise 46
- Beaker of disinfectant

In recent years, a number of miniaturized systems have been made available to microbiologists for the rapid identification of enteric bacteria. One such system is the Enterotube II, available from Roche Diagnostics. The Enterotube II is a self-contained, sterile, compartmentalized plastic tube containing 12 different media and an enclosed inoculating wire (see Appendix B). This system permits the inoculation of all media and the performance of 15 standard biochemical tests using a single bacterial colony. The results are tabulated, and a computer coded system is used to identify the unknown enteric bacterium.

RAPID IDENTIFICATION OF ENTERIC BACTERIA

PURPOSE: to rapidly identify an enteric bacterial species using 15 standard biochemical tests done in one procedure.

In this exercise, the Enterotube II will be used to learn the name of an isolated enteric organism.

PROCEDURE

1. Choose a plate from Exercise 46 that displays well-isolated colonies, and select one colony for identification.
   - Obtain one Enterotube II.
2. Remove the caps from the Enterotube II.
   - Note that the straight wire is to be used to pick up the bacteria, while the bent end is the handle.
   - Without flaming the wire, touch the end to a well-isolated colony of bacteria from a plate prepared in Exercise 46.
   - Avoid touching the agar; try to touch the bacteria only.
3. While gently twisting the wire, pull it through all 12 compartments of the Enterotube II.
   - Reinsert the wire into the first four compartments (glucose, lysine, ornithine, and H_2S/indole), thereby creating an anaerobic environment.
   - Break the needle by bending it at the notch, and discard the handle in the beaker of disinfectant.
   - Replace the caps loosely on both ends of the tube.

4. Strip off the blue tape to expose the holes in the Enterotube II and provide aerobic conditions.
   - Slide the clear plastic band over the glucose compartment to contain any wax that may otherwise escape due to excessive gas production.
   - Incubate the tube lying on its flat surface at 37°C for 18 to 24 hours, and then refrigerate it until it is to be observed.
5. Examine the Enterotube II and interpret all the results except the indole and Voges-Proskauer (VP) tests by referring to TABLE 49.1.
   - Enter your results in TABLE 49.2 in the Results section.
   - Complete the indole test by injecting two drops of Kovac's reagent from a syringe through the plastic band. (Alternately, a small hole may be melted in the plastic with a warm inoculating needle.)
   - Read the results in one minute.
   - Complete the VP test by adding two drops of Barr A reagent and two drops of Barr B reagent. A positive test is indicated by development of a red color in 20 minutes.
   - Add your results to Table 49.2.
6. Attempt to identify the unknown bacterium by the computer coding system on a report sheet available from the instructor. The instructor will provide directions for the system to determine the "ID value" to identify the genus and/or species of the organism.
   - Return the tubes for autoclaving at the conclusion of the exercise.

TABLE 49.1 Interpretation of Reactions in the Enterotube II

Test	Principle	Positive Rx	Negative Rx
GLU	Acid from glucose	Yellow	Red
GAS	Gas from glucose	Bubbles	No bubbles
LYS	Lysine decarboxylated	Purple	Yellow
ORN	Ornithine decarboxylated	Purple	Yellow
H_2S	Hydrogen sulfide produced	Black	No change
IND	Indole produced	Red	No change
ADON	Adonitol fermented	Yellow	Red
LAC	Lactose fermented	Yellow	Red
ARAB	Arabinose fermented	Yellow	Red
SORB	Sorbitol fermented	Yellow	Red
VP	Acetylmethylcarbinol produced	Red	No change
DUL	Dulcitol fermented	Yellow	Green
PA	Phenylalanine deaminated	Gray/Black	Green
UREA	Urea digested	Pink	No change
CIT	Citrate utilized	Blue	Green

Name: ________________ Date: ________ Section: ________

The Enteric Bacteria: Rapid Identification of Enteric Bacteria

EXERCISE RESULTS 49

RESULTS

RAPID IDENTIFICATION OF ENTERIC BACTERIA

TABLE 49.2 Reactions in Enterotube II

Glu	Gas	Lys	Orn	H_2S	Ind	Adon	Lac	Arab	Sorb	Vp	Dul	Pa	Urea	Cit

ID value:

____

____

Name of enteric bacterium:

____

____

Observations and Conclusions

____

____

____

____

____

Questions

1. With the Enterotube II system, why does the first value have to be at least 2?

2. Why do some species of bacteria have more than one ID value in the Enterotube II system?

The Genus *Bacillus*

EXERCISE

50

Materials and Equipment

- Nutrient agar plates
- Starch agar plates
- Iodine solution
- Hydrogen peroxide (3%)
- Tubes of motility test agar
- Malachite green
- Safranin
- Heated water bath
- Soil samples and test tubes
- Gram stain reagents

The genus ***Bacillus*** includes over three dozen species, many of which have significance in industrial and medical microbiology. For example, dried *Bacillus subtilis* is used as a source of enzymes in enzyme detergents, and *B. polymyxa* produces the antibiotic polymyxin. Also, *Geobacillus* (formerly *Bacillus*) *stearothermophilus* is used to test the effectiveness of sterilization, and *B. anthracis* is the agent of anthrax, a serious disease of animals—and a health threat to humans, as evidenced by the anthrax-contaminated letter incident in October 2001.

The organisms in this genus are gram-positive, sporeforming rods that generally produce catalase. Most are motile, and many species digest starch and DNA. In this exercise, *Bacillus* species will be isolated from the soil and their presence confirmed by noting typical characteristics of the organisms. Initially, the other soilborne aerobic organisms will be destroyed with heat so as to select out *Bacillus* spores.

THE GENUS *BACILLUS*

PURPOSE: to isolate and identify *Bacillus* species from soil samples.

PROCEDURE

1. Heat a water bath to approximately 80°C.
   - Place a generous pinch of rich soil in several milliliters of water in a test tube, and immerse the tube in the hot water for approximately 10 minutes, then remove it and let the liquid cool.
   - This procedure will destroy most organisms other than bacterial endospores and will stimulate the spores to germinate.
2. Select or prepare one nutrient agar plate and one starch agar plate.
   - Label each on the bottom side with your name, the date, the name of the medium, and the designation "soil sample."

When streaking the plate, avoid exposure to the air, since *Bacillus* spores can easily enter the plate and introduce errors.

3. Aseptically obtain a loopful of the cooled soil sample, and streak the nutrient agar plate according to the streak-plate technique, as explained in Exercise 5.
   - Repeat with the starch agar plate, and incubate both plates in the inverted position at 37°C. No more than 24 hours of incubation will be necessary.
   - Refrigerate the plates until the next laboratory period.
4. Observe the plates for uneven, white, spreading colonies, which may contain *Bacillus* species.
   - **Catalase Test.** Test for catalase production by placing drops of hydrogen peroxide onto the colonies on nutrient agar and noting the formation of bubbles of oxygen gas.
   - **Starch Test.** Many *Bacillus* species produce the enzyme amylase that is able to digest starch. The ability to digest starch can be ascertained by adding iodine to colonies growing on the starch agar. If amylase was produced, the agar around the colonies will be digested so there will be nothing with which the iodine can react. The result will be a clear area around these colonies. If the agar is black around the colonies, the starch was not digested; that is no amylase was produced. Enter representations of the plates in the Results section.
   - **Presence of Endospores.** Isolate on nutrient agar slants samples of colonies thought to contain *Bacillus* species and stain according to the method described in Exercise 16 to determine whether endospores are present.
   - **Gram Stain.** Gram stains also may be prepared.
5. When *Bacillus* species have been identified by the above tests, **motility** may be determined by inoculating motility test agar with samples, according to the procedure described in Exercise 18.
   - Observe the motility test agar for evidence of motility, and enter this information in the Results section.
6. Since *Bacillus* species often produce antibiotics, the isolates may also be tested for this ability in the following way:
   - Inoculate a nutrient agar plate with a single streak across the center of the plate (see Exercise 97, Figure 97.1).
   - A test organism supplied by the instructor is then inoculated at a right angle to the original streak, and the plate may be incubated at 37°C for 24 to 48 hours.
   - Examine the antibiotic plate and note whether the growth of the test organism has been inhibited in the region of the *Bacillus* species. Growth inhibition provides evidence of antibiotic production.
7. Prepare representations of the test results and draw appropriate observations from the results obtained in this exercise.

Name: ______________________ Date: __________ Section: __________

The Genus *Bacillus*

EXERCISE RESULTS **50**

RESULTS

BACILLUS SPECIES FROM SOIL

Nutrient Agar Plate

Starch Agar Plate

Source: ______________________ ______________________

Stained Smears of *Bacillus* Species

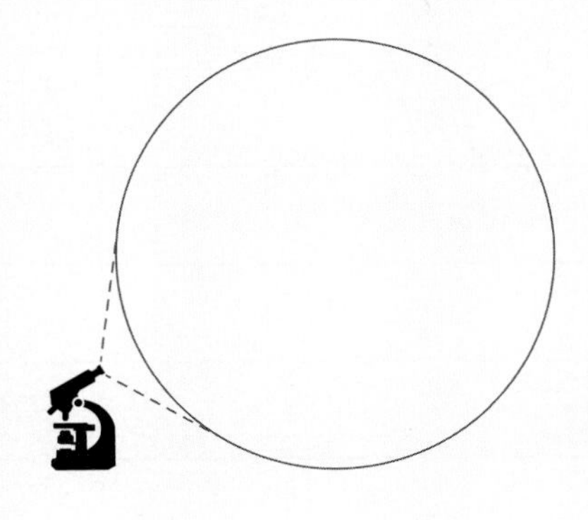

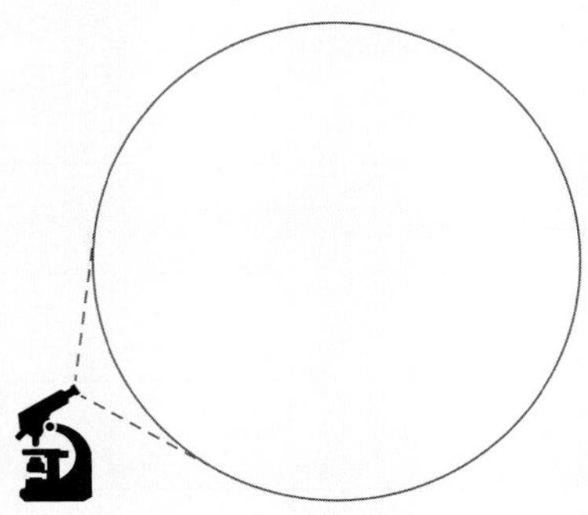

Source: ______________________ ______________________

Magnif.: ______________________ ______________________

Source: ____________________ ____________________

Magnif.: ____________________ ____________________

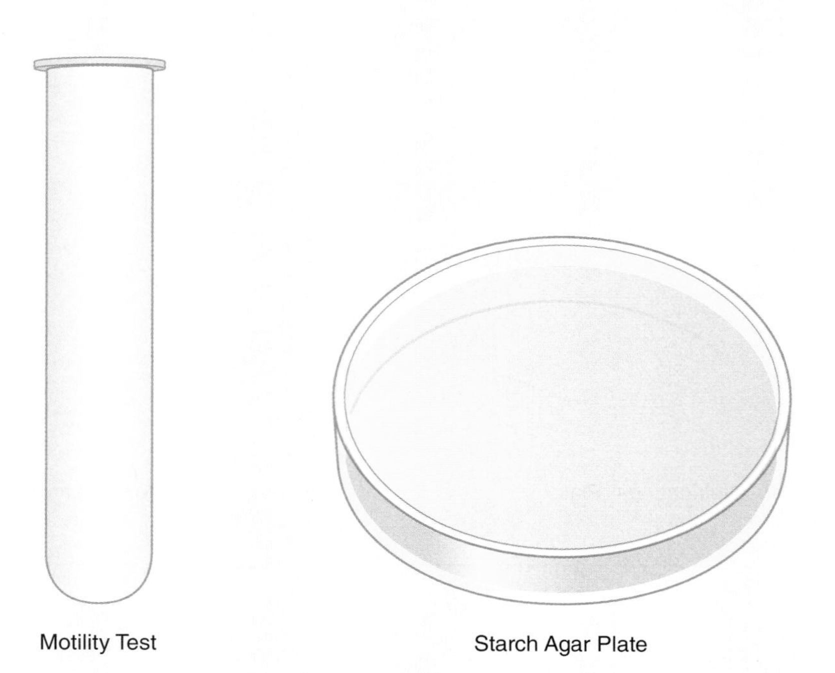

Motility Test

Starch Agar Plate

Observations and Conclusions

____

____

____

____

____

Questions

1. Summarize the biochemical characteristics of members of the genus *Bacillus*.

2. What might result if the soil sample were not heated prior to the attempt to isolate *Bacillus* species?

3. Why are microscopic observations of particular value in the identification of *Bacillus* species?

4. Given the choice, might it be more advisable to use catalase production or starch digestion as an identifying test for *Bacillus* species? Why?

5. Name several instances in which members of the genus *Bacillus* make practical contributions to human welfare.

The Genus *Clostridium*

EXERCISE

51

Materials and Equipment

- Rich organic soil samples and test tubes
- Tubes of thioglycolate medium
- Tubes of skim milk medium
- Deep tubes of brain heart infusion agar
- Tubes of cooked meat medium in screw-capped tubes
- Plates of sulfite polymyxin sulfadiazine (SPS) agar
- Plates of blood agar
- GasPak anaerobic system
- Gram stain reagents

Members of the genus ***Clostridium*** play an important role in human disease as the causes of botulism, food poisoning, tetanus, and gas gangrene. Over 100 species of *Clostridium* are recognized, most of which are widely distributed in the environment and are harmless to humans. In the soil, *Clostridium* species decay organic matter and release the carbon back to the environment. They also fix atmospheric nitrogen and make it available to plants in usable forms. In this exercise, soil will be used as a source of clostridia.

Clostridium species are gram-positive, spore-forming bacilli that grow only under anaerobic conditions. Such conditions may be found in the dead tissue of a wound or in a vacuum-sealed can of contaminated food, and the toxins produced in these environments may lead to serious disease. Anaerobic conditions also exist in sanitary landfills, swamps, and marshes, and clostridia thrive in these places as well. In this exercise, we shall employ anaerobic techniques in the laboratory to isolate the clostridia.

THE GENUS *CLOSTRIDIUM*

PURPOSE: to isolate *Clostridium* species from the soil.

PROCEDURE

1. Take a generous pinch of rich soil and suspend it in a small amount of water in a test tube. Mud is useful because it contains little oxygen in its water-soaked soil. This material will be used as a source of *Clostridium* species. The instructor may designate that soil be used directly without making a suspension.

Cover any open wounds when working with fresh soil because clostridial pathogens may be present.

2. Select a tube of **thioglycolate medium**, label it, and inoculate it with several loopfuls from the soil suspension. A few grains of soil may also be an acceptable inoculum. Thioglycolate medium contains sodium thioglycolate, a reducing agent that removes oxygen from the environment. It also contains methylene blue, an indicator that shows the aerobic layer at the surface of the medium. Below this layer, conditions are considered anaerobic.
   - Incubate the tube at 37°C for 24 to 48 hours.
   - Refer to Step 8 for possible results.
3. Obtain a tube of skim milk medium, label it, and inoculate it with several grains of soil or several loopfuls from the soil suspension. Skim milk is a rich medium with a broad variety of nutrients that support many bacteria. Certain *Clostridium* species vigorously ferment the lactose in milk and produce large amounts of acid and CO_2 gas. The acid clots the milk protein, and the gas shreds the clot, sometimes causing it to break apart. This reaction is called a **stormy fermentation**. Incubate the tube at 37°C for 24 to 48 hours, preferably in an autoclavable container to catch any contents that spill out of the tube.
4. Select a tube of cooked meat medium and inoculate it with several grains of soil or several loopfuls from the soil suspension. The medium should have been recently boiled to drive off the oxygen, and agitation of the medium should be kept to a minimum. Tightly sealed screw caps also help maintain anaerobic conditions. *Clostridium* species will digest the meat particles to a fine sediment. An odor of hydrogen sulfide and a blackening of the medium will also be apparent after incubation.
   - Incubate the tubes at 37°C for 24 to 48 hours. Refer to Step 8.
5. Select a tube of melted brain heart infusion agar from the water bath and use it quickly to avoid premature hardening. Inoculate the tube with several loopfuls of the soil suspension or with a few grains of soil, at the instructor's direction. Now mix the soil with the agar by rolling the tube between the palms of the hands, and let the medium harden with the tube standing in the upright position. *Clostridium* species will be trapped deep in the agar as it hardens, and as anaerobic conditions develop they will proliferate and produce their gases.
   - Incubate the tube at 37°C for 48 hours.
   - An alternative procedure is to perform the pour plate technique as explained in Exercise 6, except that the tubes of agar are poured into sterile culture dishes. This technique allows dilution of the soil suspension and gives *Clostridium* species more opportunity to form wide-spaced colonies than in the agar tubes.
6. Obtain plates of **sulfite polymyxin sulfadiazine (SPS) agar** and **blood agar**, label them on the bottom sides, and inoculate the plates by streaking them for isolated colonies from the soil suspension, as outlined in Exercise 5. SPS agar contains sodium sulfite, which is reduced by certain clostridia to form sulfide. The sulfide causes colonies to become black. It also contains polymyxin and sulfadiazine, two antimicrobial agents that inhibit other organisms. The blood in blood agar is hemolyzed by clostridia to form a characteristic double zone of hemolysis: an inner zone of beta-hemolysis and an outer area of alpha-hemolysis.
   - Incubate the plates in the GasPak apparatus at 37°C for 24 to 48 hours.
7. The GasPak apparatus, shown in FIGURE 51.1, contains a foil envelope that generates hydrogen and CO_2 gases when 10 ml of water is added to the envelope. A palladium catalyst in the lid combines the hydrogen with available oxygen to form water, thereby establishing anaerobic conditions. The CO_2 creates an environment favorable to the anaerobes. A pad of methylene blue in the jar indicates when anaerobic conditions have developed, because methylene blue is colorless in the absence of oxygen but blue in its presence. Refer to Step 8 for results.

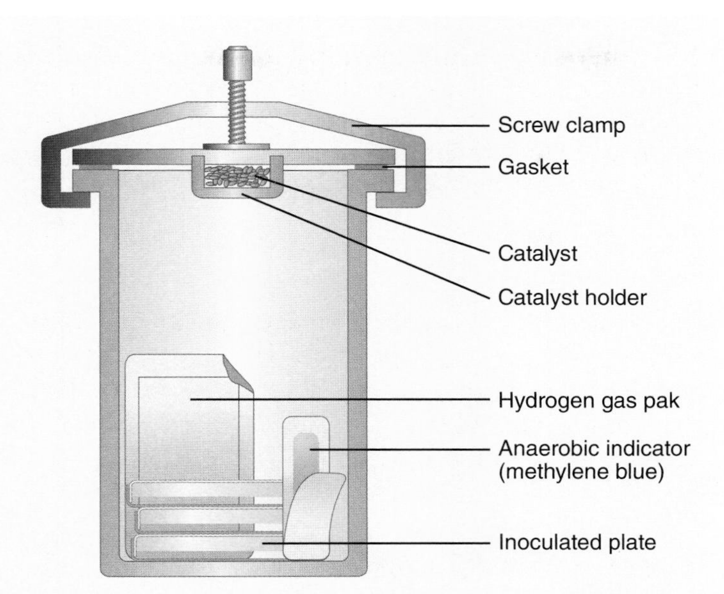

FIGURE 51.1 The GasPak system for cultivating anaerobic bacteria.

8. Examine the various laboratory media for evidence of *Clostridium* species, and enter your data in the Results section. The thioglycolate medium will show cloudiness below the methylene blue indicator, and a Gram-stained smear of a sample may reveal gram-positive, sporeforming rods.

- Examine the skim milk tube for evidence of stormy fermentation, and prepare stains.
- Note the type and color of the sediment in the cooked meat medium, as well as its odor, and take samples for staining.
- Observe the brain heart infusion tubes for blackening of the medium and colony formation deep in the agar. Samples may be obtained by melting the agar with a warm inoculating needle inserted down to the desired level.
- Examine the SPS agar plate for black colonies, which may represent *Clostridium* species, and the blood agar plate for a double zone of hemolysis. Smears should be made of appropriate colonies for Gram staining and spore staining.
- Place representations in the Results section.

Name: ____________________ Date: __________ Section: __________

The Genus *Clostridium*

EXERCISE RESULTS **51**

RESULTS

CLOSTRIDIUM SPECIES FROM THE SOIL

	Thioglycolate Medium	Skim Milk Medium	Cooked Meat Medium	Brain Heart Infusion Agar
Organism:	______	______	______	______
Test.:	______	______	______	______

SPS Agar

Blood Agar

Stained Smears of *Clostridium* Species

Source: ______________ ______________

Magnif.: ______________ ______________

Source: ______________ ______________

Magnif.: ______________ ______________

Source: ______________ ______________

Magnif.: ______________ ______________

Observations and Conclusions

Questions

1. Evaluate the different anaerobic cultivation methods used in this exercise, and explain the advantages of each.

2. Why was it unnecessary to heat the soil suspension before using it as a source of *Clostridium* spores?

3. Indicate how oxygen is reduced in the environment in the test media used in this exercise.

4. Growth of *Clostridium* species is generally accompanied by foul odors and blackening of the medium. What are the biochemical bases for these reactions?

5. Why is the spore stain technique valuable in the identification of *Clostridium* species isolated in anaerobic test media?

The Genus *Lactobacillus*: Isolation of Lactobacilli

EXERCISE 52

Materials and Equipment

- Sterile Petri dishes
- Deep tubes of melted Rogosa SL agar
- Paraffin in individual pieces
- Sterile test tubes (preferably screw-capped)
- Tubes of sterile water (5 ml)
- Yogurt samples in test tubes
- Gram stain reagents

Members of the genus ***Lactobacillus*** vary from long and slender rods to short coccobacillary forms. The bacteria are gram-positive in young, active cultures but become gram-negative with age. Some lactobacilli exhibit granulation on staining. Most are anaerobic or facultative organisms.

The lactobacilli occupy a prominent position in dairy microbiology as producers of such products as yogurt and acidophilus milk. They also may contribute to dental caries, and their number in the mouth sometimes indicates an individual's predisposition to tooth decay. In this exercise, lactobacilli will be isolated from saliva and from a dairy product.

ISOLATION OF LACTOBACILLI

PURPOSE: to isolate and identify *Lactobacillus* species from mixed population sources.

Species of *Lactobacillus* can be selected out of a mixed population by cultivating them in a medium that has a low pH, to inhibit most other organisms while encouraging the proliferation of lactobacilli. In this exercise, Rogosa SL agar (pH 6.1) is used, and lactobacilli are identified by staining after growth has taken place. An alternative medium for lactobacilli is Bacto Lactobacilli MRS medium.

PROCEDURE

1. To collect saliva, place a piece of paraffin under the tongue and allow it to soften for several minutes.
   - Chew the paraffin, moving it from side to side in the mouth.
   - As saliva accumulates, collect a few milliliters in a sterile test tube, preferably screw-capped.
   - Shake the tube vigorously to disperse the organisms present.

Before placing paraffin into your mouth, be sure it is the correct one designated for use in this exercise.

- Obtain a yogurt sample and inoculate two loopfuls into a tube of sterile water.
- Mix thoroughly.

2. Obtain two deep tubes of melted **Rogosa SL agar** and two sterile culture dishes.
    - Label one dish for "saliva" and one for "yogurt."
    - Include your name, the date, and the name of the medium.
3. Aseptically obtain a loopful of saliva, and inoculate it into a deep tube of melted Rogosa SL agar.
    - Tap the bottom of the tube to thoroughly mix the contents, then pour the agar into the culture dish labeled "saliva."
    - Aseptically obtain a loopful of the diluted yogurt, inoculate the second deep tube, and pour the agar into the dish labeled "yogurt."
    - Permit the agar to harden thoroughly.
    - Invert the plates and incubate them at 37°C for 24 to 48 hours.
    - A sample of the diluted yogurt should be Gram stained to locate long, large gram-positive rods, which are *Lactobacillus* species.
4. Observe the plates for evidence of well-isolated colonies of lactobacilli. Because lactobacilli prefer an anaerobic environment, there will be many subsurface colonies.
    - Use a warm inoculating needle to reach down to these subsurface colonies.
    - Prepare smears for Gram staining, and search for gram-positive rods similar to those seen in the yogurt sample.
    - Enter your results and observations in the Results section.

Name: ______________________ Date: __________ Section: __________

The Genus *Lactobacillus*: Isolation of Lactobacilli

EXERCISE RESULTS **52**

RESULTS

OBSERVATIONS OF LACTOBACILLI

Rogosa SL Agar Plates

Source: ______________________ ______________________

Stained Smears of Lactobacilli

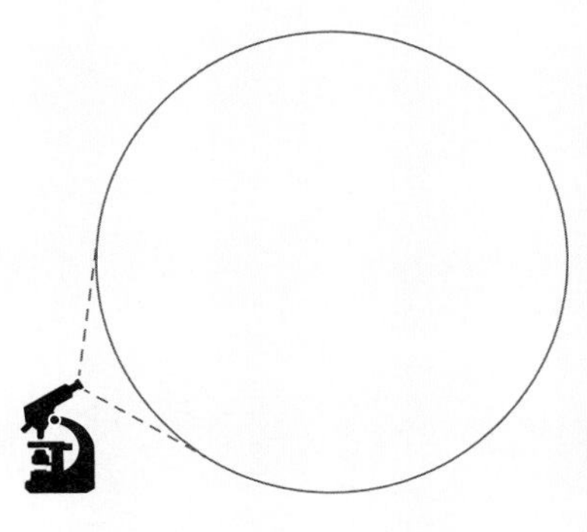

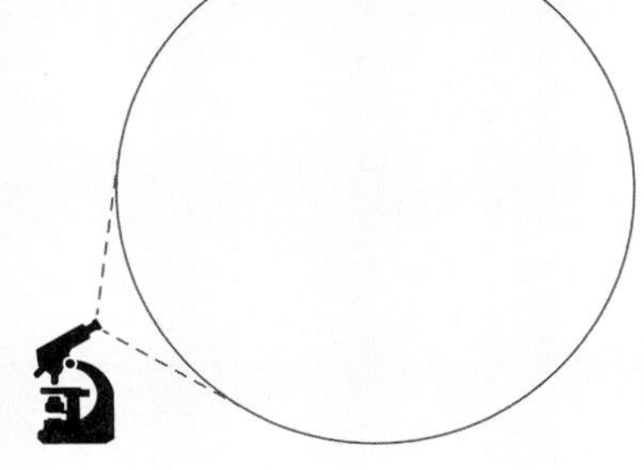

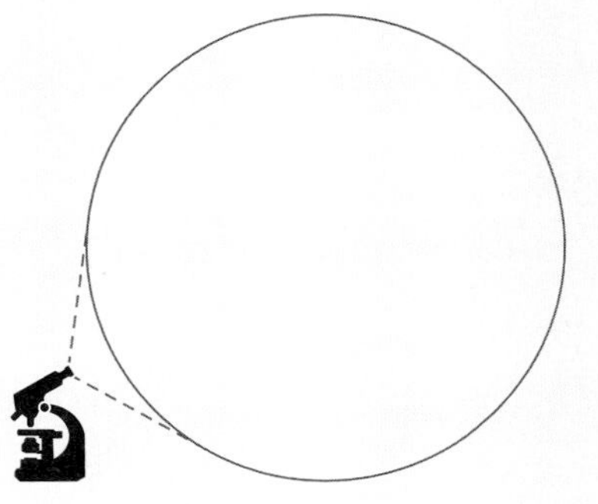

Source: ______________ ______________ ______________

Magnif.: ______________ ______________ ______________

Observations and Conclusions

Questions

1. Discuss the importance of the lactobacilli in the dairy industry.

2. Why is Rogosa SL agar of particular value in the isolation of lactobacilli?

The Genus *Lactobacillus*: Caries Susceptibility Test

EXERCISE 53

Materials and Equipment

- Tubes of Snyder test agar
- Saliva samples
- 1-ml pipettes and mechanical pipetters

A significant correlation has been noted between the amount of *Lactobacillus* in the saliva and an individual's susceptibility to dental caries. An estimate of the *Lactobacillus* population may be obtained by incubating saliva samples in a selective medium and noting the rate of acid production as it relates to incubation time. The procedure is known as the **Snyder test** after its developer, Marshall L. Snyder.

The caries susceptibility test is based on the growth of lactobacilli in **Snyder test agar**, a medium that contains glucose and the pH indicator bromcresol green. The indicator is green at a pH of 4.8 and yellow at pH 4.4 and below. As lactobacilli grow and multiply, the acid they produce causes the pH to drop and the color to change. By referring to a standard table (TABLE 53.1), an estimate of the *Lactobacillus* population may be made, and caries susceptibility may be determined.

CARIES SUSCEPTIBILITY TEST

PURPOSE: to estimate a person's susceptibility to dental caries.

In this exercise, lactobacilli will be isolated from saliva and a caries susceptibility test will be performed.

TABLE 53.1 Caries Susceptibility and Snyder Test Results

Caries Susceptibility	Time for Medium to Turn Yellow		
	24 hr	48 hr	72 hr
Marked	+		
Moderate	–	+	
Slight	–	–	+
Not susceptible	–	–	–

PROCEDURE

1. The saliva sample used in Exercise 52 may be used for this test. If unavailable, refer to Part A of that exercise and collect a sample of saliva.
2. Obtain two tubes of melted Snyder test agar maintained at 45° to 50°C.
   - Each tube should be labeled in the prescribed manner, noting one as "experimental" and the second as a "control."
3. Using a mechanical pipetter, pipette 0.2 ml of saliva into the experimental tube, and mix the contents well by tapping the bottom of the tube vigorously and rolling it between the palms of the hands. The control tube should remain uninoculated.
4. Allow the media to harden, and then incubate the tubes at 37°C.
5. Examine the experimental tube at 24-hour intervals to determine whether the medium in the experimental tube has become yellow in comparison to the control tube.
   - Refer to Table 53.1, and indicate the degree of caries susceptibility in the appropriate space in the Results section.
   - Gram stains may be prepared from colonies within the agar to observe the lactobacilli.
   - A warm inoculating needle should be used to melt the agar down to the level of the growth. Note that fungi may grow in the medium due to its acidity.
6. Conduct a survey of fellow students, and attempt to correlate the degree of caries susceptibility with the amount of dental work the student has had. Tabulate the data in TABLE 53.2 in the Results section.

Name: ______________________ Date: __________ Section: __________

The Genus *Lactobacillus*: Caries Susceptibility Test

EXERCISE RESULTS 53

RESULTS

CARIES SUSCEPTIBILITY TEST

Caries susceptibility:

____

____

TABLE 53.2 Survey of Caries Susceptibility		
Student	**Caries Susceptibility**	**Amount of Dental Work**
1.		
2.		
3.		
4.		
5.		
6.		

Observations and Conclusions

____

____

____

____

____

Questions

1. How might a person's recent eating habits affect the outcome of the caries susceptibility test?

2. Why are lactobacilli often found in environments that favor the growth of fungi?

3. How well did the results of the Snyder agar test correlate with caries susceptibility in the people surveyed in this exercise?

Identification of a Bacterial Unknown

PART IX

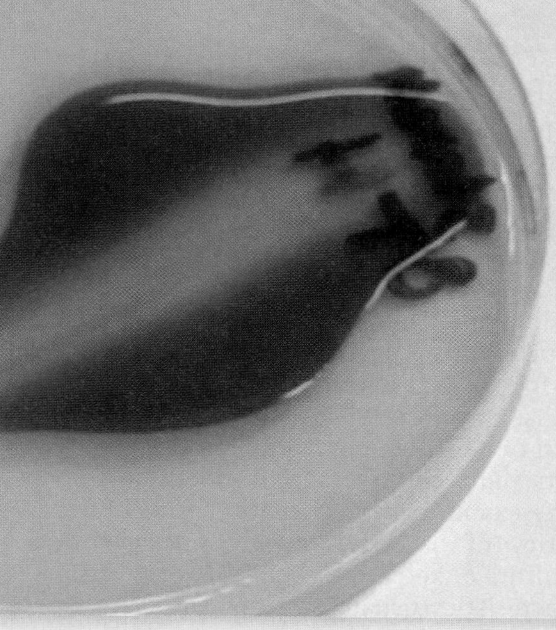

Exercises

54. Bacterial structural characteristics
55. Bacterial culture characteristics
56. Biochemical characteristics of bacteria: Carbohydrate fermentation
57. Biochemical characteristics of bacteria: Starch hydrolysis
58. Biochemical characteristics of bacteria: Catalase production
59. Biochemical characteristics of bacteria: DNA Hydrolysis
60. Biochemical characteristics of bacteria: Hydrogen sulfide production
61. Biochemical characteristics of bacteria: IMViC series
62. Biochemical characteristics of bacteria: Urea hydrolysis
63. Biochemical characteristics of bacteria: Fat (triglyceride) hydrolysis
64. Biochemical characteristics of bacteria: Casein hydrolysis

Learning Objectives

When you have completed the exercises in Part IX, you will be able to:

- Carry out a complete cultural and biochemical characterization of a bacterial unknown to species identification.
- Distinguish specific and distinctive culture characteristics needed to help identify a bacterial species.
- Perform a series of biochemical tests and interpret the outcomes of those tests as diagnostic keys to bacterial species identification.

- Biochemical Tests: Carbohydrate Fermentation
- Biochemical Tests: Starch Hydrolysis Test
- Biochemical Tests: Catalase Test
- Biochemical Tests: Hydrogen Sulfide Test
- IMViC Tests (Indole Test; Methyl Red Test; Voges-Proskauer Test; Citrate Test)
- Urease Test

One of the major responsibilities of the microbiologist is the identification of unknown bacterial organisms. The physician, for example, may request that the clinical lab microbiologist determine which organism is present in a patient's blood, or a quality control microbiologist may wish to learn the identity of a contaminant in a food sample. A drug company microbiologist may wish to isolate and identify bacteria from the soil during the search for new antibiotics, or a water quality microbiologist may desire to identify the bacteria in water as part of a regular monitoring program.

Once in pure culture, species identification can begin by examining the structural and culture characteristics of bacteria growing on agar or in broth culture (Exercises 54 and 55). However, these characteristics usually are not sufficient to accurately identify the bacterial species in pure culture (TABLE IX.1).

Rather, specific biochemical characteristics have to be gathered about the bacteria in pure culture. Such biochemical tests (Exercises 56–64) are relatively simple analyses that visually detect the presence or absence of a substrate, or the presence or absence of an end product, of an enzymatic reaction or metabolic pathway (Table IX.1). Such biochemical tests often are critical in species identification. With this information and understanding, TABLE IX.2 lists some of the structural, culture, and biochemical characteristics of 11 bacterial species.

The exercises in this section give you the opportunity to identify an unknown bacterial organism. Use the Lab Report - Identification of a Bacterial Unknown to record your findings as you complete the exercises. Your instructor may ask you to submit the final report.

TABLE IX.1 Examination Procedures and Identification Keys

Examination Procedure	Identification Keys
Structural and Culture Characteristics	■ Gram staining ■ Temperature preference ■ Motility ■ Spores ■ Oxygen requirement ■ Growth form on agar ■ Capsule ■ Growth in broth ■ Pigmentation
Biochemical Characteristics	■ Sugar fermentations ■ Indole production ■ Starch hydrolysis ■ Urea hydrolysis ■ Catalase production ■ Fat (triglyceride) hydrolysis ■ DNA digestion ■ Casein hydrolysis ■ H_2S gas production ■ Other special tests

TABLE IX.2 Some Characteristics of Selected Bacterial Species

CHARACTERISTIC*	E. coli	S. epidermidis	B. subtilis	A. faecalis	C. freundi	E. aerogenes	M. luteus	P. fluorescens	N. subflava	M. smegmatis	S. marcescens
Temperature preference	37°C	37°C	37°C	25°C	37°C	37°C	25°C	25°C	37°C	37°C	25°C
Cell morphology	rod	coccus	rod	rod	rod	rod	coccus	rod	coccus	rod	rod
Cell arrangement	single	cluster	single	single	single	single	packet	single	pair	single	single
Gram reaction	–	+	+	–	–	–	+	–	–	§	–
Spores	–	–	+	–	–	–	–	–	–	–	–
Motility	±	–	+	+	+	+	–	+	–	–	+
Pigment formation	–	–	–	–	–	–	+	±	±	–	+
Glucose fermentation	AG	A	A	–	AG	AG	–	AG	A	–	A
Lactose fermentation	AG	A	–	–	AG	AG	–	AG	–	–	–
Sucrose fermentation	A	A	A	–	AG	AG	–	±	±	–	A
Maltose fermentation	AG	A	A	–	AG	±	–	–	A	–	A
Starch digestion	–	–	+	–	–	–	–	–	±	–	–
Catalase production	+	+	+	+	+	+	+	+	+	+	+
DNA digestion	–	–	+	–	–	–	–	–	–	–	+
H2S production	–	–	–	–	+	–	–	–	+	–	–

(*continued*)

TABLE IX.2 Some Characteristics of Selected Bacterial Species (Continued)

CHARACTERISTIC*	E. coli	S. epidermidis	B. subtilis	A. faecalis	C. freundi	E. aerogenes	M. luteus	P. fluorescens	N. subflava	M. smegmatis	S. marcescens
Indole production	+	–	–	–	–	–	–	+	–	–	–
Urea digestion	–	+	–	–	+	±	+	–	–	+	–
Lipid digestion	–	+	–	–	–	–	–	+	–	–	–

* Growth characteristics on agar and broth as listed in Bergey's Manual
§ Gram reaction does not apply; organism is acid-fast
± Some strains give positive reaction; other strains give negative reaction
A = acid production
G = gas production
Your instructor may ask you to submit a compiled lab report on the bacterial unknown at the end of this entire section. Be sure to record your findings as you complete your tests. On the Lab Report, "reaction" refers to what you observe as a result of the test. "Result" indicates whether the test gave a positive (+) or negative (–) result.

Name: ______________________ Date: __________ Section: __________

Identification of a Bacterial Unknown

EXERCISE RESULTS

LAB REPORT

Student Name	Unknown Code

Structural Characteristics

Gram Stain	Cell Shape	Cell Size
Cell Arrangement		Motility
Endospores	Capsule	

Culture Characteristics

Appearance of Nutrient Broth		
Better Growth Temperature		
Oxygen Requirement		
Colony Appearance on Agar		
Form:	Margin:	Elevation:
Pigmentation		

Biochemical Characteristics

Carbohydrate Fermentation					
Acid Production			Carbon Dioxide Gas		
Sugar	**Color**	**Result**	**Sugar**	**Bubble**	**Result**
Glucose			Glucose		
Lactose			Lactose		
Sucrose			Sucrose		
Maltose			Maltose		

IMViC Tests					
Test	**Color**	**Result**	**Test**	**Color**	**Result**
Indole			Methyl Red		
Citrate			Voges-Proskauer		

Hydrogen Sulfide		Urease		Oxidase	
Reaction	**Result**	**Color**	**Result**	**Reaction**	**Result**

Starch		Catalase		Nitrate	
Reaction	**Result**	**Reaction**	**Result**	**Reaction**	**Result**

Other Tests

Test Reaction and Result

Based on the above tests, I believe the unknown bacterial species is:

Bacterial Structural Characteristics

EXERCISE

54

Materials and Equipment

- Broth cultures of unknown bacteria identified by code

Identification procedures for bacteria take into account a broad variety of factors. These factors include structural characteristics such as shape, size, arrangement of cells, their response to staining reactions, and the presence of spores, capsules, and flagella (motility). In this exercise, these characteristics will be used to help identify an unknown bacterium. All the procedures used in this exercise are described in previous exercises and shown in microbiology videos.

BACTERIAL STRUCTURAL CHARACTERISTICS

PURPOSE: to identify structural characteristics of an unknown bacterium.

This exercise relies on procedures from previous exercises. Be sure to review relevant microbiology videos to see how procedures are performed.

PROCEDURE

1. You will be provided with a broth culture of an unknown bacterium identified only by a code number or letter, or both. The object of this exercise will be to determine the bacterial characteristics of your unknown.
2. Prepare an air-dried, heat-fixed smear of a sample from the broth culture and stain it by using the Gram stain technique (Exercise 15).
   - From this smear, determine the shape, relative size, cell arrangement, and Gram reaction of the bacterium.
   - Record your results in the Results section.
3. Depending on the Gram stain result, determine whether the bacteria are sporeformers by preparing smears and staining them for spores following the technique described in Exercise 16.
   - Record your results in the Results section.
4. Determine whether the bacteria are motile by preparing hanging drop preparations and/or performing the motility test agar technique described in Exercise 11.
   - Record your results in the Results section.

To avoid confusion and possible contamination, always label plates and tubes with your name, the date, the name of the medium, and the name of the organism or the source of the inoculum.

5. Capsule formation may be determined by reference to Exercise 17.
   - Record your results in the Results section
6. From the information obtained, you should be able to narrow your choices considerably. At this point you should consider which biochemical tests will be most helpful in confirming your suspicion of the organism's identity. You may decide to perform all the available biochemical tests for a complete verification, or you may select certain tests to perform. Consult with the instructor if you need assistance.

Name: ______________________ Date: __________ Section: __________

Bacterial Structural Characteristics

EXERCISE RESULTS 54

RESULTS

BACTERIAL STRUCTURAL CHARACTERISTICS

Code: ______________

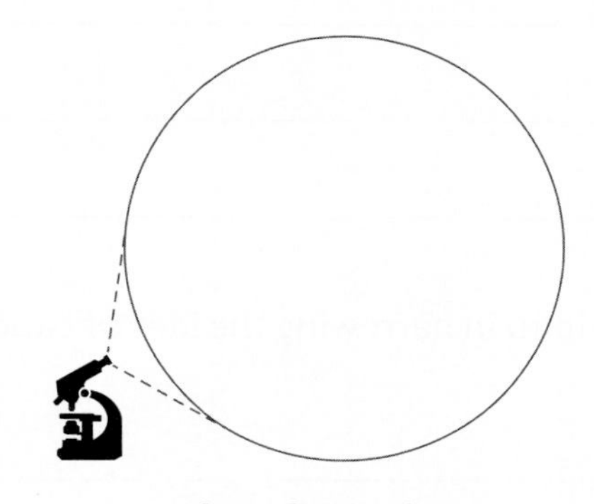
Gram-Stained Smear

Microscopic Observations
Shape of cells:
Relative size:
Cell arrangement:
Gram reaction:
Presence of spores:
Motility:
Capsule:

Enter your results in the table below and keep track of data to enter in the final lab report for this section.

Structural Characteristics		
Gram Stain	Cell Shape	Cell Size
Cell Arrangement		Motility
Endospores	Capsule	

Observations and Conclusions

Questions

1. Describe several instances, other than those in the introduction to this exercise, where identification of a bacterial unknown is practical and necessary.

2. Which structural characteristics in this exercise were most helpful in narrowing the identification of the unknown bacterium? Which tests were least helpful? Why?

3. Before doing the spore stain, what information from the Gram stain results may assist you in a preliminary determination for the presence of spores?

4. What structural characteristic can be assumed with a positive motility test?

____

____

5. What other bacterial structural characteristics could be useful with the light microscope in eventual identification of species?

____

____

____

Bacterial Culture Characteristics

EXERCISE 55

Materials and Equipment

- Nutrient agar plates, or materials for their preparation
- Nutrient broth tubes and agar slants
- Thioglycolate medium
- Incubators at 25°C and 37°C

Different species of bacteria exhibit different growth characteristics in culture media. Because these characteristics are genetically determined, they remain constant for an organism, and bacteriologists therefore can use them as identifying markers.

In this exercise, you will observe an unknown bacterium that has been cultivated in solid and liquid media, and you will note the distinctive cultural differences that exist among bacteria.

BACTERIAL CULTURE CHARACTERISTICS

PURPOSE: to identify specific bacterial culture characteristics on agar and in broth.

PROCEDURE

I. Temperature Preference

1. Obtain two nutrient agar slants, and inoculate each with the unknown bacterium; that is, the broth culture used in Exercise 54.
   - Incubate one slant at 37°C for 48 hours, and place the other one at 25°C or at room temperature.
   - From the growth, determine the better growth temperature of your unknown. One slant culture should be maintained as a stock culture to be used as the need arises, while the other slant culture will be a working culture from which samples may be taken for analysis.
2. In the Results section (TABLE 55.1), record which tube exhibits better growth. Note: there may be little difference between the tubes in which case the organism grows equally well at both temperatures.

II. Colony Appearance on Agar

1. Obtain one nutrient agar plate, or prepare it at the instructor's direction.
   - Using a wax marking pencil or felt marker, label the bottom of the plate with your name and the date.

To avoid confusion and possible contamination, always label plates and tubes with your name, the date, the name of the medium, and the name of the organism or the source of the inoculum.

2. Aseptically inoculate a section of each plate with a loopful of unknown from the working broth tube.
   - Make a streak plate as described in Exercise 5.
3. Invert the plate and incubate it at room temperature. The instructor may designate another incubation time. The plates may then be refrigerated until the next laboratory period.
4. Examine the nutrient agar plate for the quantity and quality of growth, and prepare carefully labeled representations of the plate in the Results section.
   - In TABLE 55.2 of the Results section, indicate the colony appearance with regard to form, margin, and elevation (Figure 5.3).
   - Evaluate the **pigmentation** of the colonies by indicating whether pigment is absent or present and, if present, its color.

III. Appearance in Nutrient Broth

1. Obtain a tube of sterile nutrient broth, and label the tube with your name, the date, and 25°C or an incubation temperature designated by the instructor.
2. Aseptically inoculate the broth tube with a loopful of your unknown organisms.

To avoid spills, limit the times you need to reach for something by gathering all your materials in front of you before beginning your work.

   - To ensure a successful inoculation, extend the bacteriological loop below the surface of the broth and shake the loop lightly to distribute the cells.
3. Incubate the tube at the designated temperature for the time indicated by the instructor, and then refrigerate the tubes until observed.
4. Examine the nutrient broth tube, recording the amount and type of growth in the Results section and in TABLE 55.2.
   - Note the surface of the broth, and compare it to the surfaces shown in FIGURE 55.1.

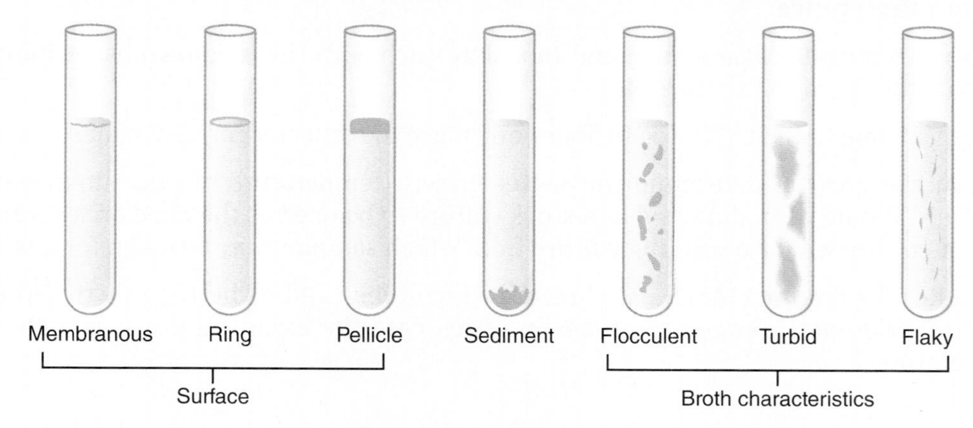

FIGURE 55.1 Characteristics of bacterial growth in broth media.

- Determine the characteristics of the broth as turbid (cloudy), granular (contains small particles), flocculent (contains small masses), flaky (contains large particles), or clear.
- Describe the type of sediment at the bottom of the tube by tapping the bottom gently and observing whether the sediment is granular, flaky, viscid (clumps together), or absent.
- Note whether the growth is throughout the tube or only in certain areas such as the surface or bottom sediment.
- Note these characteristics in your drawing.

IV. Oxygen Gas Requirement

1. Obtain a thioglycolate medium tube and label it as in (III.1).
   - Aseptically inoculate the medium with a loopful of your unknown organism (Exercise 54).
   - Move the loop up and down *gently* a few times to mix the bacteria. Then, replace the cap and gently rotate the tube 8 to 10 times between the palms of your hands .
   - Incubate the tube at room temp for 24 to 48 hours.
2. Turbid growth at the top of the tube indicates aerobic growth, while growth at the bottom indicates anaerobic growth. Growth throughout the tube indicates a facultative type of growth.
3. Record the growth results for oxygen in the Results section.

Name: ______________________ Date: ____________ Section: ____________

Bacterial Culture Characteristics

EXERCISE RESULTS **55**

RESULTS

BACTERIAL CULTURE CHARACTERISTICS

I. Temperature Preference

TABLE 55.1 Temperature Preference on Agar

	Amount	
	25°C	37°C
Unknown		

II. Colony Appearance on Agar

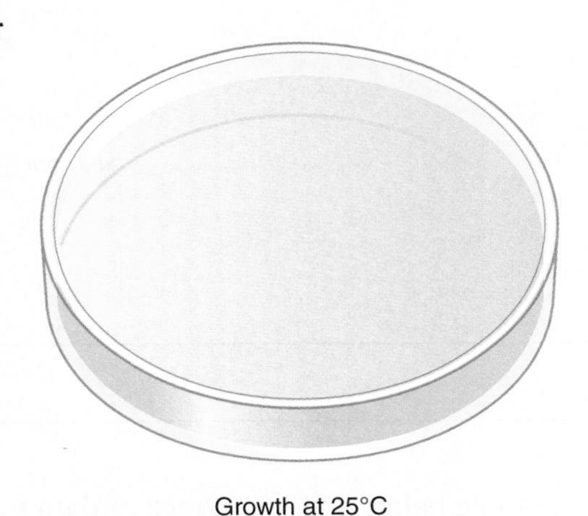

Growth at 25°C

TABLE 55.2 Summary of Colony Appearance and Pigmentation on Nutrient Agar Plate

	Colony Appearance on Agar			
	Form	Margin	Elevation	Pigmentation
Unknown				

Abbreviations used:

___

___

___

III/IV. Appearance in Nutrient Broth and Oxygen Gas Requirement

Incubation temperature: ___

Nutrient broth

Thioglycolate medium

Abbreviations used:

___

___

In addition, enter your results in the table below and keep track of data to enter in final lab report for this section.

Culture Characteristics		
Appearance of Nutrient Broth		
Better Growth Temperature		
Oxygen Requirement		
Colony Appearance on Agar		
Form:	Margin:	Elevation:
Pigmentation		

Observations and Conclusions

Questions

1. After the incubation period, why is it best to refrigerate culture plates and tubes rather than storing them at room temperature?

2. How can you account for the observation that a mass or colony of bacteria shows pigmentation while the individual cells have transparent cytoplasm?

3. Describe the factors that may alter the expression of the cultural characteristics of bacteria.

4. Determine which culture characteristics noted in this exercise are more distinctive than others.

____

____

____

5. Suppose your unknown prefers to grow at 37°C, yet both slants show similar growth. Provide some reasons to account for this observation.

____

____

____

Biochemical Characteristics of Bacteria: Carbohydrate Fermentation

EXERCISE 56

Materials and Equipment

- Selected bacterial species
- Tubes of phenol red with glucose, lactose, sucrose, maltose, and other carbohydrates as specified by the instructor

- Biochemical Tests: Carbohydrate Fermentation Test

Bacteria exhibit a wide variety of biochemical characteristics reflecting the many aspects of **metabolism** taking place within the cell. During metabolism, a broad diversity of **substrates** are used, and an equally broad series of **end products** are formed. This chemistry is catalyzed by **enzyme systems** that are genetically determined and specific for an individual species.

Biochemical characteristics serve as a fingerprint for a bacterial species. By using a series of tests such as those outlined in the next few exercises (Exercises 56-64), the laboratory technologist can learn the identity of an unknown organism and assist a physician in the diagnosis of disease. Moreover, biochemical activities reflect the role played by many organisms in the environment because they show how organic substances are degraded and how certain products are formed.

The biochemical tests performed in these exercises consist of the inoculation of a bacterial species into an organic compound, followed by an incubation period and a determination of the end products. Standard results for each organism usually are listed in *Bergey's Manual of Systematic Bacteriology*. Together with morphological, staining, immunological, and other observations, biochemical tests occupy a prominent position in the diagnostic laboratory.

CARBOHYDRATE FERMENTATION

> PURPOSE: to identify whether a bacterial species can ferment a sugar into acid or acid and gas products.

Fermentation is a type of microbial metabolism in which an organic intermediary molecule serves as an electron acceptor. The chemistry of the process also is a key element in the production of many fermented foods.

The fermentation of carbohydrates by bacteria results in various end products, among which are acid and carbon dioxide gas (CO_2). Certain bacteria produce both acid and gas from a particular carbohydrate; others produce only acid; and still others produce neither acid nor gas. In this exercise, bacteria will be inoculated into different carbohydrate broths, and the end products of fermentation will be recorded.

The media used in the fermentation tests consist of nutrient broth supplemented with 0.5% of the fermentable carbohydrate. The tubes also contain **phenol red indicator**, which is red in neutral or basic solution and yellow in acid solution. The final component is an inverted vial called a **Durham tube** to trap any CO_2 gas evolved during the fermentation.

Before starting this exercise, watch the related **Microbiology Video: Biochemical Tests: Carbohydrate Fermentation Test** to understand how these tests are performed.

Biochemical Tests: Carbohydrate Fermentation Test

PROCEDURE

1. Select tubes of the various carbohydrate media available and label them with your name, the date, the name of the medium, and the organism assigned. A slide label may be used at the instructor's suggestion. Be careful not to place the label in your mouth to moisten it.
2. Inoculate each of the carbohydrate broths as follows:
    - Aseptically obtain a loopful of bacteria in the usual manner, and replace the culture tube in the rack.
    - Take the carbohydrate broth tube, remove the cap, and flame the neck (FIGURE 56.1).
    - Place the loop containing bacteria below the broth surface while shaking it lightly (FIGURE 56.2).

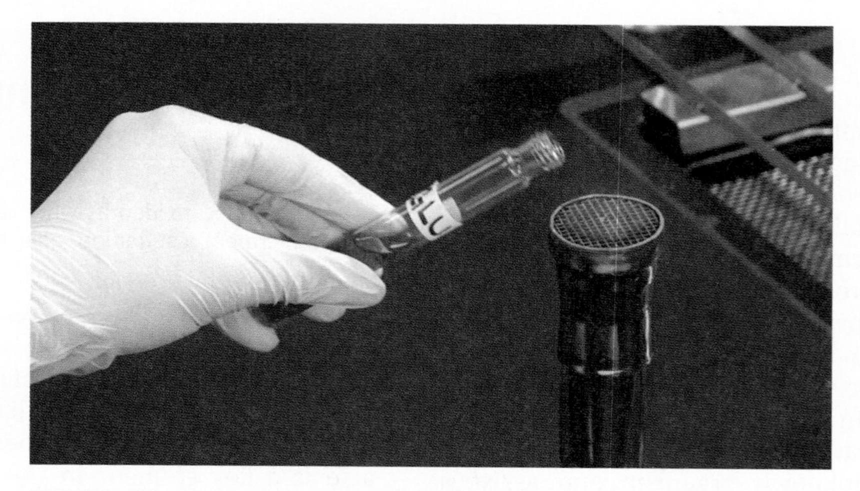

FIGURE 56.1 Always flame the neck of the tube to ensure an aseptic transfer.

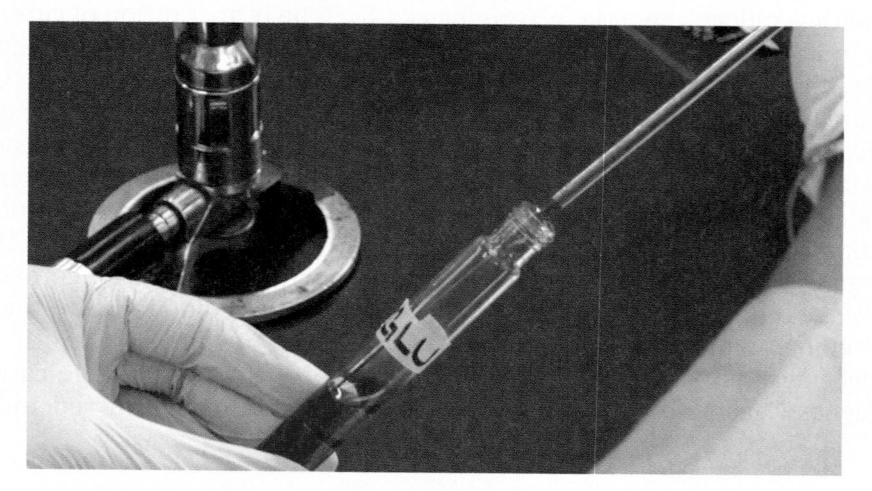

FIGURE 56.2 Place the loop containing bacteria below the broth surface while shaking it lightly.

- Reflame the neck, replace the cap, and return the tube to its rack.
- Flame the loop.
- Repeat the inoculation process for the remaining carbohydrate broth tubes.

3. Incubate the tubes for 24 to 48 hours at 37°C or as directed by the instructor. A set of uninoculated control tubes should be included.

To clean a culture spill, surround the area with the available disinfectant and gradually work inward until the disinfectant covers the spilled material. Wait for 10 minutes, then wipe up the material with paper toweling. If possible, place the toweling in a biohazard container and set it aside for autoclaving.

- At the end of the incubation period, refrigerate the tubes until the next laboratory session to retard additional growth.

4. Observe each tube for color change to yellow, indicating acid production (FIGURE 56.3).

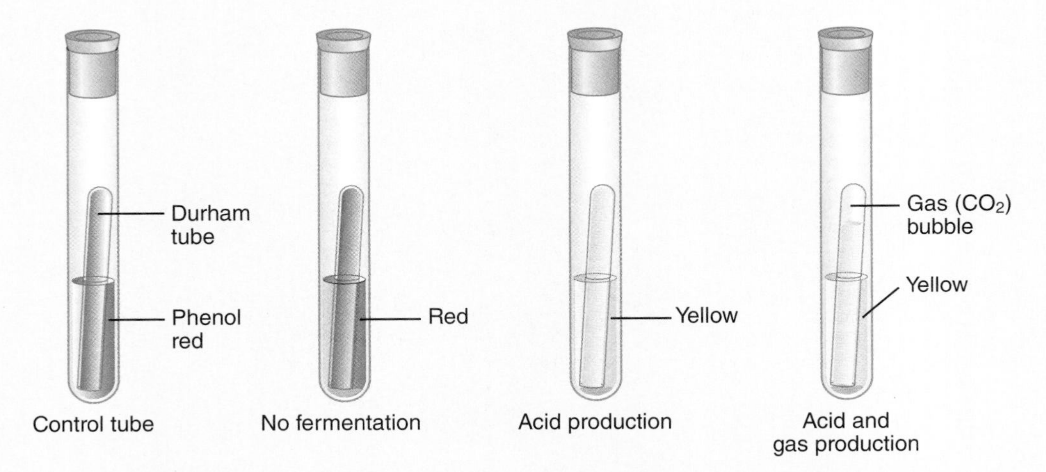

FIGURE 56.3 A control tube and three experimental tubes showing possible results of the carbohydrate fermentation tests.

- Examine the inverted vial in each tube for the presence of a gas bubble.
- In TABLE 56.1 of the Results section, enter an "A" for acid production and a "G" for gas production. The broth must have turned yellow to be positive for acid production. A red or orange tube is considered negative for acid production. Use a minus sign (–) to indicate that neither acid nor gas was produced.
- Prepare a labeled diagram of the tubes (i.e., "a talking picture") in the appropriate spaces of the Results section to illustrate your results.
- Add explanatory notes to the diagram.

Name: ______________________ Date: __________ Section: __________

Biochemical Characteristics of Bacteria: Carbohydrate Fermentation

EXERCISE RESULTS 56

RESULTS

CARBOHYDRATE FERMENTATION TEST

TABLE 56.1 Results of Biochemical Reactions

Biochemical Test	Unknown/Name of Organism			
Carbohydrate Fermentation				
Glucose				
Lactose				
Sucrose				
Maltose				

Representations of Carbohydrate Fermentation Test

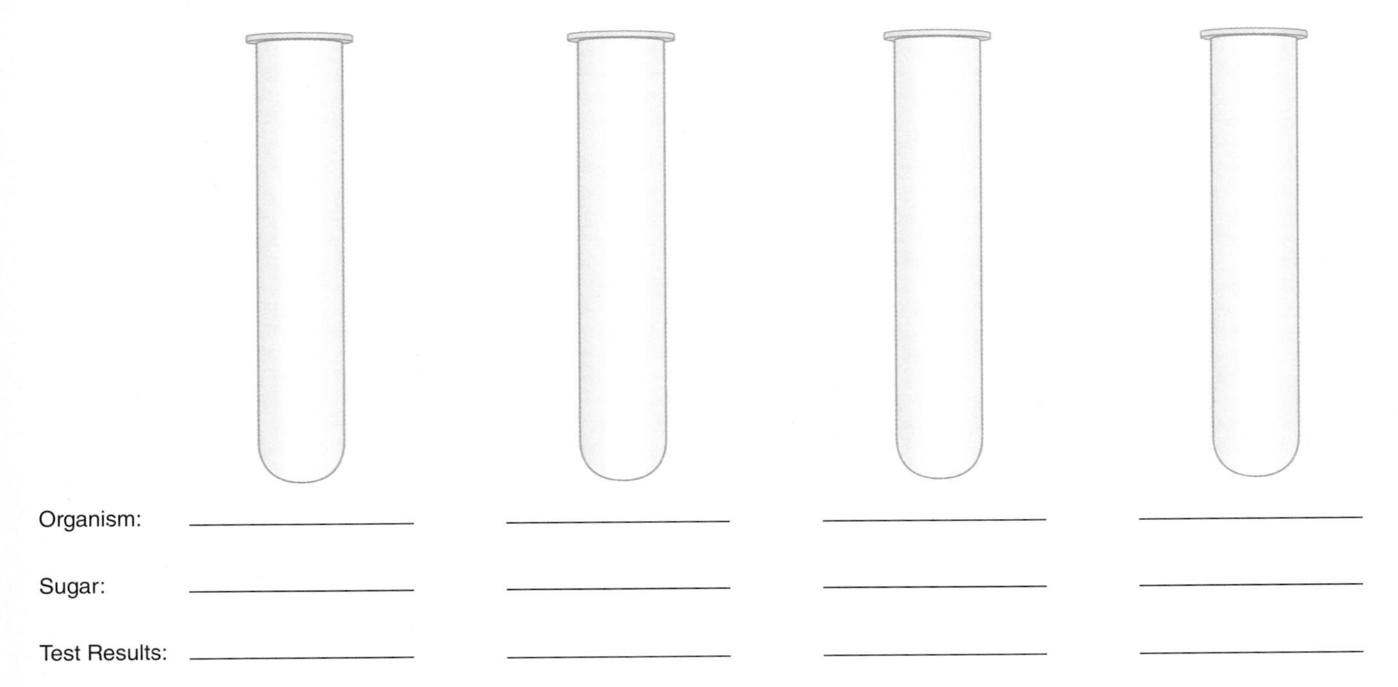

Organism: __________ __________ __________ __________

Sugar: __________ __________ __________ __________

Test Results: __________ __________ __________ __________

Observations and Conclusions

Questions

1. Is the carbohydrate fermentation test identifying the presence or absence of a substrate or a product? Explain.

Biochemical Characteristics of Bacteria: Starch Hydrolysis

EXERCISE 57

Materials and Equipment

- Selected bacterial species
- Starch agar plates, or materials for their preparation
- Iodine solution such as Gram's iodine

- Biochemical Tests: Starch Hydrolysis Test

Starch is a polysaccharide consisting of thousands of glucose molecules chemically bonded to one another. Certain bacteria produce the enzyme **amylase** and use the enzyme to biochemically break these bonds between glucose monomers. This biochemical characteristic assists identification.

STARCH HYDROLYSIS

PURPOSE: to identify if a bacterial species can digest starch.

Starch digestion is tested by inoculating bacteria onto a starch agar medium. After a suitable incubation period, **iodine** is added to determine whether the starch is still present or has been digested by secreted amylase enzymes.

Before starting this exercise, watch the related **Microbiology Video: Biochemical Tests: Starch Hydrolysis Test** to understand how this test is performed.

Biochemical Tests: Starch Hydrolysis Test

PROCEDURE

1. The instructor may demonstrate the technique for pouring starch agar plates if these have not been prepared previous to the laboratory period. One plate of starch agar will be needed.
    - Label the plate on the bottom side with a felt maker or wax marking pencil. Include the name of the medium, the date, your name, and the name of the organism or organisms.
    - Allow the medium to solidify before use.
    - The instructor may suggest dividing the plate to test several organisms.
2. Aseptically obtain a loopful of bacteria.
    - Carefully inoculate the starch agar by drawing the loop one time across the surface of the plate or as directed by the instructor. Be careful to avoid airborne contamination by lifting the lid only enough to permit entrance of the loop.
    - Apply the loop lightly to prevent tearing of the agar (FIGURE 57.1).

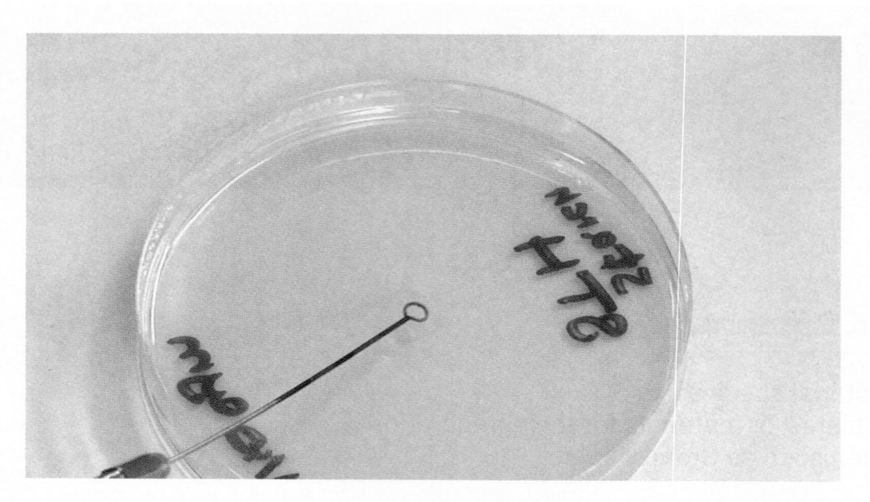

FIGURE 57.1 Use caution in applying the loop to prevent tearing of the agar.

3. Invert the plate to avoid liquid condensation on the medium, and incubate it for 24 to 48 hours at 37°C or as directed.
   - Refrigerate the plate until the next laboratory session. Include an uninoculated control plate.
4. At the next session, pour several drops of iodine solution (such as Gram's iodine) onto the surface of the starch agar.
   - Be sure the entire surface is covered (FIGURE 57.2).
   - Treat the control plate similarly.
   - Starch will normally react with iodine to form a blue-black chemical complex, and the agar medium will become blue-black if starch is present. However, if bacteria have digested the starch, none is left for the reaction, and the agar area near the streak will remain clear or may be light brown due to the iodine color. A blue-black color may still appear at the margins because enzyme activity has not reached this far. If digestion has failed to occur, the entire plate will become blue-black, similar to the control plate.

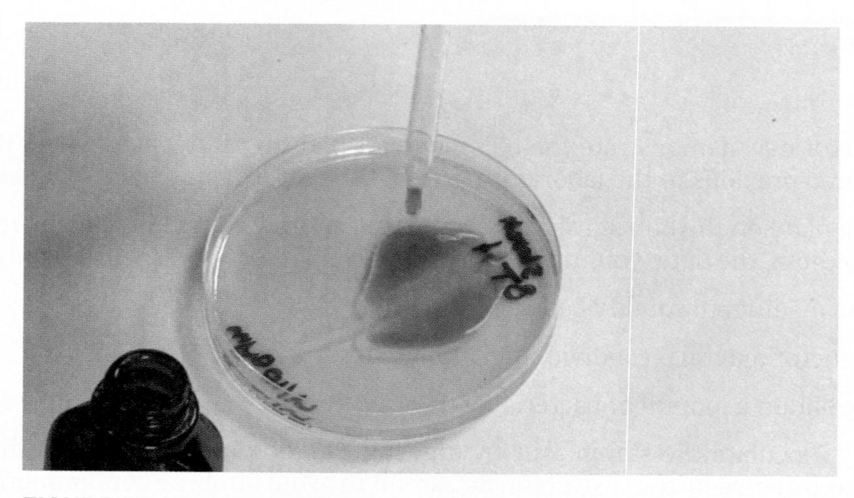

FIGURE 57.2 Cover the entire surface area with iodine solution.

- Be cautious not to confuse the white bacterial growth with starch digestion. Also, be mindful that the blue-black color will fade after several minutes, depending on how much iodine has been added.

5. Enter your results in TABLE 57.1, using a plus sign (+) to indicate starch digestion and a minus sign (−) to indicate nondigestion.
   - Include a carefully labeled representation of the plate to illustrate your results.
   - The representation should point out the area of digestion (if applicable) and the bacterial growth.
   - Include explanatory notes.

Name: ______________________ Date: __________ Section: __________

Biochemical Characteristics of Bacteria: Starch Hydrolysis

EXERCISE RESULTS 57

RESULTS

STARCH HYDROLYSIS

TABLE 57.1 Biochemical Characteristics of Bacteria				
Biochemical Test	Name of Organism			
Starch Hydrolysis				

Representations of Starch Hydrolysis Test

Organism: ______________________ ______________________

Test Results: ______________________ ______________________

Observations and Conclusions

Questions

1. Determine if this test is identifying the presence or absence of a substrate or a product.

Biochemical Characteristics of Bacteria: Catalase Production

EXERCISE 58

Materials and Equipment

- Selected bacterial species
- Nutrient agar plates, or materials for their preparation
- Hydrogen peroxide solution (3%)
- Inoculating loop

Watch This Video

- Biochemical Tests: Catalase Test

Catalase is an enzyme that breaks down hydrogen peroxide (H_2O_2) to oxygen and water:

$$2H_2O_2 \xrightarrow{\text{catalase}} 2H_2O + O_2$$

This enzyme is important in bacterial metabolism because the hydrogen peroxide produced during energy-yielding processes must be broken down. Otherwise it will accumulate and kill the cells. Hydrogen peroxide is generally a poor antiseptic because it is rapidly degraded by bacterial and tissue catalase.

CATALASE PRODUCTION

PURPOSE: to identify a bacterial species as a catalase producer.

Production of catalase may be tested by growing bacteria on a nutrient medium and then adding hydrogen peroxide.

Before starting this exercise, watch the related **Microbiology Video: Biochemical Tests: Catalase Test** to understand how this test is performed.

Biochemical Tests: Catalase Test

PROCEDURE

1. Select one nutrient agar plate or pour one plate of nutrient agar as directed by the instructor.
   - Label the bottom side of the plate as described in Exercise 57, Step 1, using your name, the date, the medium name, and the organism's name.
   - Wait for the medium to harden before using it, and divide the plate into sections at the instructor's direction.

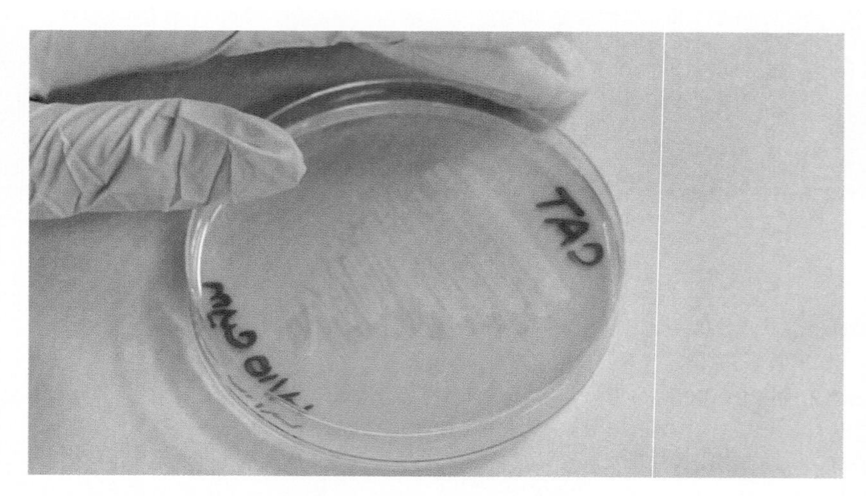

FIGURE 58.1 Draw the bacteria across the plate in a zig-zag motion with an inoculating loop.

2. Obtain a loopful of bacteria, and aseptically inoculate the nutrient agar by drawing the loopful of bacteria across the surface of the plate, or as directed by the instructor (FIGURE 58.1).
   - Be careful to avoid airborne contamination.
   - The loop should not cut into the agar.
3. Incubate the plate in the inverted position (to keep any condensation on the lid) for 24 to 48 hours at 37°C or as directed.
4. At the next period, test for the presence of catalase by placing several drops of 3% hydrogen peroxide directly onto the bacterial growth (FIGURE 58.2). If catalase has been produced during incubation, the enzyme will break down the hydrogen peroxide. Oxygen bubbles will appear, and the area will effervesce. It may take a few moments for this to become apparent. If no bubbles appear, catalase was not produced by the bacterium tested. Be mindful that if the hydrogen peroxide remains on the plate for a few minutes, it naturally will break down and bubbles will be seen. Do not mistake this for a positive reaction.

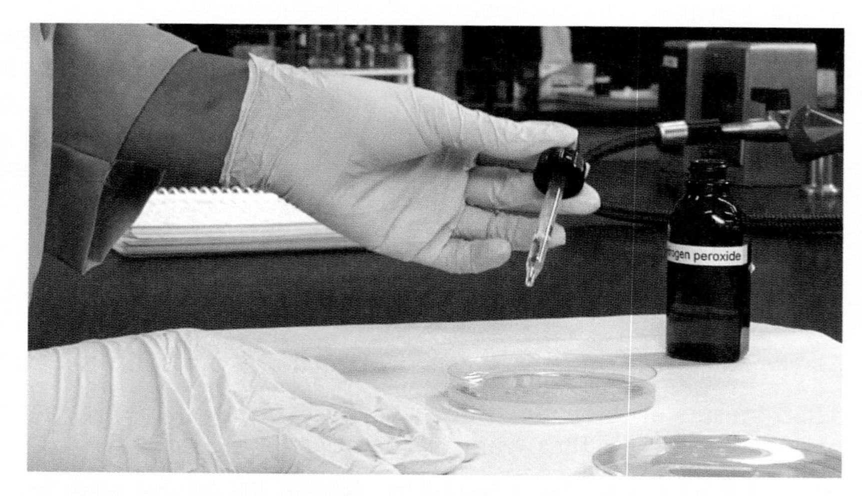

FIGURE 58.2 Place several drops of 3% hydrogen peroxide directly onto the bacterial growth.

5. An alternate way of testing for catalase is as follows:
   - Fill two small test tubes each with about 10 milliliters of 3% hydrogen peroxide.
   - Obtain two filter paper disks and a forceps.
   - Take one disk with the forceps, touch the disk lightly to the surface of the bacteria on the agar plate, and then place the disk into the first tube of H_2O_2.
   - As a control, take the second disk, touch it to the agar surface where there are no bacteria, then place it into the second tube of H_2O_2.
   - Both disks should sink to the bottom of the tube.
   - However, if the bacteria have produced catalase, the first tube will begin to bubble and the disk will rise to the surface as the catalase digests the hydrogen peroxide.
   - If no catalase was produced, no hydrogen peroxide breakdown will occur, no oxygen will be released, and the disk will remain at the bottom of the tube. If bacteria are available in a broth culture (as in Exercise 61), then disks may be dipped into the broth (rather than touching them to the growth on the agar) before placing them into the hydrogen peroxide tubes.
6. Enter your results in TABLE 58.1, using a plus sign (+) to indicate catalase production and a minus sign (−) to indicate nonproduction.
   - Include a representation of the plate in the Results section.

Name: ______________________ Date: __________ Section: __________

Biochemical Characteristics of Bacteria: Catalase Production

EXERCISE RESULTS **58**

RESULTS

CATALASE PRODUCTION

TABLE 58.1 Results of Biochemical Reactions

Biochemical Test	Name of Organism			
Catalase Test				

Representations of Catalase Test

Organism: ______________________ ______________________

Test Results: ______________________ ______________________

Observations and Conclusions

Questions

1. For each of the tests you performed in this exercise, determine if the test is identifying the presence or absence of a substrate or a product.

2. How might you explain the bubbles that appear when hydrogen peroxide is applied to a wound even though there may be no bacteria present?

Biochemical Characteristics of Bacteria: DNA Hydrolysis

EXERCISE 59

Materials and Equipment

- Inoculating loop
- Selected bacterial species
- Plates of DNase test agar, or materials for their preparation
- 1N hydrochloric acid

Certain bacteria produce the enzyme **DNase**, which digests **deoxyribonucleic acid** (**DNA**) into its constituent nucleotides. This ability may be used as an identifying biochemical characteristic of a bacterial species.

The DNA digestion test is performed by incubating bacteria on a medium that contains DNA (DNase test agar). Hydrochloric acid then is added to the medium to determine whether the DNA has been digested.

DNA HYDROLYSIS

PURPOSE: to identify if a bacterial species can digest DNA.

PROCEDURE

1. Select one plate of DNase test agar or pour one plate as directed by the instructor.
   - Label the bottom side of the plate as described in Exercise 57, Step 1, and divide it at the instructor's direction.
2. Aseptically inoculate the solid DNase agar by drawing a loopful of bacteria across the surface of the medium, being careful not to cut into the agar.
3. Incubate the plate in the inverted position for 24 to 48 hours at 37°C or as directed.
   - Include an uninoculated control plate.

4. Test for the digestion of DNA by flooding the plate with 1N hydrochloric acid (HCl). DNA will normally react with HCl and form a very fine precipitate, which will give a cloudy appearance to the agar after several minutes. If DNA digestion has occurred, the area near the streak will remain clear. The remainder of the plate, however, becomes cloudy.
   - A dark background is recommended, and observations should be made from an angle to see the cloudiness.
   - Compare your results with the control plate containing no bacteria.
5. Enter your results in TABLE 59.1, using (+) for DNA digestion and (–) to indicate nondigestion.
   - Enter a labeled diagram in the appropriate space of the Results section to illustrate your results.

Name: ______________________ Date: __________ Section: __________

Biochemical Characteristics of Bacteria: DNA Hydrolysis

EXERCISE RESULTS 59

RESULTS

DNA DIGESTION

TABLE 59.1 Results of Biochemical Reactions

Biochemical Test	Name of Organism			
DNA Hydrolysis				

Representations of DNA Hydrolysis Test

Organism: ______________________ ______________________

Test Results: ______________________ ______________________

Observations and Conclusions

Questions

1. Determine if this test is identifying the presence or absence of a substrate or a product.

Biochemical Characteristics of Bacteria: Hydrogen Sulfide Production

EXERCISE 60

Materials and Equipment

- Inoculating needle
- Selected bacterial species
- Tubes of SIM medium

- Biochemical Tests: Hydrogen Sulfide Test

Certain bacteria have the ability to produce the enzyme **cysteine desulfurase**. This enzyme breaks down the amino acid cysteine to form alanine and hydrogen sulfide (H_2S) gas. If iron ions are present, the ions will react with H_2S to form iron sulfide. The latter precipitates and causes the medium to become black:

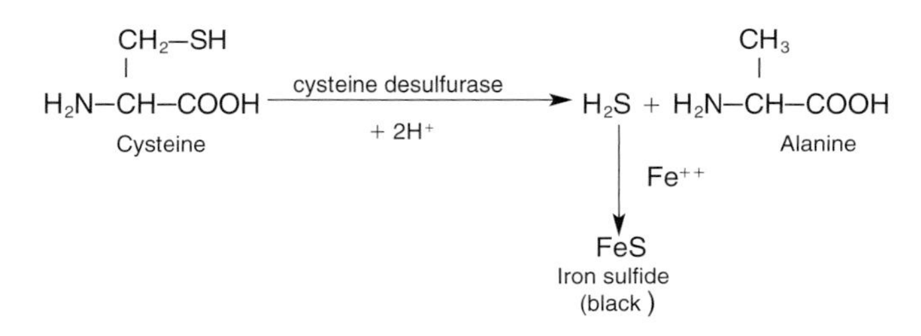

Hydrogen sulfide has significance in the food industry because it leads to the black rot of eggs, with the characteristic rotten egg smell. Anaerobic muds also become black with hydrogen sulfide and other sulfide precipitates.

HYDROGEN SULFIDE PRODUCTION

PURPOSE: to identify if a bacterial species can produce H_2S gas.

The medium commonly used in this test is called **SIM (Sulfide Indole Motility) agar**. It contains cysteine and iron ions. The medium is semisolid, and many bacteria will grow in it, but only those that produce hydrogen sulfide will cause it to darken. Indole production and motility also may be tested in the medium.

Before starting this exercise, watch the related **Microbiology Video: Biochemical Tests: Hydrogen Sulfide Test** to understand how these tests are performed.

Biochemical Tests: Hydrogen Sulfide Test

PROCEDURE

1. Select a tube of **SIM medium** and label it in the prescribed manner, with your name, the date, the organism's name, and the medium name.
2. Since the medium is semisolid, a straight wire inoculating needle will be used instead of a loop (FIGURE 60.1).

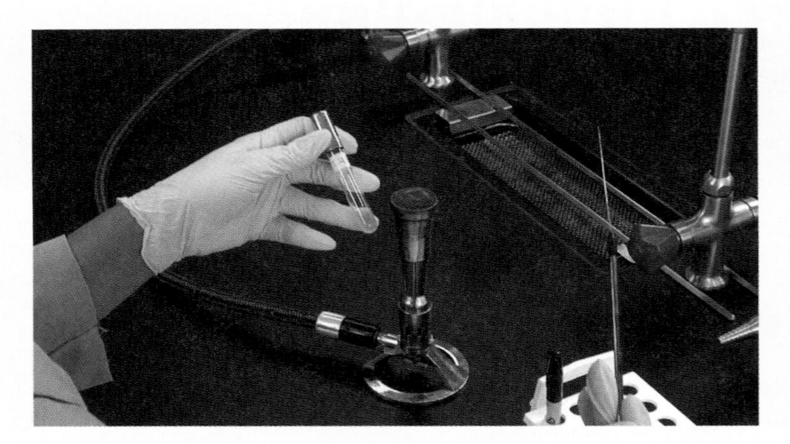

FIGURE 60.1 Use an inoculating needle and obtain the bacteria through aseptic technique.

- Using an inoculating needle, obtain a sample of bacteria, then inoculate the cells (stab the needle) into the center of the SIM agar about three-fourths the way down.
- Be sure to withdraw the needle along the same line it entered.

3. Incubate the tube for 24 to 48 hours at 37°C or as directed (FIGURE 60.2).
   - Then refrigerate the tube until observed.
   - An uninoculated control tube should be included.
4. At the next session, observe the SIM tube for evidence of hydrogen sulfide production. This is manifested by a blackening along the line of the inoculation and, possibly, in the entire tube.
   - If the tube has remained the same color as the control tube, hydrogen sulfide production has not taken place.
   - Bacteria should be seen along the inoculation line even though H_2S production has not occurred.
5. Enter your results in TABLE 60.1, using (+) for H_2S production and (−) for nonproduction.
   - Include a labeled representation of the tube in the appropriate space to complete your results.
   - An explanatory note is worthwhile.

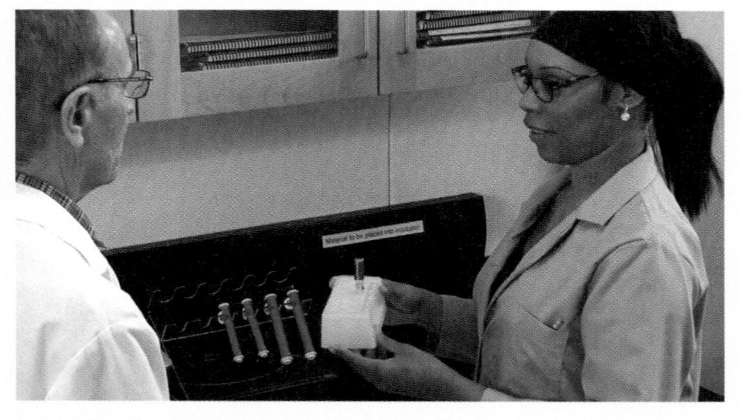

FIGURE 60.2 Incubate the tube for 24 to 48 hours at 37°C.

Name: ______________________ Date: __________ Section: __________

Biochemical Characteristics of Bacteria: Hydrogen Sulfide Production

EXERCISE RESULTS 60

RESULTS

HYDROGEN SULFIDE PRODUCTION

TABLE 60.1 Results of Biochemical Reactions

Biochemical Test	Name of Organism			
Hydrogen Sulfide Production				

Representations of Hydrogen Sulfide Test

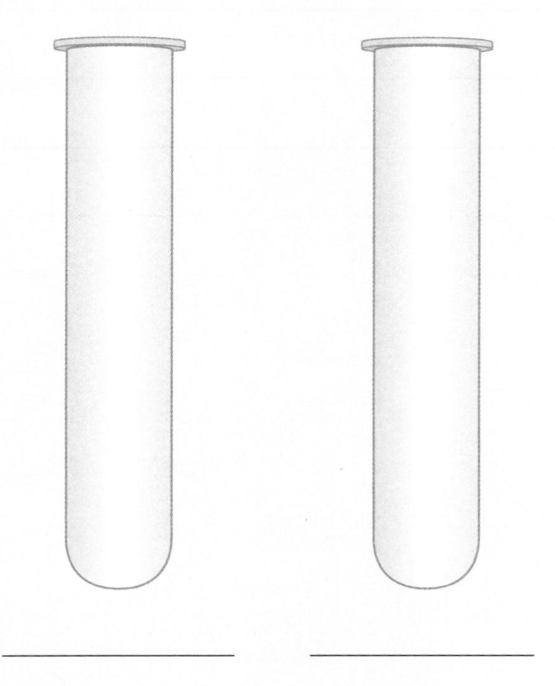

Organism: ______________ ______________

Test Results: ______________ ______________

Observations and Conclusions

Questions

1. For each of the tests you performed in this exercise, determine if the test is identifying the presence or absence of a substrate or a product.

2. Explain the practical significance of the production of hydrogen sulfide by bacteria.

Biochemical Characteristics of Bacteria: IMViC Series

EXERCISE 61

Materials and Equipment

- Inoculating loop and/or inoculating needle

Watch These Videos

- IMViC: Indole Test
- IMViC: Methyl Red Test
- IMViC: Voges-Proskauer Test
- IMViC: Citrate Test

The IMViC tests (indole, methyl red, Voges-Proskauer, and citrate), described in Exercise 48, are used in the preliminary identification of an organism. The indole test identifies the presence of the enzyme tryptophanase, which converts tryptophan to indole. The methyl red test produces mixed acids, resulting in an acidic pH and a positive methyl red reaction. The VP test produces a neutral product that is detected via potassium hydroxide and alpha-naphthol. The citrate test is designed to identify whether an organism can use citrate as its only carbon source.

IMViC SERIES

PURPOSE: to identify if a bacterial species can digest tryptophan.

Before starting this exercise, you may want to review the related **Microbiology Videos: Indole Test, Methyl Red Test, Voges-Proskauer Test,** and **Citrate Test** to understand how these tests are performed.

IMViC: Indole Test

IMViC: Methyl Red Test

IMViC: Voges-Proskauer Test

IMViC: Citrate Test

PROCEDURE

1. Follow the procedures outlined in Exercise 48.
2. Record your results in TABLE 61.1.
   - Include a labeled representation of the tube in the Results section, being sure to point out the important features of the test.

Name: ______________________ Date: __________ Section: __________

Biochemical Characteristics of Bacteria: IMViC Series

EXERCISE RESULTS 61

RESULTS

IMViC SERIES

TABLE 61.1 Results of Biochemical Reactions

Biochemical Test	Name of Organism			
Indole				
Methyl Red				
Voges-Proskauer				
Citrate				

Representations of IMViC Tests

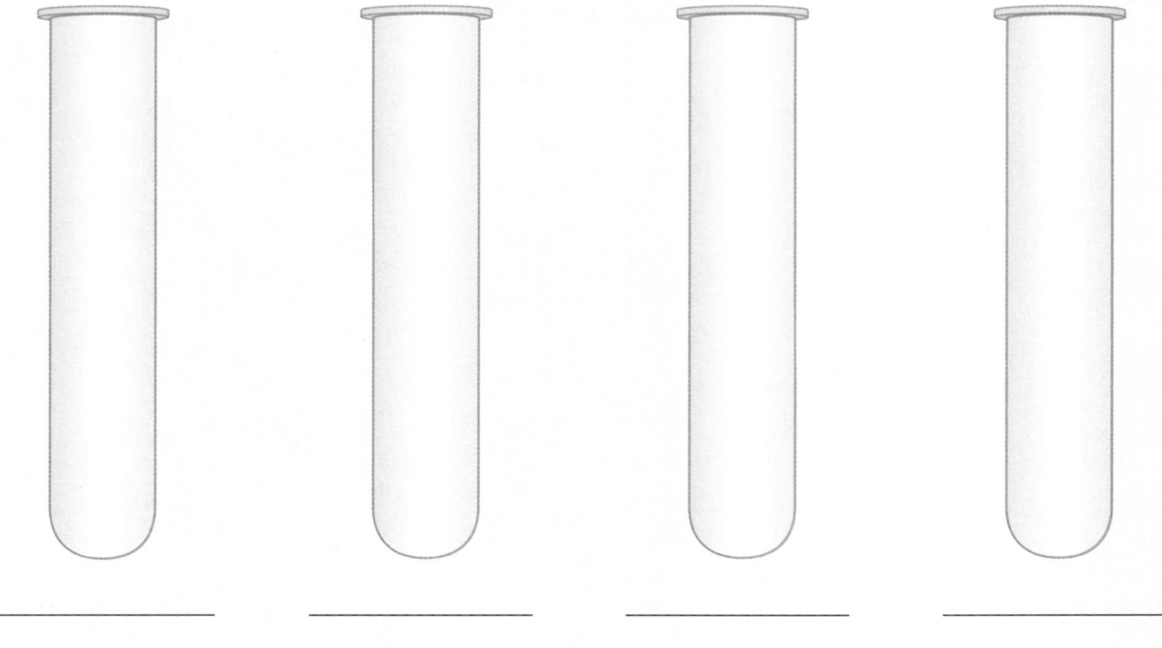

Organism: __________ __________ __________ __________

IMViC Test: __________ __________ __________ __________

Test Results: __________ __________ __________ __________

Observations and Conclusions

____

____

____

____

____

Questions

1. For each of the tests you performed in this exercise, determine if the test is identifying the presence or absence of a substrate or a product.

   ____

   ____

   ____

Biochemical Characteristics of Bacteria: Urea Hydrolysis

EXERCISE 62

Materials and Equipment

- Inoculating loop
- Selected bacterial species
- Tubes of urea agar

Watch This Video

- Urease Test

Urea is an end product of protein metabolism in bacteria. Certain bacteria produce the enzyme **urease**, which breaks down the urea into ammonia and carbon dioxide:

$$H_2N{-}C({=}O){-}NH_2 \;(\text{Urea}) \xrightarrow[2\,H_2O]{\text{urease}} 2\,NH_3 \;(\text{Ammonia}) + CO_2 \;(\text{Carbon dioxide}) + H_2O \;(\text{Water})$$

Urea hydrolysis is important in nitrogen cycles in the soil because urea is a major component of animal urine, and bacteria are responsible for breaking it down to release the nitrogen for reuse. The production of urease is a distinguishing characteristic of certain bacteria.

UREA HYDROLYSIS

PURPOSE: to identify if a bacterial species can digest urea.

In the test for urea digestion, bacteria are inoculated to an agar slant that contains urea and **phenol red indicator**. If the urea is digested during incubation, the ammonia will raise the pH of the medium and cause the indicator (and the agar) to become deep fuchsia or purple.

Before beginning this exercise, watch **Microbiology Video: Biochemical Tests: Urease Test** to understand how this test is performed.

Urease Test

PROCEDURE

1. Obtain a tube of urea agar, and label it in the prescribed manner (FIGURE 62.1).
2. Aseptically streak the urea agar slant with a loopful of bacteria (FIGURE 62.2), and incubate the tube for 24 to 48 hours at 37°C or as directed.
   - Include an uninoculated control tube for comparison purposes.

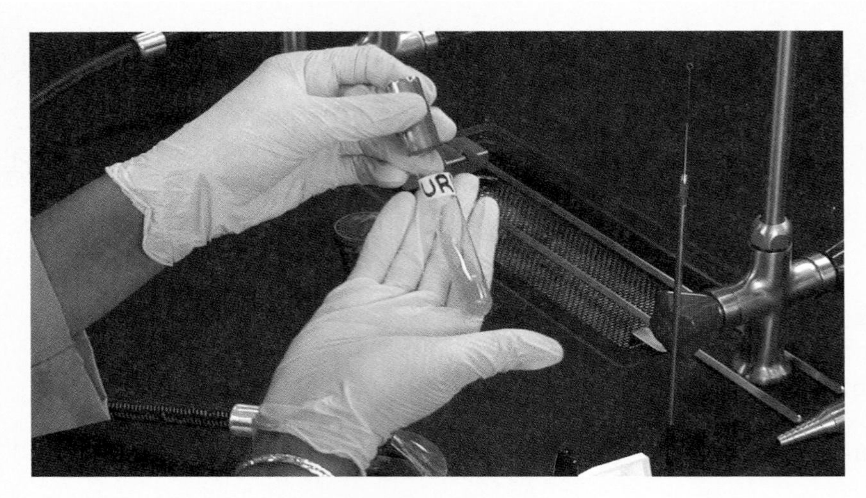

FIGURE 62.1 Label the tube of urea broth.

3. Compare the inoculated tube with the control tube, and note whether the agar color has become a deeper fuchsia or purple. This color change indicates urea digestion and ammonia production. If the color has not changed, then urea has not been hydrolyzed.
4. Enter your results in TABLE 62.1, using (+) to indicate urea digestion and (−) for nondigestion.
   - Include a labeled representation of the tube and a note of explanation to complete your results.

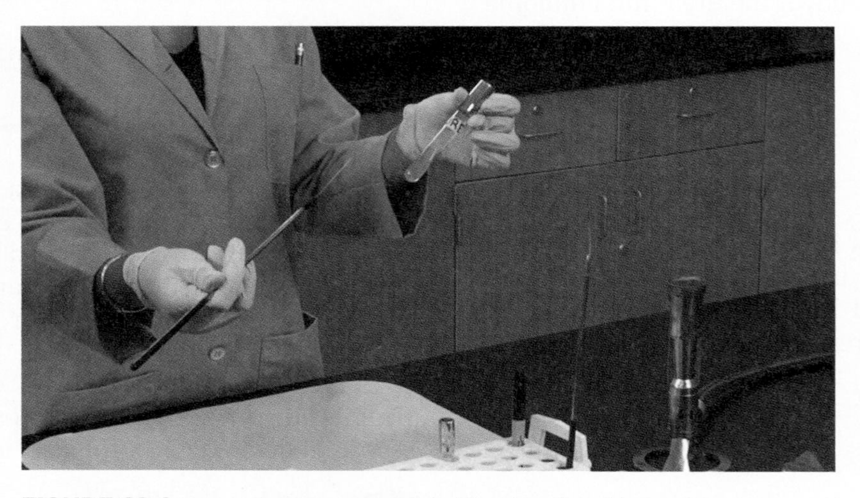

FIGURE 62.2 Inoculate the agar slant with a loopful of bacteria.

Name: ______________________ Date: __________ Section: __________

Biochemical Characteristics of Bacteria: Urea Hydrolysis

EXERCISE RESULTS **62**

RESULTS

UREA HYDROLYSIS

TABLE 62.1 Results of Biochemical Reactions

Biochemical Test	Name of Organism			
Urea Hydrolysis				

Representations of Urease Test

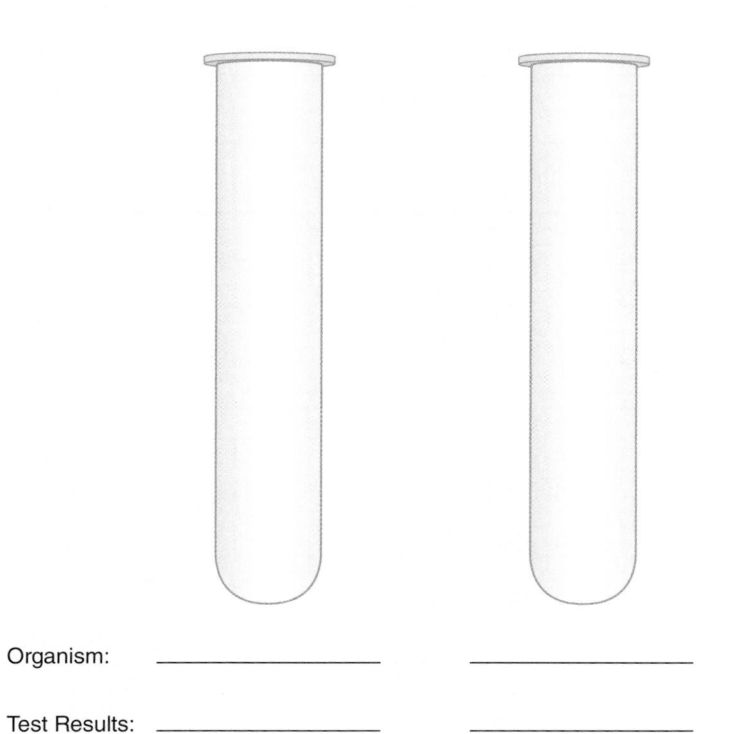

Observations and Conclusions

Questions

1. Determine if this test is identifying the presence or absence of a substrate or a product.

2. Why is it important to use pure cultures when performing the biochemical tests on bacteria?

Biochemical Characteristics of Bacteria: Fat (Triglyceride) Hydrolysis

EXERCISE 63

Materials and Equipment

- Inoculating loop
- Selected bacterial species
- Plates of spirit blue agar, or materials for their preparation

Certain bacteria secrete enzymes called **lipases**. In the environment, these enzymes hydrolyze the linkages in fats (triglycerides), separating the glycerol molecule from its fatty acid molecules. The glycerol and fatty acid molecules resulting from the digestion are transported into the bacterial cytoplasm, where they are used as the building blocks for new cell structures and as energy sources.

Lipid-digesting bacteria may be found in environments containing animal and vegetable fats, such as in fatty foods. Here they cause rancidity, a type of food spoilage in which foul odors and flavors develop because of the fatty acids released from the lipids. Butter, vegetable oil, and fish with high lipid content are particularly susceptible to rancidity.

LIPID DIGESTION

PURPOSE: to identify if a bacterial species can digest triglycerides.

Lipid digestion by bacteria is tested in a special medium called **spirit blue agar**. This medium is rich in triglycerides. It also contains a pH indicator called spirit blue. The indicator is pale lavender in a neutral environment, but it becomes a deep blue in the presence of acids. Thus, a color change to deep blue will occur in the medium if the bacteria are able to digest the lipids and as fatty acids accumulate and the pH lowers. The ability to digest lipids is an identifying characteristic of a bacterial species.

PROCEDURE

1. Select one plate of spirit blue agar or pour one plate as directed by the instructor.
   - Label the bottom side of the plate with your name, the name of the medium, the date, and the name of the organism tested.
2. After the agar has hardened, inoculate the medium by aseptically drawing a loopful of bacteria across the surface.
   - Try to avoid cutting into the agar as you streak the plate.
   - Several bacteria may be tested on the plate at the direction of the instructor.

3. Invert the plate to avoid formation of moisture on the agar surface, and incubate it for 24 to 48 hours at 37°C or as directed by the instructor.
    - An uninoculated control plate should be included for comparison.
4. At the conclusion of the incubation period, examine the plate and note the area near the bacterial growth. A dark blue area indicates that lipid hydrolysis has taken place.
    - If the area remains the original color or becomes a pale, lighter blue, then lipid digestion has not occurred.
    - Your results may be compared to the control plate containing no bacteria.
5. Enter your results in TABLE 63.1, using (+) for lipid digestion and (–) to indicate nondigestion.
    - Include a labeled diagram of the plate in the Results section to illustrate your results.

Name: ______________________ Date: __________ Section: __________

Biochemical Characteristics of Bacteria: Fat (Triglyceride) Hydrolysis

EXERCISE RESULTS 63

RESULTS

FAT (TRIGLYCERIDE) HYDROLYSIS

TABLE 63.1 Results of Biochemical Reactions

Biochemical Test	Name of Organism			
Triglyceride Hydrolysis				

Representations of Fat (Triglyceride) Hydrolysis

Organism: ______________________ ______________________

Test Results: ______________________ ______________________

Observations and Conclusions

____

____

____

____

____

Questions

1. Determine if this test is identifying the presence or absence of a substrate or a product.

   ____

   ____

   ____

Biochemical Characteristics of Bacteria: Casein Hydrolysis

EXERCISE 64

Materials and Equipment

- Inoculating loop
- Selected bacterial species
- Plates of skim milk agar, or materials for their preparation

Casein is the major protein in milk, comprising about 85% of the total protein content. Many species of bacteria have the ability to digest casein by producing enzymes called proteinases, specifically one called **caseinase**. Secreted into the environment, caseinases digest casein into smaller peptides and amino acids. The bacteria then import these products and use them in their metabolism to produce unique bacterial proteins. Other bacteria are unable to produce proteinases, and casein digestion does not take place.

CASEIN HYDROLYSIS

PURPOSE: to identify if a bacterial species can digest casein.

The medium used to test casein digestion is **skim milk agar**. This medium contains skim milk, where casein exists as very large, insoluble molecules. Indeed, the suspended particles of casein give milk its cloudiness and white color.

When casein is digested by bacterial proteinases, it reverts to soluble peptides and amino acids. As the casein particles disappear from the milk, the agar becomes clear. Thus, it is possible to determine whether a bacterium digests the casein by taking note of the area near the bacterial growth after a suitable incubation period. This characteristic may be useful in identification of an unknown bacterial species.

When you are finished with this exercise, fill out all the tables in the Results section, based on your work in Exercises 54 through 64. Complete the Test Results section and identify the bacterial species.

PROCEDURE

1. Obtain one plate of skim milk agar for this exercise, or pour one plate at the instructor's direction.
   - The bottom side of the plate should be labeled in the usual manner. It can be divided to test several bacteria at the direction of the instructor.
2. After the agar has solidified, inoculate the surface by drawing a loopful of bacteria across the surface.
   - An inoculating loop is recommended.
   - Streak lightly to avoid cutting into the agar.

3. Place the plate in the inverted position and incubate it for 24 to 48 hours at 37°C, or as the instructor directs.
   - Include an uninoculated control plate for comparison.
4. Observe the plate after the incubation period and note the area immediately adjacent to the streak of bacterial growth.
   - If the area has remained cloudy, you may conclude that casein digestion has not occurred.
   - However, if the area near the streak has become clear, then digestion has taken place.
   - Note that various bacteria will cause various degrees of digestion, and the clearing may be larger for one bacterial species than for another.
   - Make a comparison with the control plate containing no bacteria.
5. Enter results for the test in TABLE 64.1 using (+) for case– in digestion and (–) to indicate nondigestion.
   - Use a carefully labeled diagram to illustrate your results in the appropriate space of the Results section.

Name: ______________________ Date: __________ Section: __________

Biochemical Characteristics of Bacteria: Casein Hydrolysis

EXERCISE RESULTS 64

RESULTS

CASEIN HYDROLYSIS

TABLE 64.1	Results of Biochemical Reactions			
Biochemical Test	Name of Organism			
Casein Hydrolysis				

Representations of Casein Hydrolysis Test

Organism: ______________________ ______________________

Test Results: ______________________ ______________________

Observations and Conclusions

Questions

1. Determine if this test is identifying the presence or absence of a substrate or a product.

2. Explain why biochemical characteristics serve as a fingerprint for a bacterial species.

Bacterial Genetics

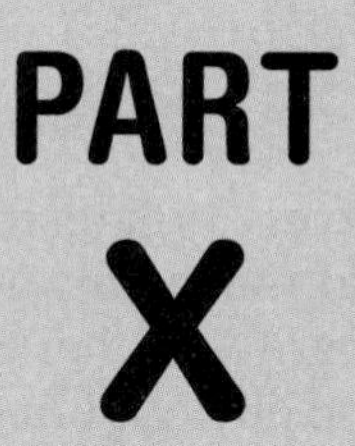

PART X

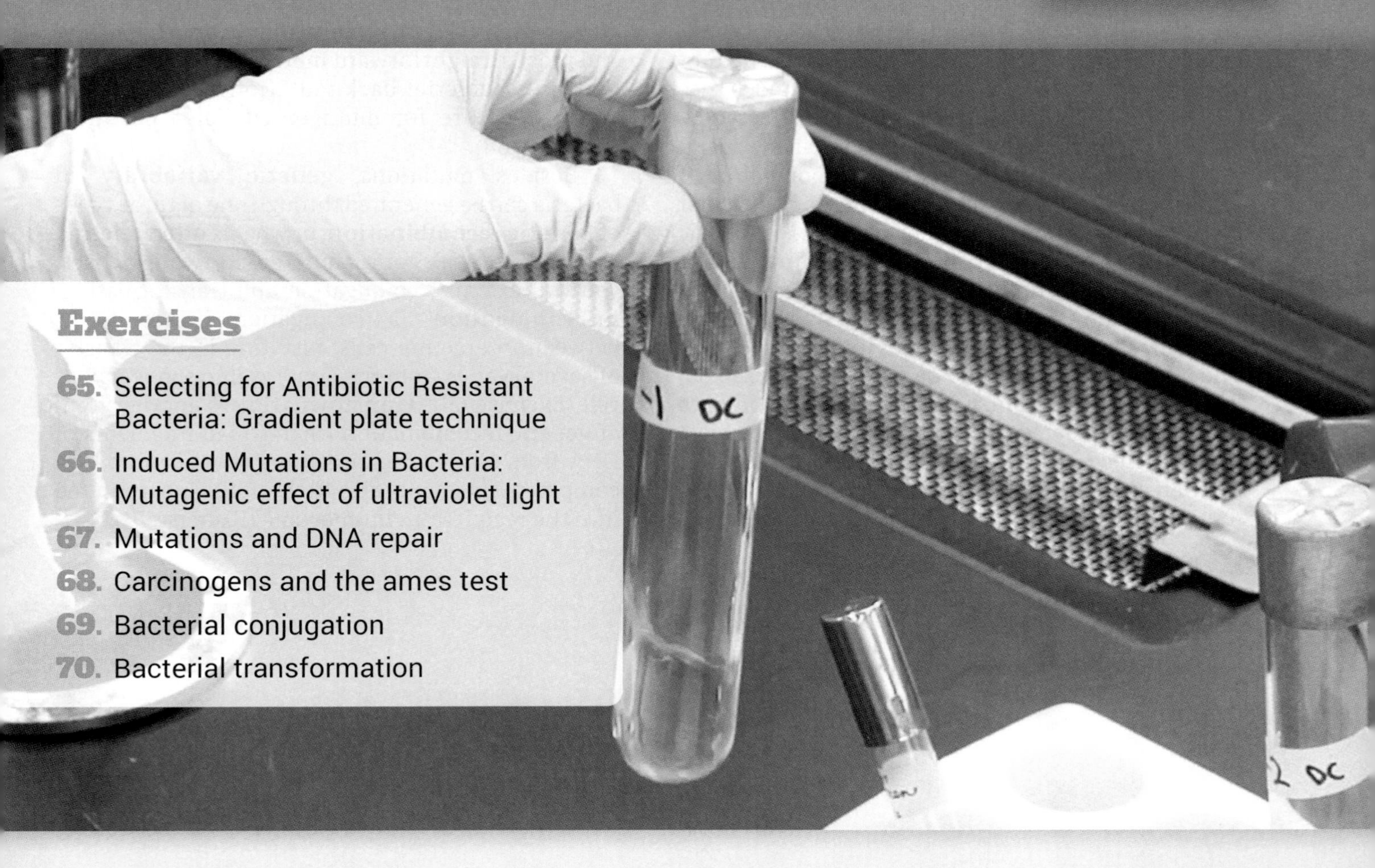

Exercises

65. Selecting for Antibiotic Resistant Bacteria: Gradient plate technique
66. Induced Mutations in Bacteria: Mutagenic effect of ultraviolet light
67. Mutations and DNA repair
68. Carcinogens and the ames test
69. Bacterial conjugation
70. Bacterial transformation

Learning Objectives

When you have completed the exercises in Part X, you will be able to:

- Carry out a gradient plate technique to detect antibiotic-resistant mutants in a bacterial population.
- Prepare an experiment to examine the effect of UV light as a physical mutagen.
- Test an organism's ability to carry out DNA repair following a mutational event such as that caused by UV light.
- Complete an Ames test to screen for chemical mutagens (carcinogens).
- Demonstrate genetic recombination through the process of conjugation.
- Visualize genetic recombination through transformation by using a fluorescent plasmid.

Every organism, from bacteria to humans, is subject to **mutation** during the life of the organism. These heritable changes to genes in a cell's DNA may cause no obvious effect or they may alter the characteristics or functioning of the organism for better or worse. Although mutation rates usually are very rare, because many bacteria can multiply at record rates, mutations crop up quite often. Indeed, the field of **bacterial genetics** has prospered and much has been learned about mutations by studying bacteria.

Mutations can be detected in bacteria in several ways. Mutations that already exist in a population of bacteria can be detected by selecting for particular types of mutational consequences (e.g., antibiotic resistance). A **gradient plate technique** is one such selection technique (Exercise 65). Agents that cause mutations are called **mutagens**. Some are physical agents, such as **ultraviolet (UV) light**. The effects of UV light can be studied (Exercise 66) by selecting for mutants that have lost the ability to produce a product, such as a pigment. Being subject to the effects of mutagens, most bacteria have mechanisms to repair damaged DNA, including mutational damage caused by UV light. Fairly simple experiments can be designed to assay pigment production and DNA repair (Exercise 67).

Many chemicals, both natural and chemically synthesized, are mutagens and have been found to be **carcinogens** (cancer-causing agents) in humans. Therefore, screening new chemicals for mutagenicity is extremely important. The best and most straightforward method is the Ames test that uses bacterial back mutations in *Salmonella* as the measure for mutagenicity of a chemical (Exercise 68).

Besides mutations, **genetic variability** in bacteria can be generated through one of three types of **genetic recombination**, or what is often referred to as **horizontal gene transfer**. In two exercises, you will examine conjugation and transformation. In **conjugation**, a conjugation bridge forms between interacting cells, resulting in the transfer of several genes from the donor cell to the recipient cell (Exercise 69). **Transformation** is another form of genetic recombination where small fragments of DNA from a lysed donor cell may be taken up by a competent recipient cell and the genes incorporated into the recipient's chromosome (Exercise 70).

Selecting for Antibiotic Resistant Bacteria: Gradient Plate Technique

EXERCISE

65

Materials and Equipment

- Broth cultures of *Escherichia coli*
- Solution of 0.01% tetracycline or other selected antibiotic
- Sterile culture dishes
- Deep tubes of liquid nutrient agar
- Sterile swabs and disinfectant solution
- 1-ml pipettes

Mutations are permanent and inheritable changes in the genome of an organism (or virus). If the change occurs in the nitrogenous base sequence of a gene, the base change may lead to a miscoded messenger RNA molecule, resulting in the insertion of one or more incorrect amino acids into the protein during translation. Some mutations have little effect, but other mutations significantly change the functions of the cell or organism.

Spontaneous mutations are those that take place in nature without human intervention or an identifiable cause. In the normal course of events, cells arising from a spontaneous mutation are masked by normal cells. Should a selecting agent be introduced into the environment, however, the mutant cells will survive, multiply, and emerge. This exercise will demonstrate how bacteria resulting from a spontaneous mutation can be selected out from a mixed population of bacteria. In this case, the mutant bacteria will be a strain that displays antibiotic resistance.

GRADIENT PLATE TECHNIQUE

PURPOSE: to identify *E. coli* bacteria resistant to an antibiotic.

Spontaneous mutations are estimated to occur at a rate of one in every billion or so replications of a bacterium. This statistic implies that a mutant may exist in every billion bacteria. By culturing the bacteria on a selecting agent, the mutant can be culled from the population and encouraged to multiply and form a colony. The **gradient plate technique** will be used to demonstrate that antibiotic-resistant mutants already exist in a laboratory culture of ***Escherichia coli***. In this technique, a gradual increase in antibiotic concentration is established, and **antibiotic-resistant cells** are selected out in high-concentration areas.

PROCEDURE

1. Obtain a sterile culture dish, and label it with your name and the date.
2. Aseptically pour one deep tube of liquid nutrient agar into the dish. Using a suitable object such as a pencil, immediately place the plate on an angle, as illustrated in FIGURE 65.1A.
3. Allow the agar to harden on a slant for several minutes, being certain that the agar covers the surface on the low side of the slant.
4. After the agar has hardened, place the plate flat on the laboratory bench.
5. Obtain a second tube of liquid nutrient agar, a pipette, and a sample of tetracycline or other designated antibiotic.
    - Pipette 0.1 ml of the antibiotic solution into the nutrient agar tube, and mix the solution well by vigorously tapping the bottom of the tube.
    - Pour the contents of the tube into the plate, as indicated in FIGURE 65.1B. A gradient of drug will be formed as the antibiotic diffuses throughout the medium.
    - When the top layer of agar has hardened, label the bottom of the plate, indicating the low and high concentrations of the drug.
6. Using a sterile swab, obtain a sample of the broth culture of *Escherichia coli*, and prepare a lawn of bacteria over the entire surface of the plate, as indicated in **Figure 66.1**.
    - Deposit the contaminated swab in the disinfectant as directed by the instructor.
7. Invert the plate and incubate it at 37°C for 24 to 48 hours. Refrigerate the plate until observed.
8. Observe the plate and note how the amount of bacterial growth diminishes as the concentration of antibiotic increases. This illustrates the killing power of the drug.
    - Examine the region of high drug concentration and locate isolated colonies of bacteria. These colonies have resulted from the multiplication of drug-resistant mutants present in the original culture tube.
    - Realize the drug was not the mutagen, which cannot be identified since the mutants already existed in the population.

When pipetting, always use a mechanical pipetter, and avoid putting anything in your mouth.

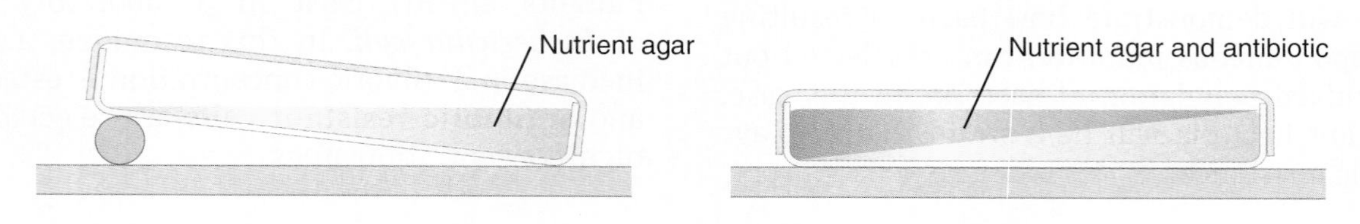

A Liquid nutrient agar is poured into a sterile culture dish and the dish is placed at an angle. The agar solidifies.

B Liquid nutrient agar with antibiotic is poured over the first layer. It solidifies and forms a gradient of antibiotic.

FIGURE 65.1 Preparation of an antibiotic gradient in an agar plate.

9. Draw a labeled representation of the plate in the Results section, indicating the colonies that contain drug-resistant mutants.
   - Enter your observations concerning the success of the technique in the space provided.
10. At the direction of the instructor, samples of mutant bacteria may be isolated on nutrient agar slants and tested for higher drug resistance by repeating the experiment with an increased level of drug in the gradient. On occasion, the gradient plate technique will not display colonies of mutants, possibly because there were none in the original culture.

Quick Procedure

Gradient Plate Technique

1. Prepare a gradient of drug on an agar plate.
2. Inoculate the gradient plate with bacteria.
3. Incubate.
4. Observe for mutant bacteria in high-drug areas.

Name: ______________________ Date: __________ Section: __________

Selecting for Antibiotic Resistant Bacteria: Gradient Plate Technique

EXERCISE RESULTS **65**

RESULTS

GRADIENT PLATE TECHNIQUE

Organism: ______________________

Observations and Conclusions

____

____

____

____

____

Questions

1. How might antibiotic-resistant bacteria have arisen by spontaneous mutation in the natural environment?

2. Suppose an organism were unable to synthesize an essential nutrient for its growth. How could the gradient plate technique be used to detect a mutant organism that was able to synthesize the nutrient?

3. Is the gradient plate technique used in this exercise an example of a positive selection method or a negative selection method? Explain.

Induced Mutations in Bacteria: Mutagenic Effect of Ultraviolet Light

EXERCISE 66

Materials and Equipment

- Broth cultures of *Serratia marcescens*
- Ultraviolet lamps (germicidal quality)
- Nutrient agar plates, or materials for their preparation
- Sterile cotton swabs and disinfectant solution

Two general types of mutations are induced mutations and spontaneous mutations. **Induced mutations** are those in which the cause can be identified. Occasionally these mutations occur by accident, but more often they result from planned laboratory experiments where bacteria are subjected to chemical or physical agents. These agents are referred to as **mutagens**. In this exercise, **ultraviolet (UV) light** will be the mutagen used to induce a mutation in bacteria. We shall see how a mutation changes a pigment-producing strain of bacteria into one that cannot produce pigment.

MUTAGENIC EFFECT OF ULTRAVIOLET LIGHT

PURPOSE: to examine the effect of UV light exposure on bacterial pigment production.

Mutations resulting from treatment with UV light are among the best understood. Ultraviolet light is a form of energy not perceived by the human eye. When this energy is absorbed by a bacterium's DNA, it induces adjacent thymine or cytosine molecules to link together, thus disrupting the genetic code. The gram-negative rod ***Serratia marcescens*** normally produces a bright red pigment at 25°C. However, it often loses this ability after undergoing treatment with UV light, and mutant strains appear white. Depending on the degree of exposure to UV light, the disruption of the genetic code may even lead to cell death. In this exercise, we shall observe how mutation results in both of these effects.

PROCEDURE

1. Select or prepare five plates of nutrient agar, and label them on the bottom side with your name and the date. Choose four plates and label one each with the designations "2 seconds," "5 seconds," "10 seconds," and "20 seconds." These times will indicate the time of exposure of the bacteria to UV light. The fifth plate will be the control, so label it "0 seconds," as it will not be exposed to UV light.

Be careful not to look directly into the UV light, since damage to the eye may occur.

2. Dip a sterile cotton swab into a broth culture of *Serratia marcescens*, and prepare a lawn of bacteria by swabbing the surface of the nutrient agar in three directions, as shown in FIGURE 66.1. Be sure to cover all parts. Then make a final swab around the perimeter.
   - A separate cotton swab should be used for each plate (including the control), and precautions should be taken to avoid airborne contamination by lifting the lid only enough to permit access (FIGURE 66.2).
   - When the swabbing is complete, the swabs should be deposited in the beaker of disinfectant as directed by the instructor.
3. Turn on the UV light and allow it to warm up for a few minutes. Be careful not to look into the light because ultraviolet energy may damage the retina of the eye. The wavelength of the energy should be about 265 nm, and the light should be about 6 inches away from the plates.

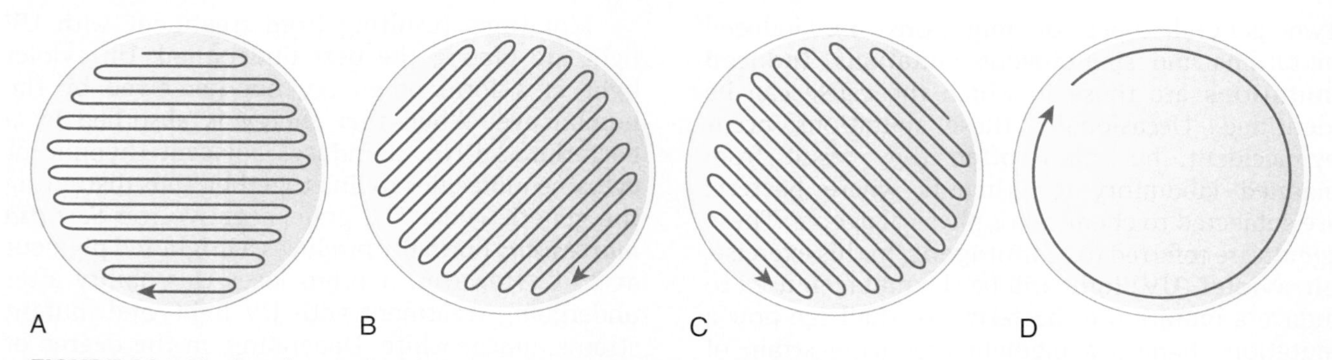

FIGURE 66.1 The "3+1" procedure for making a lawn of bacteria on a plate of medium.

FIGURE 66.2 Method of swabbing the plate of agar to avoid airborne contamination.

4. Remove the cover of the first plate, and expose the bacteria to UV light for the time designated on the label.
   - Replace the cover quickly.
   - Expose the remaining three plates for the designated time periods.
5. Incubate the plates in the inverted position in a dark environment at 25°C or room temperature for 3 to 5 days. A dark environment such as a box is essential because visible light may reverse the mutations. The fifth, inoculated but unexposed plate, also should be incubated as a control.
6. Observe the plates, and note the presence of colonies of mutated and normal *Serratia marcescens*. Mutated *S. marcescens* forms white, nonpigmented colonies, while normal *S. marcescens* appears as red colonies.
   - In TABLE 66.1 in the Results section, record the number of pigmented and nonpigmented colonies in each plate.
   - In your explanation, state whether the increasing time of exposure affected the number of mutant colonies that formed and, if so, how substantial the effect was.
   - Draw appropriate conclusions from the exercise.
7. Note the reduction in the number of colonies in all plates by comparing the experimental plates to the control plate. This reduction arises from the formation of lethal mutants.
   - In the appropriate space, write an explanation of the effect of UV irradiation on bacterial death.
   - Examine several red colonies for pie-shaped sectors of white growth. These sectors represent mutant cells that expressed the mutation after the colony began to develop.
   - In the Results section, illustrate one plate that contains colonies of mutated bacteria.
   - Comment on the ability of UV light to cause mutation in bacteria.
8. Save these plates to compare with Exercise 67.

Quick Procedure

Inducing Mutations with UV Light

1. Inoculate an agar plate with pigment-producing bacteria.
2. Subject the bacteria to ultraviolet light for various time periods.
3. Incubate.
4. Observe for colonies containing mutated bacteria.

Name: ______________________ Date: __________ Section: __________

Induced Mutations in Bacteria: Mutagenic Effect of Ultraviolet Light

EXERCISE RESULTS 66

RESULTS

MUTAGENIC EFFECT OF ULTRAVIOLET LIGHT

TABLE 66.1 Effect of UV Light Exposure on Pigment Production

Exposure to Ultraviolet Light	Number of Colonies of *Serratia marcescens*	
	Pigmented	Nonpigmented
0 seconds		
2 seconds		
5 seconds		
10 seconds		
20 seconds		

Irradiated Colonies of *Serratia marcescens*

Exposure Time: __________ __________ __________

Exposure Time: ____________________ ____________________

Effect of increasing exposure time on number of mutants:

Effect of irradiation on cell death:

Questions

1. What method might be used to determine whether an induced mutation was a temporary or permanent change in the organism?

Mutations and DNA Repair

EXERCISE

67

Materials and Equipment

- Broth cultures of *Serratia marcescens*
- UV lamps (germicidal quality)
- Nutrient agar plates, or materials for their preparation
- Sterile cotton swabs and disinfectant solution
- Quebec colony counter

In Exercise 66, the damage caused by UV light was determined by examining the UV light-exposed plates after they had been incubated in the dark for 3 to 5 days. Many bacteria have **DNA repair enzymes** capable of repairing alterations and distortions to the DNA caused by spontaneous or induced mutations.

DNA REPAIR OF UV LIGHT DAMAGE

PURPOSE: to examine the ability of bacterial cells to repair DNA damage.

One way bacteria attempt to repair UV-induced damage to DNA is to separate the adjacent thymine or cytosine molecules that were linked by UV light. This repair or "reactivation" is accomplished in the presence of visible light, which gives the repair enzyme, called **photolyase**, the energy to separate the linked bases. The DNA repair process is called **photoreactivation**. In this exercise, we shall see if UV damage can be reversed by this repair process.

PROCEDURE

1. As you did in Exercise 66, select or prepare five plates of nutrient agar, and label them on the bottom side with your name and the date. Choose four plates and label one each with the designations "2 seconds," "5 seconds," "10 seconds," and "20 seconds." These times will indicate the time of exposure of the bacteria to UV light. Again, the fifth plate will be the control ("0 seconds" exposure to UV light).
2. Dip a sterile cotton swab into a broth culture of *Serratia marcescens*, and prepare a lawn of bacteria by swabbing the surface of the nutrient agar in three directions as described in Exercise 66. Be sure to cover all parts. Then make a final swab around the perimeter.
   - A separate cotton swab should be used for each plate, and precautions should be taken to avoid airborne contamination by lifting the lid only enough to permit access.

- When the swabbing is complete, the swabs should be deposited in the beaker of disinfectant as directed by the instructor.

3. Turn on the UV light and allow it to warm up for a few minutes. The wavelength of the energy should be about 265 nm, and the light should be about 6 inches away from the plates.

Be careful not to look directly into the UV light, since damage to the eye may occur.

4. Remove the cover of the first plate, and expose the bacteria to UV light for the time designated on the label.
   - Replace the cover quickly.
   - Expose the remaining three plates for the designated time periods.
5. After exposing the plates to UV light, incubate the plates in the inverted position in a brightly lit environment for 3–5 days, including the unexposed control plate.
6. Observe the plates, and note the presence of colonies of mutated and normal *Serratia marcescens*.
7. Now, compare similarly exposed plates and controls between those incubated in the dark (Exercise 66) and those incubated in the light. Using a **Quebec colony counter** (see Exercise 81) or dissecting microscope, determine which plates (dark or light incubated) contain (a) more pigmented colonies and (b) more total colonies. Record your observations in TABLE 67.1 of the Results section and draw appropriate conclusions concerning the activity of a photoreactivation process.

Quick Procedure

DNA Repair of UV Light Damage

1. Inoculate an agar plate with pigment-producing bacteria.
2. Subject the bacteria to ultraviolet light for various time periods.
3. Incubate.
4. Observe for colonies containing mutated bacteria.

Name: ______________________ Date: __________ Section: __________

Mutations and DNA Repair

EXERCISE RESULTS 67

RESULTS

DNA REPAIR OF UV LIGHT DAMAGE

TABLE 67.1 The Effects of Visible Light on the Reversal of UV Light Damage

Exposure to UV Light	Number of Colonies	
	UV Light - Visible Light	UV Light + Visible Light
Control		
2 sec		
5 sec		
10 sec		
20 sec		

Observations and Conclusions

____

____

____

____

____

Questions

1. Explain how visible light can repair DNA damage caused by UV light.

Carcinogens and the Ames Test

EXERCISE
68

Materials and Equipment

- Nutrient broth cultures of *Salmonella typhimurium*, strain TA 1538 or TA 98
- Minimal agar plates
- 2-ml top agar tubes
- Sterile histidine solution
- Commercial hair dyes or other chemicals to screen
- 1-ml sterile pipettes
- Sterile disks with and without 4-NOPD
- Sterile forceps

In today's increasingly industrialized society, we are exposed to an ever-larger variety of man-made chemicals that are liberated into the air, water, or soil. We potentially may breathe, ingest, or in some way be exposed to air pollutants from automobiles and industry, pesticides, food additives, animal hormones, cigarette smoke, and many other chemicals yet identified. Research has shown that many of these chemicals may pose a significant hazard to our health through their induction of cancer. Many of these chemicals are mutagens that are **carcinogens**; that is, cancer-causing agents. It is imperative, therefore, that chemicals that may act as mutagens be identified. Carrying out such determinations on human subjects is socially unacceptable. Because some 90 percent of human carcinogens also induce mutations in bacteria, in the early 1970s, Bruce Ames developed a procedure using bacteria as test organisms to rapidly screen for possible mutagens (and therefore possible carcinogens) in humans.

THE AMES TEST

PURPOSE: to screen potential chemicals as mutagens and therefore cancer-causing agents in humans.

The **Ames test** is a simple and inexpensive procedure that uses ***Salmonella typhimurium*** as the bacterial test organism. A mutant strain that is unable to synthesize the essential amino acid histidine (his$^-$) is used. This strain will not grow on a growth medium lacking histidine unless a back mutation occurs that restores the bacterium's ability to synthesize histidine (his+) has occurred. In the Ames test, one assumes any such back mutation was the result of the chemical being tested.

In the standard Ames test, a liver homogenate is used as a source of activating enzymes for the chemical tested. This is because in the human body the mutagenic effect of some chemicals only results after liver enzymes convert a nonmutagen into a mutagen. In the following procedure, you will perform a modified Ames test without the need for a liver homogenate. Molten minimal agar containing *S. typhimurium* and a trace of histidine is poured on four minimal agar plates. The trace amount of histidine is needed to allow the bacteria to undergo the several cell divisions necessary for mutation to occur. Then, a disk containing a known carcinogen, **4-nitro-o-phenylenediamine** (**4-NOPD**), is placed in the center of one plate, the **positive control**. The second plate, the **negative control**, is not exposed to 4-NOPD. The two other plates are used to screen two other "test chemicals" of interest. Diffusion of the known and test chemicals from the disk establish a concentration gradient of the chemical in the dish.

PROCEDURE

1. Obtain the two test chemicals that will be screened by the Ames test.
   - Obtain four minimal agar plates and label one as "positive control" and another as "negative control."
   - Label the two other plates with the names of the test chemicals to be screened.
   - Put your name on the bottom of all four plates.
2. Obtain 4 tubes of molten top agar at 45°C.
   - Aseptically, add 0.2 ml of the sterile histidine solution to each tube.
   - Then add 0.1 ml of the *S. typhimurium* broth culture.
   - Quickly mix the tubes by rotating each between the palms of your hands and then pour each tube onto the surface of a minimal agar plate. Allow the top agar to solidify.
3. With sterile forceps, obtain a disk containing the known mutagen 4-NOPD and place the disk flat in the center of the "positive control" plate.
   - Gently press down the disk to make sure it is in tight contact with the agar surface.

Caution. Since you are working with a known mutagen and possible mutagens, wear gloves when carrying out Steps 3–5. Dispose of materials as directed by your instructor.

4. Take a plain sterile disk (without 4-NOPD) and place the disk on the "negative control" plate.
5. For each of the two plates that will be used for the other chemicals to be screened, obtain a sterile disk and, with the sterile forceps, dip half the disk into one of the test chemicals (see Exercise 28).
   - Remove any excess liquid by touching the disk to the side of the container.
   - Place the disk on the center of the plate and tap down the disk.
   - Repeat for the second chemical.
6. Incubate the four plates in the inverted position at 37°C for 24 to 48 hours in the dark.
7. Following incubation, observe the four plates and count the number of colonies on each plate.
   - Record the number in TABLE 68.1 of the Results section. The number of induced mutations caused by 4-NOPD can be determined by subtracting the number of colonies on the negative control plate from the number of colonies on the positive control plate.
   - Record this number in the table.
   - Remember, each colony represents a his^- to his+ revertant (back-mutation). A positive result, indicating some degree of mutagenicity, is indicated by the increased number of induced mutations (colonies) on the 4-NOPD plate as compared to the number of spontaneous revertants (due to spontaneous back-mutations) on the negative control plate.
8. For the two test chemicals, the number of induced mutations can be determined in a similar fashion by again subtracting the number of colonies on the negative control plate from those on each of the test chemical plates. Record these numbers in Table 68.1.

9. The relative degree of mutagenicity is determined by examining the number of induced mutations (colonies). If there are fewer than 10 colonies, the mutagenicity is negligible (–). If between 10 and 100, mutagenicity is slight (+); between 100 and 500 is moderate (++); and if over 500, mutagenicity is strong (+++).
   - Record the degree of mutagenicity in Table 68.1.
   - Draw appropriate conclusions concerning the mutagenicity of the test chemicals.
10. When this exercise is complete, **do not** dispose of any materials in the usual manner. Some of the plates contain mutagens and live bacteria. Your instructor will inform you as to the proper disposal process.

Quick Procedure

The Ames Test

1. Mix *S. typhimurium* and sterile histidine solution.
2. Add 4-NOPD to the positive control plate
3. Add test chemicals to the other plates
4. Incubate.
5. Observe for colonies containing mutated bacteria.

Name: ______________________ Date: __________ Section: __________

Carcinogens and the Ames Test

EXERCISE RESULTS 68

RESULTS

THE AMES TEST

TABLE 68.1 The Mutagenicity of Known and Test Chemicals

Test Chemical	Number of Colonies	Number of Induced Mutations	Degree of Mutagenicity (−, +, ++, +++)
Negative control			−
Positive control (4-NOPD)			
Test chemical: ______			
Test chemical: ______			

Observations and Conclusions

____

____

____

____

____

Questions

1. Identify the advantages and disadvantages of using bacteria instead of laboratory test animals to screen for chemicals that act as mutagens and carcinogens in humans.

Bacterial Conjugation

EXERCISE 69

Materials and Equipment

- 12-hour nutrient broth cultures of F^- *E. coli* strain [thr^-, leu^-, thi^-, and Str-r] (ATCC e23724)
- 12-hour nutrient broth cultures of Hfr *E. coli* strain [thr^+, leu^+, thi^+, and Str-s] (ATCC e 23740)
- Minimal medium agar plates with streptomycin and thiamine
- Beakers with 95% ethyl alcohol
- Bent (L-shaped) glass rods
- 1-ml sterile pipettes and mechanical pipetting devices
- Sterile 13- × 100-mm test tubes
- Waterbath at 37°C

Genetic variability in eukaryotes involves processes occurring during sexual reproduction. **Meiosis** itself helps "shuffle" the genes donated to an egg or female gamete, while the process of **crossing over** (exchange of genetic material between homologous chromosomes) during meiosis I further mixes up the genes. However, in prokaryotic organisms, there is no sexual reproduction. Yet genetic variability occurs to maintain the evolutionary success within and between species. In prokaryotes, **horizontal (lateral) gene transfer** (**HGT**) can occur between cells within the same generation.

One mechanism driving genetic variability in bacteria is **conjugation**. This form of HGT involves the one-way transfer of genetic material from a donor cell to a recipient cell. Transfer requires a physical contact between the two cells through the formation of a conjugation bridge. Cells having an **F factor** represent donor cells because they have the ability to produce the conjugation bridge. If the F factor is extrachromosomal (a **plasmid**), the donor cells are designated as **F^+**. If the F factor becomes incorporated into the bacterial chromosome of the donor, the resulting cells are designated **Hfr**, for **high-frequency recombinants**. Cells missing the F factor are the recipients of the conjugation process and are designated as **F^-**. In F^+ cells, a copy of the plasmid usually is completely transferred to an **F^-** recipient. The result is that the **F^-** cell now has an F plasmid, so it has become an F^+ cell. In Hfr cells, there is a transfer of chromosomal genes, but the conjugation bridge usually breaks before a complete copy of the donor chromosome can be transferred to an **F^-** cell. A few chromosomal genes may be transferred, so the recipient is referred to as an **F^- recombinant**.

BACTERIAL CONJUGATION

PURPOSE: to demonstrate bacterial genetic recombination through conjugation and the formation of recombinant cells.

In this experiment, you will prepare three cultures of *Escherichia coli*: one containing an Hfr strain sensitive to the antibiotic streptomycin (Str-s);

another containing an **F⁻** strain that is resistant to streptomycin (Str-r) but requires the amino acids threonine (thr) and leucine (leu) to build proteins, and the vitamin thiamine (thi); and a third containing a mixture of the Hfr and F⁻ strains. Following a short incubation period, isolation of only the F⁻ recombinants will be performed by plating the mixed culture on a minimal medium containing streptomycin and thiamine. A genetic map indicating the origin and direction of transfer and the sites of the relevant marker genes is provided in FIGURE 69.1.

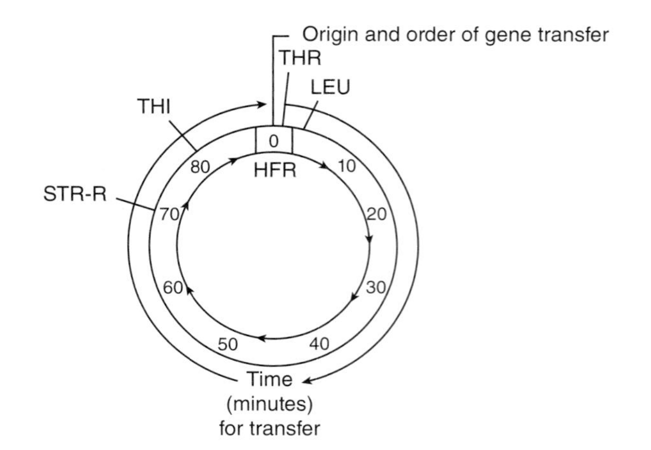

FIGURE 69.1 The genetic map for the *E. coli* HFR strain.

PROCEDURE

1. Obtain two 1-ml sterile pipettes. Use one pipette to aseptically transfer 1 ml of the F⁻ *E. coli* broth culture and the other pipette to aseptically transfer 0.3 ml of the Hfr *E. coli* broth culture into a sterile 13- × 100-mm test tube.
2. Mix by gently rotating the culture between the palms of your hands.
3. Incubate the culture for 30 minutes at 37°C waterbath.
4. Obtain two minimal medium plus streptomycin and thiamine agar plates.
   - Label the bottom of both plates with your name and date.
   - Label one plate "Hfr" and the other "F⁻."

Only place the glass rod in the Bunsen burner flame long enough to catch the alcohol on fire. Longer exposure to the flame will melt the glass or cause the rod to break.

5. Using the spread-plate technique illustrated in FIGURE 69.2, aseptically add 0.1 ml of each *E. coli* strain to its appropriately labeled agar plate.
   - With an alcohol-dipped and flamed glass rod, spread the inoculum over the entire surface of the agar plate.
6. Following the 30-minute incubation, *vigorously* agitate the mixed culture to terminate the genetic transfer by breaking the conjugation bridge.
7. Obtain another minimal medium plus streptomycin and thiamine agar plate. Label the bottom of the plate with your name and date. Label this plate "Hfr × F⁻."
8. Aseptically add 0.1 ml of the mixed culture.
   - Spread the inoculum over the entire surface with a sterile glass rod as described in Step 5.
9. Incubate all three plates in an inverted position for 48 hours at 37°C; then refrigerate.
10. Observe the three plates for the presence of (+) or absence of (−) bacterial colonies.
    - Enter your data in the Results section (TABLE 69.1).
    - In your conclusions, write why you would expect to see colonies or see no colonies on the appropriate minimal medium plus streptomycin and thiamine agar plates.

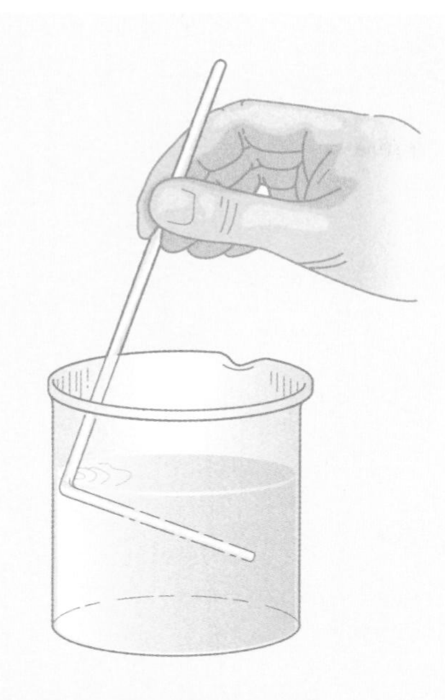

A Dip the bent glass rod into the beaker of 95% ethyl alcohol.

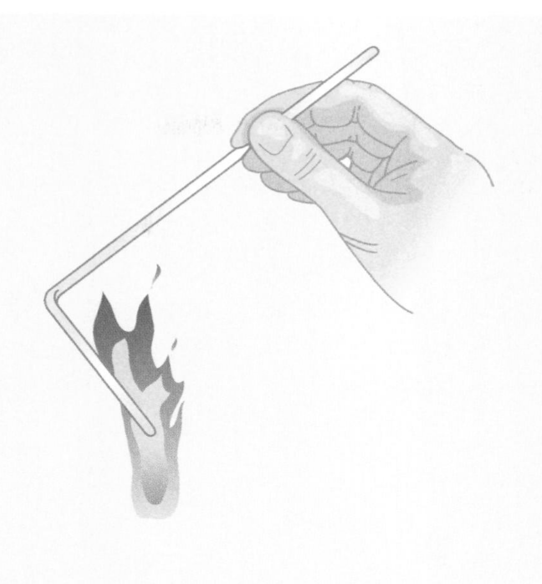

B Sterilize the glass rod by flaming with a Bunsen burner.

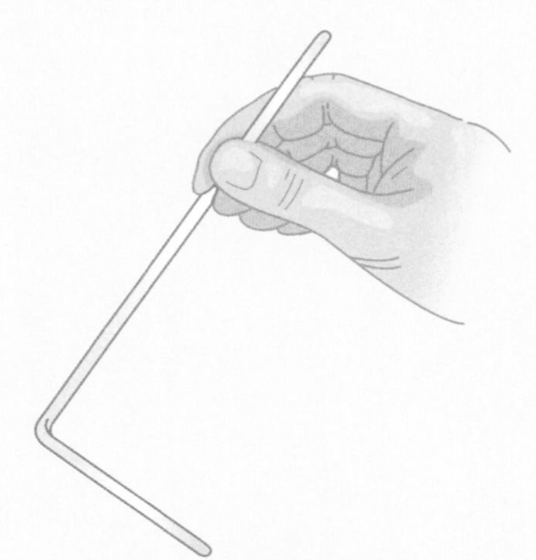

C Remove from Bunsen burner, allow flame to extinguish, and cool the glass rod.

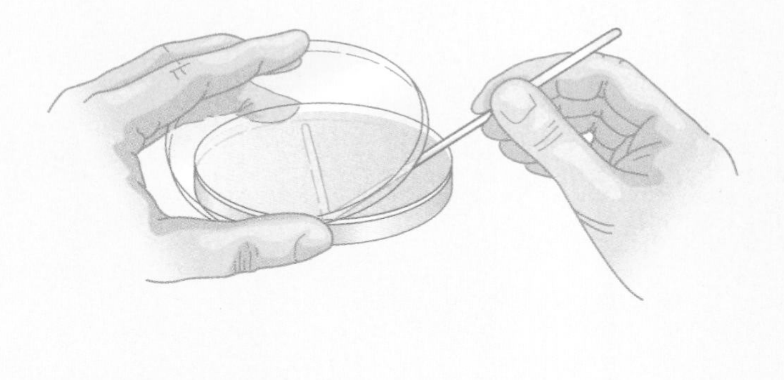

D Spread the inoculum over the agar surface by rotating the plate.

FIGURE 69.2 The Spread-Plate Technique.

Name: ______________________ Date: __________ Section: __________

Bacterial Conjugation

EXERCISE RESULTS 69

RESULTS

BACTERIAL CONJUGATION

TABLE 69.1 Conjugation in Bacteria

	Minimal Medium Plus Streptomycin and Thiamine Agar Plate		
	"Hfr"	"F^-"	"Hfr × F^-"
Colonies present (+) or absent (–)			

Observations and Conclusions

____

____

____

____

____

Questions

1. Explain why you would not expect any growth on the agar plates containing only donor or recipient cells.

____

____

____

____

2. What specific genes were transferred from donor to recipient cells?

3. Assess the importance of streptomycin as a marker in the donor and recipient cells.

4. Which plate or plates represented the controls in the conjugation experiment?

5. Describe how conjugation can be a mechanism for genetic variability in bacteria.

Bacterial Transformation

EXERCISE 70

Materials and Equipment

Note: The cultures, media, reagent, plasmid, and some of the equipment presented here are available in the pGLO Bacterial Transformation Kit (1660003EDU) available from Bio-Rad Laboratories, Hercules, California 94547 (www.bio-rad.com).

- 18- to 24-hour Luria-Bertani (LB) agar plate cultures of *Escherichia coli* HB101 K12
- LB agar plates.
- LB agar plates plus ampicillin (amp)
- LB agar plates plus ampicillin and arabinose (ara)
- LB broth tubes
- 50 mM cold $CaCl_2$ solution.
- pGLO plasmid DNA stock (0.03 μg/ml)
- Sterile plastic 13 × 100 mm test tubes or plastic 1.5-ml centrifuge tubes
- Adjustable micropipette (0.5 to 100 μl)
- Sterile plastic micropipette tips (10 to 100 μl) (or 1.0-ml graduated, individually wrapped, disposable plastic transfer pipettes)
- Disposable plastic inoculating loops (standard wire loops may be used)
- Waterbath at 42°C
- Beakers of crushed ice
- Bent glass rods
- Beakers of alcohol
- Permanent black markers
- UV lamp (hand-held black light)

Transformation is a process whereby small fragments of DNA or small plasmids are taken up by a **competent recipient cell** (a cell able to take up DNA from the environment) and are incorporated into the recipient cell's cytoplasm or genome. The genome of the recipient cell has now been modified to contain DNA with genetic characteristics of the donor cell. Naturally occurring transformations are of great interest medically because they may serve as a vehicle for genetic exchange among pathogenic organisms. Although the frequency is less than 1 percent, many pathogenic bacteria, including *Streptococcus pneumoniae, Neisseria gonorrhoeae*, and *Haemophilus influenzae* can undergo natural transformation. This suggests that the transfer of genetic material to a nonpathogenic recipient cell could make that cell pathogenic or more virulent and perhaps acquire the ability to evade a host's immune system defenses.

Because all bacteria are not naturally transformable, scientists and genetic engineers have developed laboratory methods to produce competency in those bacteria that do not undergo natural transformation. Competency can be generated by treating recipient cells with a cold calcium chloride ($CaCl_2$) solution. This permits the transfer of donor DNA across the cell membrane, allowing DNA to pass through the membrane and into the cytoplasm of the recipient cell.

With our rapidly advancing knowledge in the field of molecular genetics, it is now possible to artificially induce transformation by using plasmids as the donor DNA. **Plasmids** are small, circular pieces of extrachromosomal DNA capable of autonomous replication in the bacterial cytoplasm. The uptake of these exogenous DNA molecules may give new properties to the recipient bacterial cell, such as resistance to certain antibiotics or the ability to synthesize an amino acid.

BACTERIAL TRANSFORMATION

PURPOSE: to transform a competent ampicillin-susceptible strain of *Escherichia coli* into one that is ampicillin resistant by means of a DNA plasmid and to visualize transformed cells using a color marker gene carried in the plasmid.

In the following experiment, the ability to transform *E. coli* K12 will be examined using the plasmid pGLO (FIGURE 70.1). The plasmid contains:

- A gene encoding an enzyme that provides resistance to the antibiotic ampicillin (amp-r). This gene is a **selection marker**, meaning only transformed bacteria that express the *amp-r* gene can grow in media containing ampicillin. Untransformed *E. coli* are sensitive to ampicillin (amp-s) and will not grow in the presence of this antibiotic.

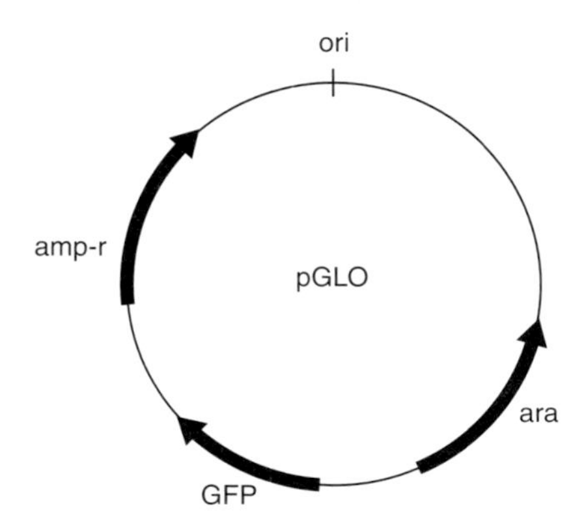

FIGURE 70.1 pGLO Plasmid.

- The gene for **green fluorescent protein (GFP)**. GFP is a protein normally found in the jellyfish *Aequorea victoria*. This protein is luminescent and allows the jellyfish to glow green in the dark. In the absence of the sugar arabinose (ara), transcription of the *GFP* gene is turned off. Only when ara is present does transcription of *GFP* occur. Thus, transformed colonies containing bacteria expressing this protein will glow a green color when observed under ultraviolet (UV) light.

PROCEDURE

I. Producing Competent Cells

1. Obtain two microcentrifuge tubes. With a permanent marker, label one tube "+PLA" (for plasmid) and another tube "−PLA" (FIGURE 70.2).
   - Label both tubes with your name.
2. Open the tubes and using a sterile transfer pipette, transfer 250 µl (0.25 ml) of cold $CaCl_2$ into each tube.
3. Place the tubes in a beaker of ice.
4. Obtain an agar plate culture of *E. coli* K12.
5. With a plastic disposable inoculating loop (or a flame-sterilized metal loop):
   a. Pick up one single bacterial colony from the agar plate and immerse the loop to the bottom of the +PLA tube.
   b. Twirl the loop to disperse the cells in the solution. There should be no visible floating chunks of bacteria.
   c. Place the tube back in the rack in the ice.

Do not eat the ice used in the laboratory.

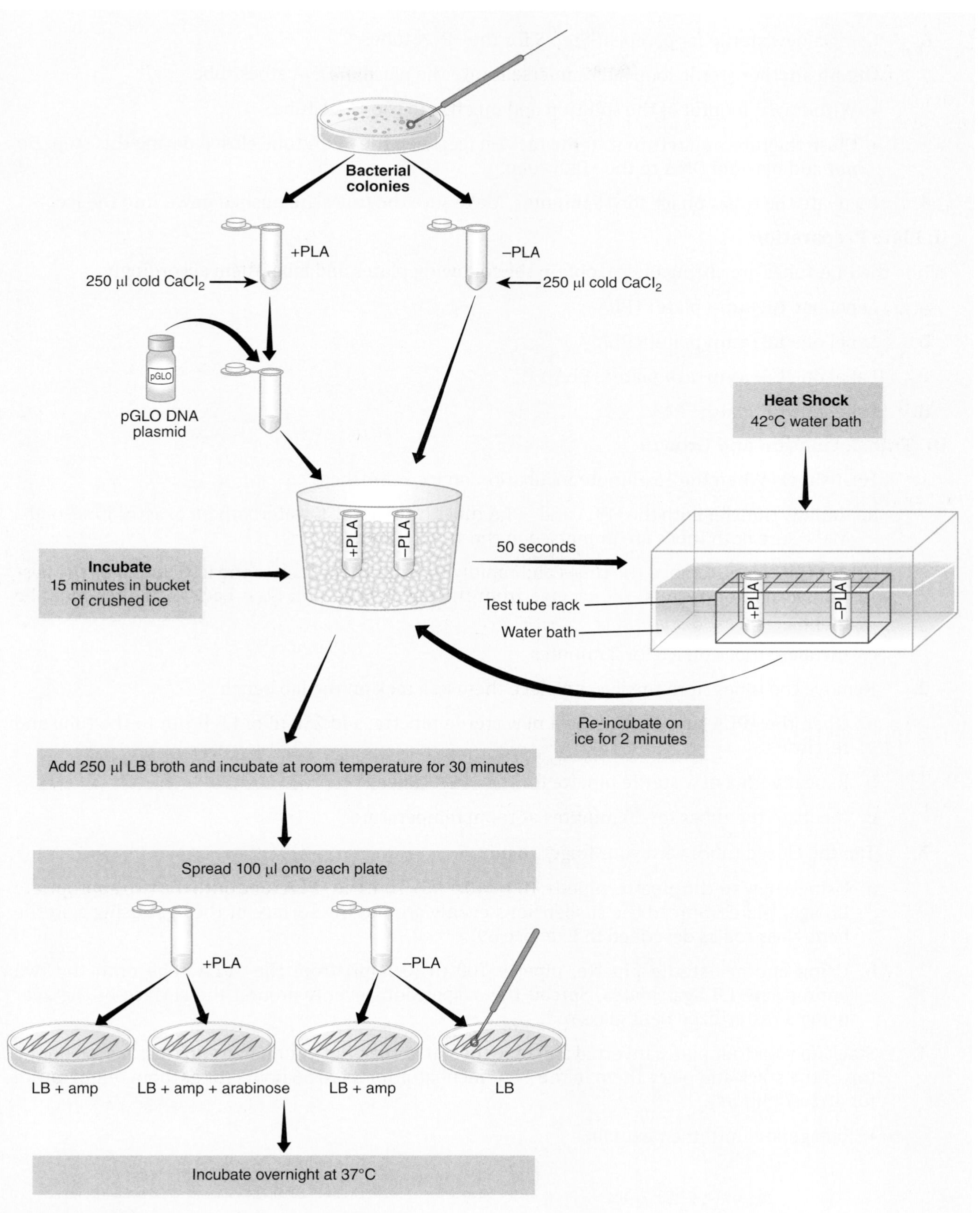

FIGURE 70.2 The Bacterial Transformation Procedure.

6. Using a new sterile loop, repeat Step 5 for the −PLA tube.

7. Obtain another sterile loop and immerse it into the plasmid DNA stock tube.
   - Withdraw a loopful of the solution and mix it into the +PLA tube.
   - Close the tube and return it to the rack on ice. Keep the −PLA tube closed during this step. ***Do not*** add plasmid DNA to the −PLA tube.

8. Incubate the tubes on ice for 15 minutes. Make sure the tubes are pushed down into the ice.

II. Plate Preparation

While the PLA tubes are sitting on ice, obtain the following plates and label them accordingly:

a. Label one **LB+amp** plate: +PLA

b. Label one **LB+amp** plate: −PLA

c. Label one **LB+amp+ara** plate: +PLA

d. Label one **LB** plate: −PLA

III. Transformation and Growth

1. Heat shock. When the 15-minute incubation on ice is finished:
   a. Rapidly transfer both the +PLA and −PLA tubes into the 42°C waterbath for exactly 50 seconds. Make sure both tubes are immersed in the warm water.
   b. At 50 seconds, remove the tubes and rapidly transfer them back to the ice. Note: For the best transformation results, the change from the ice to 42°C and then back to the ice must be rapid.
   c. Incubate tubes on ice for 2 minutes.

2. Remove the tubes from the ice and place them in a rack on the lab bench.
   a. Open the +PLA tube and, using a new sterile pipette, add 250 µl of LB broth to the tube and re-close it.
   b. Repeat with a new sterile pipette for the −PLA tube.
   c. Incubate the tubes for 30 minutes at room temperature.

3. Tap the closed tubes with your finger to mix.
   a. Using a new sterile pipette, pipette 100 µl (0.1 ml) from the +PLA tube onto the two appropriate LB agar plates. Spread the suspensions evenly around the surface of the agar using a sterile bent glass rod as described in Exercise 69.
   b. Using another sterile pipette, pipette 100 µl (0.1 ml) from the −PLA tube onto the two appropriate LB agar plates. Spread the suspensions evenly around the surface of the agar using a resterilized bent glass rod.

4. Stack up your four plates inverted and tape them together. Put your name and class period on the top of the stack and place them in the 37°C incubator for 24 to 36 hours (or at room temperature for 48 to 72 hours).
   - Refrigerate until the next lab.

IV. Observation

1. Before examining your plates, answer the pre-examination questions in the Results section.
2. Place all four LB agar plates upside down (lids down) on the lab bench.
   - Count the number of colonies on each plate and record your results in TABLE 70.1 in the Results section. Plates with more than 300 colonies should be recorded as too numerous to count (TNTC). For each plate, did transformation occur?
   - Record your results in Table 70.1.

When using a black light, do not shine the light into your eyes or those of your neighbors.

3. Using a UV-light source (hand-held black light), shine the light on each plate. On which plate(s) did you see green fluorescent bacterial colonies? Record your results in Table 70.1.

Name: ______________________ Date: __________ Section: __________

Bacterial Transformation

EXERCISE RESULTS 70

RESULTS

BACTERIAL TRANSFORMATION

Pre-Examination Questions

1. On which plate(s) are the bacterial colonies (cells) most similar to those on the original *E. coli* K12 plate used in Step I.4 of the procedure?

____

____

____

____

2. Which plate(s) should be compared to determine if:

   a. Transformation occurred?

   ____

   b. GFP is being expressed?

   ____

3. Which plate represents:

   a. The positive control?

   ____

   b. The negative control?

   ____

TABLE 70.1 Transformation in Bacteria

	+PLA		−PLA	
	LB+amp	LB+amp+ara	LB+amp	LB
Number of colonies				
Fluorescent colonies (yes/no)				

Observations and Conclusions

____

____

____

____

____

Questions

1. What is the purpose of the −PLA tube?

   ____

   ____

   ____

2. Why must the *E. coli* cells be subjected to ice followed by heat?

   ____

   ____

   ____

3. What does the term "cell competency" mean in relation to transformation?

   ____

   ____

   ____

4. Do the colonies you observe on the LB+amp plate:

   a. Represent transformed cells? Explain.

   ____

   ____

   ____

   b. Represent cells expressing the GFP protein? Explain.

   ____

   ____

   ____

5. Suppose you find a few scattered colonies on the LB+amp+ara plate that are not fluorescent.

   a. How would you explain this result?

   ____

   ____

   ____

   ____

   b. How might examination of the −PLA LB+amp agar plate support your hypothesis?

   ____

   ____

   ____

Immunology

PART XI

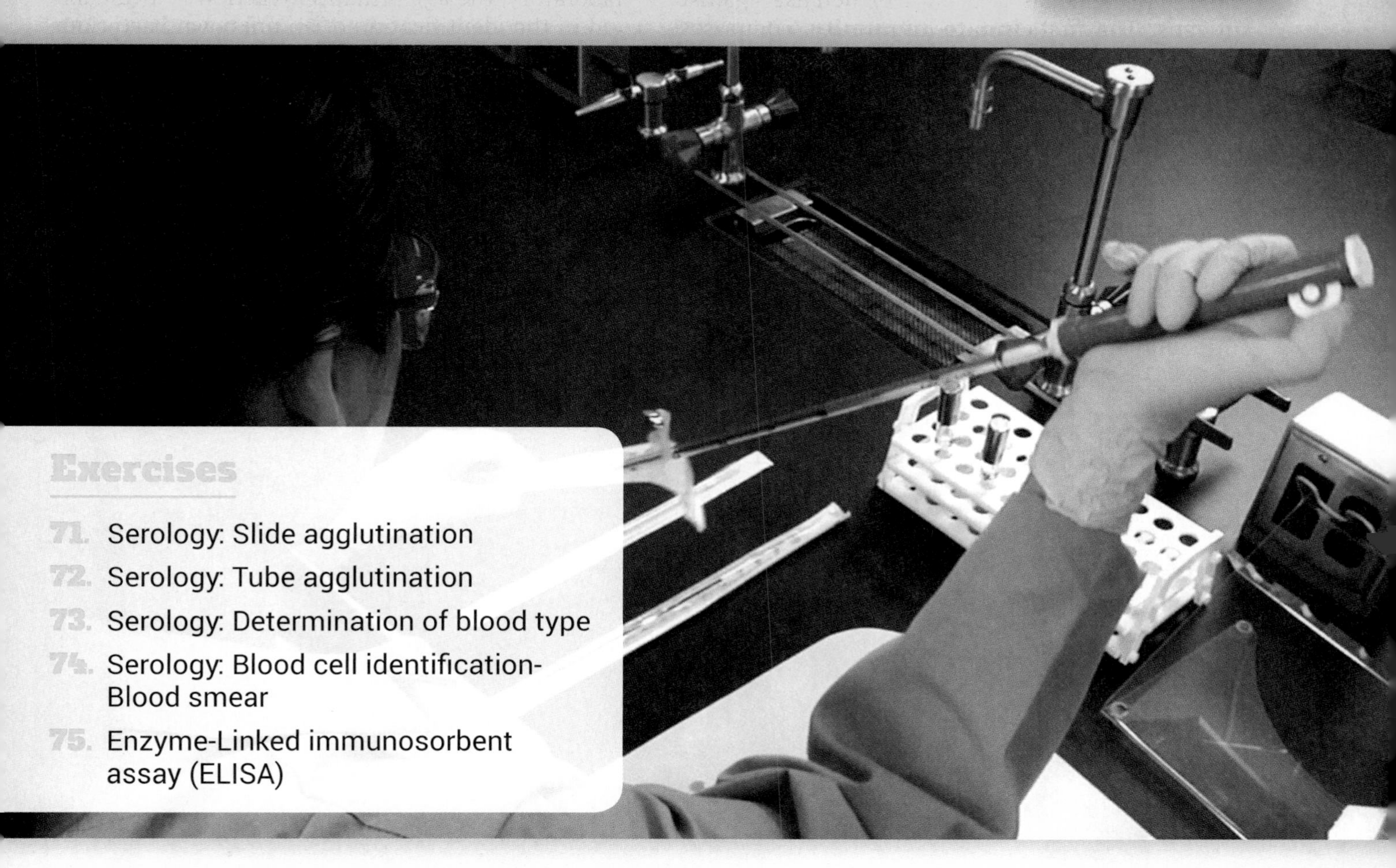

Exercises

71. Serology: Slide agglutination
72. Serology: Tube agglutination
73. Serology: Determination of blood type
74. Serology: Blood cell identification-Blood smear
75. Enzyme-Linked immunosorbent assay (ELISA)

Learning Objectives

When you have completed the exercises in Part XI, you will be able to:

- Identify bacteria by the slide agglutination technique.
- Determine an antibody titer from a serum sample.
- Carry out a hemagglutination assay to determine blood type.
- Prepare a blood smear and perform a differential white blood cell count.
- Detect and measure antibodies in a serum sample.

Immunology is the field of science that focuses on the immune system of multicellular organisms. Our (human) immune system is designed to fight off foreign substances. Those substances that provoke the immune system are known as **antigens**. The immune system has two lines of defense against antigens. The first, **innate immunity**, comprises broad defense mechanisms that are nonspecific, such as skin, fever, and natural killer cells. If infection is not rapidly contained and eliminated by innate defenses, elements of innate immunity alert and activate—**adaptive immunity**. The adaptive immune system develops a response that is specific to the invading pathogen.

Serology is a branch of immunology that deals with techniques to identify and measure antigens, and to detect serum antibodies (how many antibodies are present in the blood). **Agglutination** is a serological reaction in which antibodies react with antigens on the surface of particles and cause the particles to clump together, or agglutinate. This reaction generally results in a visible mass that can be seen with the unaided eye. In the diagnostic laboratory, the agglutination reaction is a valuable aid in the identification of an unknown bacterium, because the unknown can be pinpointed by its agglutination with known antibodies.

In this section, you practice different agglutination techniques (Exercises 71 and 72), determine the blood type of a commercially available blood substitute (Exercise 73), and observe the red and white blood cells present in a drop of blood (Exercise 74). In addition, you will have the opportunity to perform an important test to detect and measure antibodies: the enzyme-linked immunosorbent assay, or ELISA test (Exercise 75).

Serology: Slide Agglutination

EXERCISE 71

Materials and Equipment

- Inoculating loop
- Polyvalent *Salmonella* antiserum in dropper bottles (Thermo Fisher Scientific)
- Unknown broth cultures of various bacteria
- Wooden applicator sticks or toothpicks
- Beakers of disinfectant
- Saline solution in dropper bottles
- Nutrient broth culture of *Salmonella*
- Tube of nutrient broth

Although serology has become highly automated, this exercise illustrates how agglutination reactions can be studied without sophisticated or specialized equipment. In this exercise, the slide agglutination technique will be performed to learn the identity of an unknown bacterium.

SLIDE AGGLUTINATION

PURPOSE: to identify a bacterial species using a polyvalent antiserum.

The **slide agglutination technique** is used to determine which of several bacteria agglutinates a preparation of selected antibodies. Emulsions of unknown bacteria are mixed with drops of known antibodies on a microscope slide, and the mixture is observed for agglutination. This process is essentially a trial-and-error method because different combinations of bacteria and antibodies are tried until agglutination is observed. In this exercise, *Salmonella* antibodies will be combined with various bacteria to determine which, if any, is a *Salmonella* serotype.

PROCEDURE

1. Obtain a clean glass slide and, using a wax pencil, draw a circle for each of the unknown broth cultures available.
   - Aseptically deposit two loopfuls of each broth culture into their respective circles.
2. Add one drop of the *Salmonella* antiserum (a solution of *Salmonella* antibodies) to each circle.
   - Mix the bacteria and antiserum well, using a wooden applicator stick or toothpick.
   - Use different sticks for each sample, and place the sticks in the beaker of disinfectant after use.

3. As a **positive control**, take a clean glass slide, draw a circle, and mix a drop of *Salmonella* antiserum with two loopfuls of a known broth culture of *Salmonella*. An agglutination reaction should be evident.
    - As a **negative control** (to show the absence of agglutination), mix a drop of saline solution with two loopfuls of the known *Salmonella* culture in a second circle.
    - Also as a negative control, mix a drop of antiserum with two loopfuls of sterile, bacteria-free nutrient broth in a third circle.
4. Observe the various unknowns and determine which agglutinated with the *Salmonella* antiserum by comparing the mixtures to the control reactions.
    - Enter your results in TABLE 71.1 of the Results section, using (+) to indicate agglutination and (-) for nonagglutination.
    - Discard the slide in the beaker of disinfectant after use.
    - From your data, determine the identity of the unknown organism. If this is not possible, indicate why.

Name: ______________________ Date: __________ Section: __________

Serology: Slide Agglutination

EXERCISE RESULTS 71

RESULTS

SLIDE AGGLUTINATION

TABLE 71.1	Determination of Unknown Organism			
Organism	1	2	3	4
Agglutination				

Identity of unknown organism:

____

____

Observations and Conclusions

____

____

____

____

____

Questions

1. Explain the importance of positive and negative control preparations in the slide agglutination procedure.

____

____

____

____

____

Serology: Tube Agglutination

EXERCISE
72

Materials and Equipment

- Inoculating loop
- Polyvalent *Salmonella* antiserum (Thermo Fisher Scientific)
- *Salmonella* antigen solution
- Saline solution (0.85% NaCl)
- Serological tubes with racks
- 1-ml pipettes
- 5-ml pipettes
- Mechanical pipetters
- Beakers of disinfectant

In the **tube agglutination technique**, the object is to use the agglutination reaction to determine the titer of antibodies in a serum sample. The **titer** is a measure of the level of antibodies in a blood sample. It is determined by finding the most dilute concentration of antibodies that gives a detectable reaction with an antigen. Various dilutions of antibodies are prepared, after which a standard amount of antigen is added. Following an incubation period, the titer is determined.

TUBE AGGLUTINATION

PURPOSE: to determine an antibody titer in a serum sample.

This exercise will demonstrate the tube agglutination technique and the method for ascertaining the titer of antibodies.

PROCEDURE

1. Select ten clean serological tubes, number them, and set them up in a rack on the desk.
   - Obtain samples of coded *Salmonella* antiserum, *Salmonella* antigen, and saline solution with which to work.
2. Using a mechanical pipetter and a sterile 5-ml pipette, place 0.9 ml of saline solution in tube #1, and 0.5 ml of saline solution in tubes #2 through #10, as indicated in FIGURE 72.1.
3. With the mechanical pipetter and a sterile 1-ml pipette, transfer 0.1 ml of the antiserum solution from the stock supply to tube #1. This yields a 1:10 dilution of antiserum.
   - Mix the contents by drawing them up and expelling them from the pipette, and then transfer 0.5 ml from tube #1 to the saline in tube #2. This forms a 1:20 dilution, as shown in Figure 72.1.
   - Mix the contents of tube #2, and transfer 0.5 ml to tube #3, as shown in the figure.

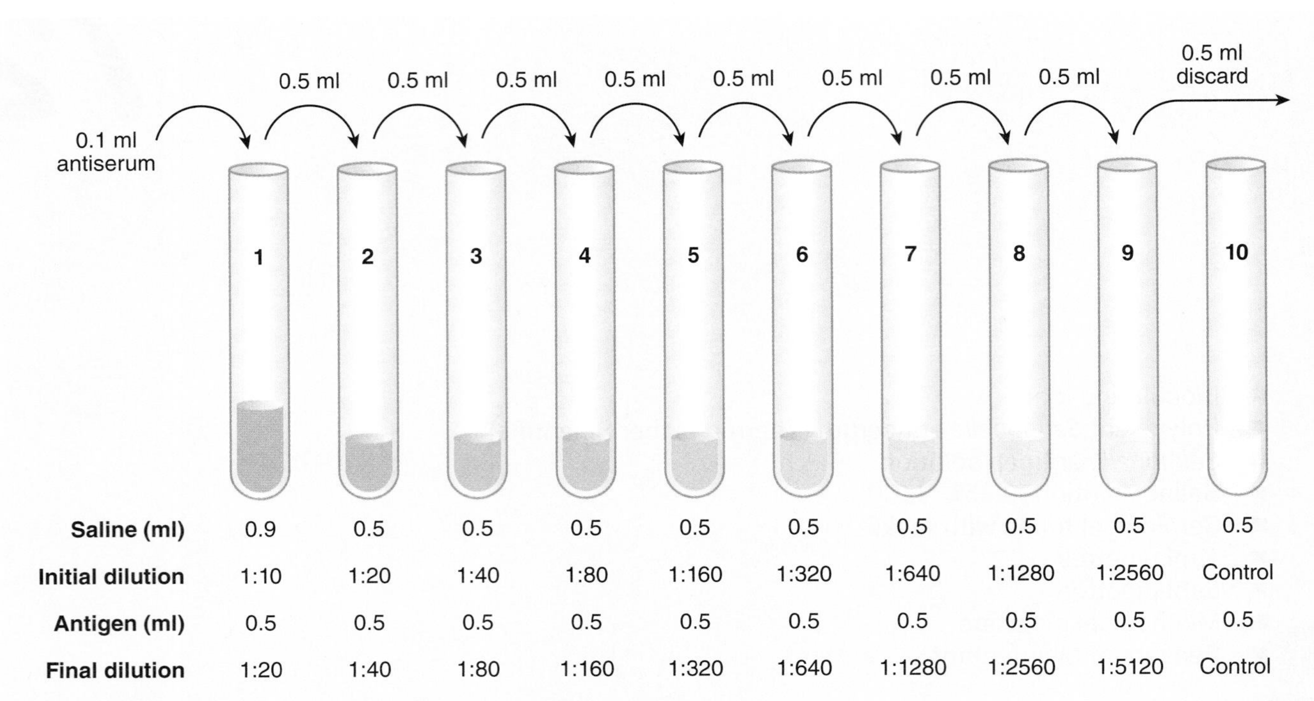

FIGURE 72.1 Preparation of a dilution series for the tube agglutination test.

- Mix and transfer 0.5 ml to tube #4, and continue this process on through to tube #9.
- Remove 0.5 ml from tube #9, and discard it into a beaker of disinfectant.
- Tube #10 will be a control tube and will not receive antiserum.
- Discard the pipette as directed by the instructor.

4. Using a mechanical pipetter and a sterile 5-ml pipette, transfer 0.5 ml of *Salmonella* antigen solution to each tube. This doubles the dilutions to the final dilutions shown in Figure 72.1.
   - Mix the contents, and set the rack aside to incubate at 37°C.
   - A water bath is helpful for the incubation step.
5. After an hour (or longer at the instructor's suggestion), observe the tubes for evidence of agglutination by tapping the bottom of the tube to suspend the sedimented material.
   - Large masses should be observed in the lower dilutions (e.g., tubes #1 and #2), and the amount of agglutinated material should decrease as the dilution becomes higher.
   - Eventually there should be a tube that shows the last evidence of agglutination. The dilution of this tube is the titer of antibodies.
   - Note that tube #10, the control tube, should show no evidence of agglutination since it does not contain anti-serum.
   - Overnight refrigeration will enhance the results.
6. Enter your results in TABLE 72.1 of the Results section, using (+++) to indicate maximum agglutination, (++) to indicate moderate agglutination, (+) to indicate minimum agglutination, and (-) to indicate no agglutination.
   - Determine the titer of antibodies in the serum sample by noting the dilution of the tube where the last evidence of agglutination appeared.
   - Write your answer in the Results section.

Name: ______________________ Date: __________ Section: __________

Serology: Tube Agglutination

EXERCISE RESULTS 72

RESULTS

TUBE AGGLUTINATION

TABLE 72.1 Determination of Titer of Antibodies

Tube number	1	2	3	4	5	6	7	8	9	10
Dilution	1: 20	1: 40	1: 80	1: 160	1: 320	1: 640	1: 1280	1: 2560	1: 5120	Control
Agglutination										

Titer of antibodies:

____

Serum sample code:

____

Observations and Conclusions

____

____

____

____

____

Questions

1. Why is 0.5 ml of material removed and discarded from tube #9 in the tube agglutination procedure?

2. What is meant by titer, and why may the titer be a significant factor in the diagnosis of disease?

Serology: Determination of Blood Type

EXERCISE

73

Materials and Equipment

- Inoculating loop and/or inoculating needle
- 70% ethyl alcohol and gauze or cotton pads
- Anti-A and anti-B typing sera
- Anti-Rh typing serum
- Toothpicks or applicator sticks
- Vials containing 0.5 ml of 0.85% sterile saline solution
- Sterile disposable lancets

Hemagglutination is a type of agglutination that involves blood cells. In microbiology, there are many clinical uses for this reaction, including the diagnostic procedures for measles, mumps, and influenza. The hemagglutination reaction also is important in blood banking to determine the blood type of an individual prior to using the blood for transfusion; that is making sure the transfusion uses a compatible blood type.

HEMAGGLUTINATION: DETERMINATION OF BLOOD TYPE

PURPOSE: to determine ABO and Rh blood types.

In this exercise, the **blood type** will be determined by hemagglutination. The blood used in this exercise will not be taken from the student unless specified by the instructor. Instead, a human blood substitute will be used. This blood substitute is commercially available and consists of animal blood or microscopic beads bonded to blood antigens. It will give reactions identical to those of human blood.

Four major blood types are recognized. **Type A** individuals possess A antigens on the red blood cells, while **type B** persons have B antigens. Both A and B antigens occur in **type AB** individuals, and neither is found on the red blood cells of **type O** persons. The antigens may be detected by reactions of the blood with antisera containing A or B antibodies, as shown in this exercise.

Another antigen, the **Rh factor**, is found in 85% of Caucasians, who are said to be Rh-positive. It is absent in the remaining 15%, who are Rh-negative. In African-American populations, the percentages are 93% positive and 7% negative. This factor is an important consideration in hemolytic disease of the newborn. Its presence also may be determined by a hemagglutination reaction.

PROCEDURE

1. Wash and dry two glass slides.
   - Divide one slide into sections "A" and "B" by marking the underside of the slide with a wax pencil or felt pen. The underside is used to prevent wax or ink from interfering with the agglutination reaction.
2. Obtain the tube of blood to be used for this exercise.
   - The instructor will specify the source of the blood.
   - The type of this blood will be determined.
   - If the student's blood is to be used, the directions for securing the blood will be explained by the instructor.
3. Place a drop of blood in sections A and B on the first slide and a drop in the center of the second slide.
   - Also, place a small drop in the vial of saline solution.
4. Place a drop of anti-A serum next to the blood in section A, and a drop of anti-B serum next to the blood in section B. These sera contain A and B antibodies, respectively.
   - Place a drop of anti-Rh serum next to the blood on the second slide. Anti-Rh serum (Rh antibodies) detects the D antigen, which causes the Rh-positive condition.
5. Mix the drops of blood and respective antibodies with clean toothpicks or applicator sticks, being sure to use a fresh one for each mixing.
   - Agglutination with anti-A and anti-B sera is observed by the presence of large, grainy, dark clumps on the slide.
   - Agglutination with anti-Rh sera is determined by rocking the slide back and forth for approximately 2 minutes over very mild heat.
   - A warming tray or warming box at 45°C may be used. Small, finer clumps will be seen on close examination.
   - It should be pointed out that determination of the Rh factor by this method is questionable at best and that further testing is necessary for a definitive conclusion.
6. Record results in the Results section using the following:

Agglutination with anti-A serum	type A blood
Agglutination with anti-B serum	type B blood
Agglutination with both anti-A and anti-B sera	type AB blood
Agglutination with neither anti-A nor anti-B sera	type O blood
Agglutination with anti-Rh serum	Rh-positive
No agglutination with anti-Rh serum	Rh-negative

7. To verify your results, repeat the above process using blood from the saline vial.
   - The test should be performed only for A and B antigens, since detection of the Rh antigen will not work by this method.
   - Cover the blood–antiserum mixtures with coverslips, and observe the presence or absence of agglutination under the lower power (10×) objective of the microscope.
   - Clumps of red blood cells indicate that agglutination occurred with the antiserum.
   - Unclumped, free-floating cells reflect nonagglutination.

Name: ______________________ Date: __________ Section: __________

Serology: Determination of Blood Type

EXERCISE RESULTS 73

RESULTS

HEMAGGLUTINATION: DETERMINATION OF BLOOD TYPE

A	B

Rh

Blood type:

____

Observations and Conclusions

____

____

____

____

____

Questions

1. Consider the blood type you have determined in the laboratory. What type or types of blood could successfully be transfused to this individual under emergency conditions? Explain carefully.

Serology: Blood Cell Identification-Blood Smear

EXERCISE

74

Materials and Equipment

- 70% ethyl alcohol and gauze or cotton pads
- Sterile disposable lancets
- Wright's stain
- Dropper bottles of distilled water
- Buffer solution (Giordano) if available

A **blood smear** provides the opportunity to view red blood cells and various types of white blood cells that function in immune system defense. Noting which cells are present in unusual numbers may provide insight to the nature of the disease. For example, the number of abnormal lymphocytes is unusually high in patients with infectious mononucleosis, and the monocyte count is elevated in cases of listeriosis.

BLOOD SMEAR

PURPOSE: to identify blood cell types and perform a differential white blood cell count from a blood smear.

In this exercise, the different blood cells will be observed, and their percentage may be determined.

PROCEDURE

1. Clean two glass slides and dry them thoroughly. One slide will be used for the blood smear, the second for spreading the blood.
    - Obtain a large drop of blood as specified by the instructor, and place it at the end of one slide.

When taking your own blood, be certain that the lancet is taken from a sealed package and is sterile.

2. Spread the drop of blood across the face of the slide, using the second slide as a spreading slide as follows:
    - Place the spreading slide at a 45° angle, and bring it back into the drop as shown in FIGURE 74.1A.
    - Allow the blood to spread out to the slide's edge, and then drag the blood the length of the slide one time only, as shown in FIGURE 74.1B.

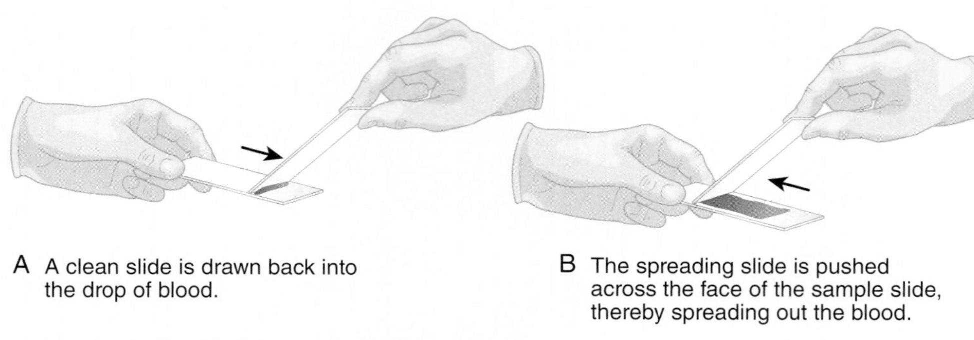

A A clean slide is drawn back into the drop of blood.

B The spreading slide is pushed across the face of the sample slide, thereby spreading out the blood.

FIGURE 74.1 Preparation of a blood smear.

- Lift the spreading slide at the end of the smear in order to "feather" the end.
- Repeat the process with a fresh drop of blood and a new slide if a successful smear is not obtained the first time.

3. Thoroughly air-dry the slide.
   - Cover the smear with **Wright's stain** while counting the number of drops added.
   - Permit the stain to remain for 2 minutes.
   - Without washing the slide, add an equal number of drops of distilled water or **buffer solution** to the slide, and mix the stain and water through gentle rocking.
   - Allow the mixture to remain an additional 2 minutes.
4. Gently wash the slide with water for 30 seconds and blot it dry.
   - Examine the slide under the oil immersion (100×) lens, beginning at the feathered end.
5. Note the presence of red blood cells, which should appear as pale orange disks if the residual stain has been washed free.
   - **Red blood cells** (**erythrocytes**) do not stain with Wright's stain, but if the dye has not been removed, the cells may appear pale to dark blue.
   - Try to locate an uncluttered area where the cells are standing alone and may be observed individually.
   - Enter a representation of several red blood cells in the Results section.
6. Note the presence of **white blood cells** (**leukocytes**), which should appear larger than red blood cells.

   Stained white blood cells will exhibit a blue nucleus in an unstained or pale blue cytoplasm.
   - **Neutrophils** are the most common white blood cells observed (FIGURE 74.2). These cells have dark blue nuclei divided into two, three, or more lobes.
   - **Lymphocytes** are the next most common white blood cell. These cells have large blue nuclei that take up almost the entire space of the cytoplasm and make the entire cell appear dark blue.
   - **Monocytes/macrophages** are relatively uncommon. They are large cells, each with a large nucleus characteristically indented on one side.

Name	Microscopic appearance	Approximate percentage of white blood cells	Description
Neutrophils		50–70	Most abundant WBCs; pale pink granules
Eosinophils		2–4	Red granules
Basophils		<1	Blue granules
Lymphocytes		20–30	Large oval or spherical nucleus; no granules
Monocytes/macrophages		5–7	Largest cell; irregular nucleus; no granules
Erythrocytes			Pale orange disks
Thrombocytes (platelets)			Fragment-like cells

FIGURE 74.2 Formed elements of stained blood smears.

- **Eosinophils** and **basophils** are the least common white blood cells. They appear as multilobed, nucleated cells with distinctive red and blue granules, respectively.
- Two types of **lymphocytes** are the **B lymphocytes** (**B cells**) and **T lymphocytes** (**T cells**), each of which has a large cell nucleus. They are important for producing antibodies (B cells) and for their cytotoxic properties (T cells).
- Small groups of cells also will be observed as clusters of fragment-like bodies. These are the **thrombocytes** (**platelets**) used in blood clotting.
- Enter representations of the types of white blood cells and the thrombocytes in the Results section.

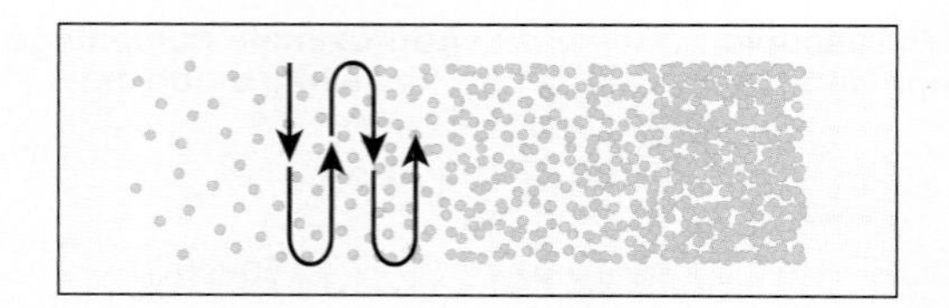

FIGURE 74.3 Counting procedure for a differential white blood cell count.

7. A **differential white blood cell count** may be obtained by counting 100 white blood cells and determining the percentage of each of the five types.
   - To perform this step, begin at the feathered end and move the smear vertically while recording each type of white blood cell as it is observed.
   - When the end of the smear is reached, move horizontally to the next plane (FIGURE 74.3) and once again move the smear vertically, continuing the tally.
   - Determine the percentage of each type of white blood cell, and enter it into TABLE 74.1 of the Results section.
   - The slide may be labeled and retained if necessary.

Name: ______________________ Date: __________ Section: __________

Serology: Blood Cell Identification-Blood Smear

EXERCISE RESULTS 74

RESULTS

BLOOD SMEAR

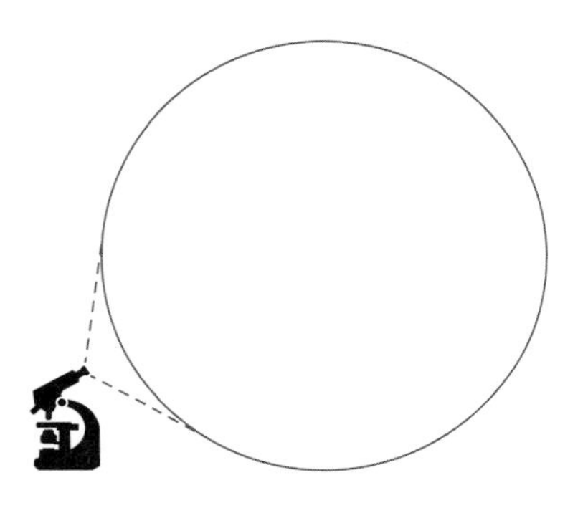

Magnif.: ______________

TABLE 74.1 Differential White Blood Cell Count					
Type of WBC	**Neutrophil**	**Basophil**	**Eosinophil**	**Lymphocyte**	**Monocyte**
Number in blood					
Percentage in blood	%	%	%	%	%

Observations and Conclusions

____

____

____

____

____

Questions

1. Name several factors that contribute to a successful blood smear and several that contribute to a poorly executed blood smear.

    ____

    ____

    ____

    ____

Enzyme-Linked Immunosorbent Assay (ELISA)

EXERCISE 75

Materials and Equipment

- Microwell plate
- Plastic tape
- Washing buffer
- Coating buffer
- Blocking buffer
- Micropipette
- Stop solution
- Alkaline phosphate substrate solution
- Antigen sample

The **enzyme-linked immunosorbent assay (ELISA)** is a procedure used frequently in laboratories to detect the presence or amount of an antigen. In this process, antigens are bound to an enzyme that is able to catalyze a reaction (FIGURE 75.1). The procedure involves looking for an interaction between an antibody and an antigen. The reaction can be agglutination, precipitation, or complement fixation. Known antibodies, labeled with enzymes to increase their sensitivity in detecting antigens, are used. A substrate is later added. A reaction is considered positive if there is a color change after addition of the substrate.

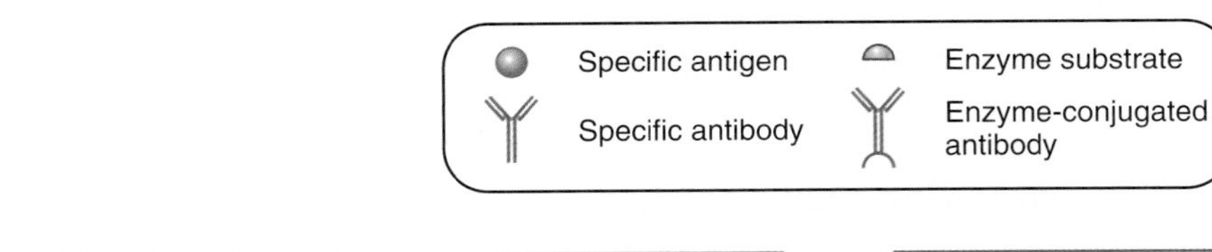

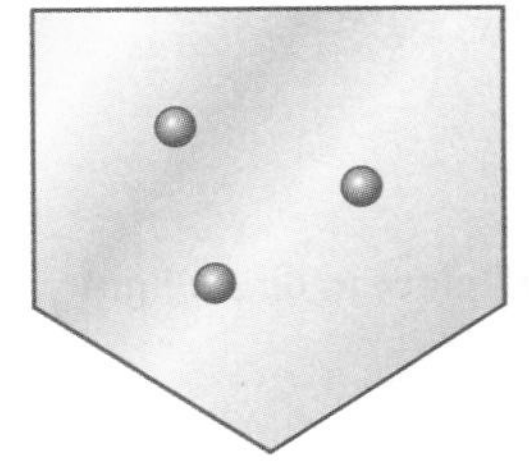

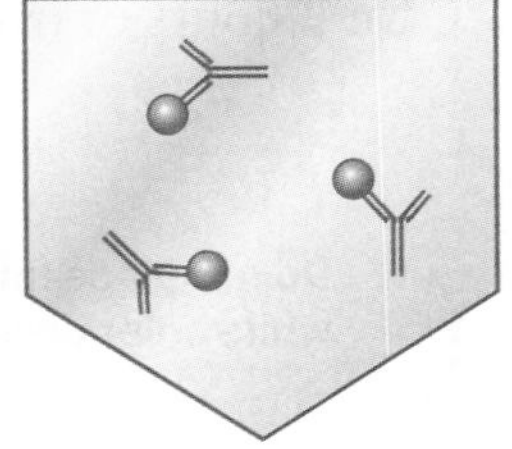

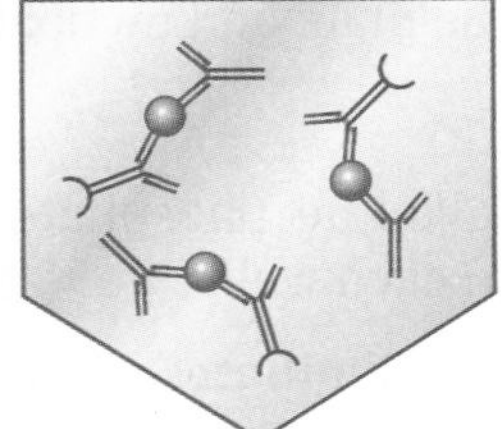

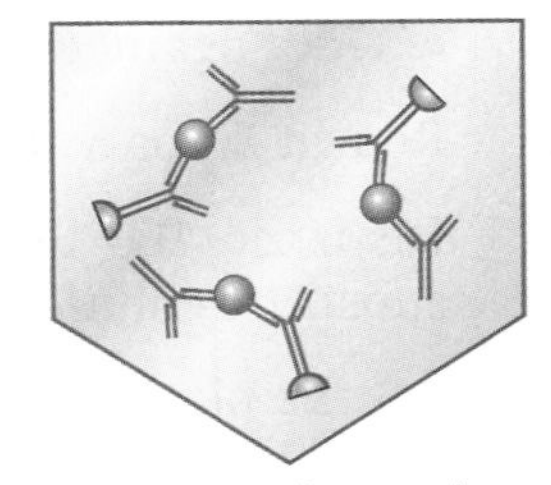

FIGURE 75.1 The enzyme-linked immunosorbent assay (ELISA) process.

An ELISA is used often in laboratories because it is a sensitive, economical procedure that is not labor intensive and does not require expensive instrumentation. Both indirect and direct ELISA techniques are useful to detect the presence of specific antibodies to a variety of infecting agents, such as hepatitis C, herpes simplex virus 1 and 2, HIV, varicella virus, *Chlamydia*, and rubella.

DIRECT ELISA

PURPOSE: to detect antigens in a sample using an ELISA procedure.

While indirect ELISA is considered more sensitive, direct ELISA is a quicker process. In this exercise, you will perform a direct ELISA.

PROCEDURE

1. Obtain a microwell and label the bottom with the date and your initials.
2. Using a sterile pipette, add 100 µL of the antigen sample to the first 12 wells.
3. With a new pipette, add 100 µL of bicarbonate/carbonate coating buffer to the same 12 wells (this immobilizes the antigen).
4. Seal the wells (plastic tape will work) and incubate overnight (24–48 hours) at 4°C.
5. After incubation, remove the tape and invert the plate over a sink to remove any liquid.
6. Fill each well with washing buffer, shake gently, and invert the plate over the sink to remove the liquid.
   - Repeat this process two more times.
   - Invert the plate over a paper towel to absorb any remaining liquid.
7. Add 100 µL of blocking buffer to each well. Again, cover the plate with plastic tape and incubate at 4°C for 2 hours or overnight.
8. After incubation, remove the tape and invert the plate over a sink to remove any liquid.
   - Repeat Step 6.
9. Add 100 µL of antibody diluted in blocking buffer to each well.
   - Once again, cover the plate and incubate at 4°C for 2 hours or overnight.
10. Repeat Steps 5 and 6.
11. Dispense 100 µL of alkaline phosphate substrate solution into each well.
12. Let the plate sit for 20–30 minutes while color develops.
13. Add 100 µL of stop solution to each well.
14. Record your observations in the Results section.

Do not place microwell plate in direct light while color develops.

Name: ______________________ Date: __________ Section: __________

Enzyme-Linked Immunosorbent Assay (ELISA)

EXERCISE RESULTS **75**

RESULTS

DIRECT ELISA

Observations and Conclusions

____

____

____

____

____

Questions

1. Which well had the most color? Why?

____

____

____

2. Why would the direct ELISA technique be used if the indirect technique has better sensitivity?

____

____

____

____

Public Health and Environmental Microbiology

PART XII

Exercises

76. A simulated epidemic: Microbial transmission via fomites

77. A microbial hunt

78. Transmission of microorganisms through toilet paper

79. Disinfection of drinking water

80. Microbiology of foods: Food preservation with salt and garlic

81. Microbiology of foods: Standard plate count

82. Microbiology of foods: Fermentation of wine and beer

83. Microbiology of milk and dairy products: Standard plate count of milk

84. Microbiology of milk and dairy products: Coliform plate count of milk

85. Microbiology of milk and dairy products: Methylene blue reduction test

86. Microbiology of milk and dairy products: Preparation of cheese

87. Microbiology of milk and dairy products: Preparation of yogurt

88. Microbiology of milk and dairy products: Natural bacterial content of milk

89. Microbiology of water: Presumptive test by the MPN method

90. Microbiology of water: Confirmed test

91. Microbiology of water: Completed test

92. Microbiology of water: Membrane filter technique

93. Microbiology of water: Preparation of a biofilm

94. Microbiology of soil: Isolation of *Rhizobium* from legume roots

95. Microbiology of soil: Ammonification of soil microorganisms

96. Microbiology of soil: Isolation of *Streptomyces* from soil

97. Microbiology of soil: Antibiotic production by *Streptomyces*

98. Microbiology of soil: Plate count of soil bacteria

99. Microbiology of soil: Microbial Ecology in the Soil—The Winogradsky Column

Learning Objectives

When you have completed the exercises in Part XII, you will be able to:

- Evaluate a simulated epidemic in terms of the nonliving source or the index patient, and explain the spread of the infectious agent.
- Assess the role of salt and other preservatives in food preservation.
- Carry out a standard plate count and coliform plate count of food and dairy products.
- Analyze the importance of the fermentation process to beverage products.
- Employ the methylene blue reduction test to determine the bacterial content of a milk sample.
- Demonstrate the role of microorganisms in such food processes as cheese and yogurt production.
- Explain the natural progression of bacterial populations under specific environmental conditions.
- Complete the series of tests (presumptive, confirmed, and completed) to detect coliform bacteria and to estimate their numbers.
- Carry out a membrane filtration procedure and using selective and differential media to determine coliform numbers in a water sample.
- Construct an apparatus to obtain and study a biofilm.
- Isolate and perform an examination of *Rhizobium* from legume roots.
- Employ the diagnostic test to detect bacterial ammonification.
- Isolate *Streptomyces* from soil and assess its capability to produce antibiotics.
- Carry out a plate count of soil bacteria.
- Construct and analyze the changing patterns of microbial succession in a Winogradsky column.

Outbreak! The word has become part of our modern lexicon of frightening words to describe a sudden and rapid spread of an infectious disease like Ebola or hepatitis A in a limited area. An **outbreak** represents a more limited form of an **epidemic**. Some outbreaks or epidemics (or even pandemics) may be the result of newly **emerging infectious diseases**, such as Zika virus, severe acute respiratory syndrome (SARS) or West Nile fever, the result of **reemerging infectious diseases**, such as malaria or tuberculosis, or from recurring infectious diseases like the flu.

Whatever the form, there are many ways a disease can be spread. It might be spread by aerosolized droplets resulting from a cough or sneeze. It may be transmitted via contaminated water or food. Many infectious diseases are carried and spread by **vectors**, which include insects like mosquitoes and fleas. Epidemics also may be spread by pathogens on **fomites**; that is, inanimate objects such as doorknobs, eating utensils, and bed linens. Even currency, including paper money and coins, can harbor and contribute to an epidemic. In fact, coins "contaminated" with yeast cells represent a good example to use in a simulated epidemic to illustrate how disease might spread in a population. Often in such epidemics, the nonliving source of the disease or the first individual (index patient) spreading the disease needs to be identified. All of these aspects can be examined through a simulated epidemic (Exercise 76).

Clearly, a priority in public health is addressing communicable disease. The field of public health grew out of the Industrial Revolution of the mid-1800s when large populations of people migrated to large cities, primarily in Europe and North America. These migrations brought new groups of people into contact with one another, altered diets, and exposed them to the effects of urbanization. Those effects included uncontrolled urban growth and indescribably grim conditions. Garbage and dead animals littered the streets; human feces and sewage stagnated in open sewers; rivers were used for washing, drinking, and excreting; and filth was rampant.

However, sanitary reformers and other budding public health advocates spoke up for effective public health measures. Working with bacteriologists, they pressed the government for the development of better methods for sewage treatment, water purification, and food preservation. Pioneers like John Snow and Hermann Biggs discovered the ability of infectious diseases to spread in numerous ways through a population, highlighting the need to continually maintain high levels of sanitation and strengthen public health measures to ensure potentially infectious agents are controlled or destroyed. In Part VIII of this lab manual, exercises examined the physical and chemical agents, and antibiotic treatments, that have been developed over the last century to limit the spread of infectious and communicable disease. The discovery of antibiotics resulted in huge gains in life expectancy in the industrialized world.

Long before the present era, people had used salt as a food preservation agent. The effectiveness of salt and the antimicrobial effects of other foods can be clearly demonstrated (Exercise 80). Food preservation also advanced by the establishment of **standard plate counts** to monitor contamination in foods and the addition of preservatives to other foods to prevent microbial growth (Exercise 81).

However, not all microorganisms are bad. For example, yeasts have been used for millennia in the fermentation of beer (Exercise 82). Microorganisms are also critical to the production of milk and other dairy products. Standard plate counts, coliform plate counts, and procedures such as the methylene blue reduction test are used to monitor the natural microorganisms in milk and any unnatural ones that should contaminate it (Exercises 83–85). Other microbes also are involved in the production of an assortment of foods, including cheeses and yogurt (Exercises 86–88).

The purity of water also is reflected in public health measures. **Coliform bacteria** can contaminate water and make it unfit for human consumption. A series of **water quality tests** can be performed to detect coliform bacteria and to estimate their numbers if present (Exercises 89–92). Aquatic environments are also home to many nonpathogenic microorganisms and these have recently become of great interest to microbiologists because these organisms usually exist as a **biofilm** (Exercise 93).

Plate counts demonstrate that the soil is almost bursting with bacteria (Exercise 98). Some play a dominant role in numerous **biogeochemical** cycles and forge links between what is useless and what is useful to other living organisms. For example, ***Rhizobium*** and other soil bacteria are essential for the conversion of inert nitrogen gas into ammonia and nitrates, useful products for plants and other organisms (Exercise 94). Other soil bacteria and fungi recycle carbon, while several genera of ***Streptomyces*** are involved in antibiotic production (Exercises 96–98). Importantly, the populations of bacteria in the environment can fluctuate over time as the environment changes. This can be elegantly demonstrated by constructing and observing a **Winogradsky column** (Exercise 98).

A Simulated Epidemic: Microbial Transmission via Fomites

EXERCISE 76

Materials and Equipment

- Several washed coins
- Suspension of yeast cells (commercial yeast such as "active dry" yeast or yeast cake, or from a broth culture of *Saccharomyces cerevisiae*)
- Sterile cotton swabs and forceps
- Sterile saline for moistening swabs
- Tubes of hydrogen peroxide (H_2O_2)

Epidemic diseases (or **epidemics**) break out suddenly within a population. Often such epidemics are caused by microorganisms transmitted by lifeless objects known as **fomites**. It often happens that several people handling the same fomite come in contact with the same organism, or one person transmits a disease organism to several others via a fomite. For example, if a dollar bill were contaminated with staphylococci (such as from an abscess), the staphylococci could be transmitted to other people handling that dollar bill. An open wound such as an abrasion or pimple could then provide the staphylococci with an entryway to deeper tissues.

To locate the responsible fomite, public health officials commonly take swab samples from various objects and environmental surfaces and cultivate any microorganisms collected on the swabs. Usually they wait for the microorganisms to grow and form colonies (and hope they have used the correct growth media). In some cases, however, a marker enzyme is available. In this exercise, the enzyme **catalase** is used as a marker. Catalase breaks down hydrogen peroxide (H_2O_2) into water and oxygen, and causes the solution to bubble as the oxygen gas is released (see Exercise 58). Catalase is produced by a wide variety of microorganisms, including the harmless yeast cells used in brewing and baking (Exercise 22). Thus, the enzyme can be used to track yeast cells and demonstrate how a true pathogen could move through a set of individuals.

MICROBIAL TRANSMISSION VIA FOMITES

PURPOSE: to simulate the spread of a disease-causing organism by fomites and to identify the index patient.

Fomites are a widely known mechanism for transmitting microorganisms during epidemics. Items like bed linens, for example, can transmit intestinal microorganisms such as salmonellae and amoebic cysts, and contaminated syringes may be the source of bloodborne viruses, including hepatitis B and C. In this exercise, we shall demonstrate how yeast cells are transmitted among various individuals by contaminated coins, and we shall use catalase as the experimental marker for the yeast cells' presence.

PROCEDURE

I. Simulation of Disease Spread

1. The fomites used in this exercise will be a set of coins.
    - To prepare for the exercise, wash several coins with soap and water until they are clean; dry the coins with paper toweling before using them.
2. To demonstrate that a contaminated fomite can transmit microorganisms to a series of individuals, proceed as follows:
    - Using forceps, dip a coin into a heavy suspension of yeast cells (*Saccharomyces cerevisiae*).
    - Place the coin on a clean paper towel to dry. (Drying may be hastened by placing the paper toweling on a warming tray.)
    - Now, one student should take the contaminated coin and rub it between the thumb and forefinger to simulate handling the coin.
    - The coin should then be passed to a series of fellow students, who will also handle the coin before passing it on.
    - After the coin has been passed to a designated number of individuals, it should be returned to the instructor.

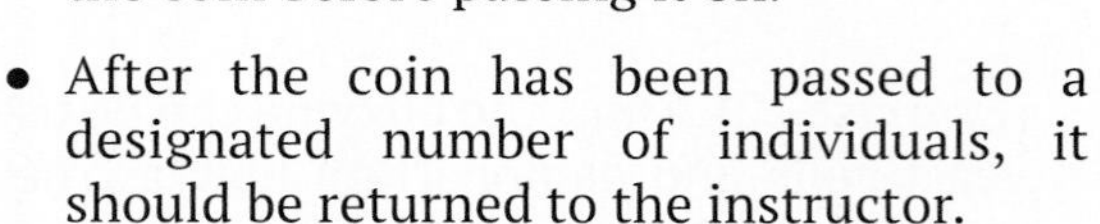

The yeast culture used in this exercise should be freshly made. If it has remained for too many days, it may have become contaminated with bacteria and be potentially dangerous.

3. Having simulated an epidemic, the object will be to verify which individuals have been contaminated.
    - To do this, obtain a sterile cotton swab and tubes of hydrogen peroxide and sterile saline.
    - Moisten the swab with sterile saline and swab the fingertips of a person who handled the coin.
    - Then drop the swab into a tube of hydrogen peroxide.
4. Observe the hydrogen peroxide for signs of oxygen release, signified by bubbling.
    - If yeast cells were present on the subject's fingertips, they will release catalase, which breaks down H_2O_2 to water and oxygen gas (**FIGURE 76.1**).
    - The bubbling will be vigorous if many yeast cells are recovered and mild if there are few yeast cells. Thus, you may estimate the relative concentration of microorganisms "infecting" the subject.
    - A negative control tube should be prepared using a swab sample taken from a known uncontaminated person.
    - A positive control tube should contain a swab that has been dipped into the yeast suspension.
    - Note your observations in the Results section, and draw your conclusions on the importance of the fomite in microbial transmission.

II. Identification of the Index Patient

1. It often happens that one individual is the index patient in an epidemic. The **index patient** is the first person to transmit microorganisms to other individuals.

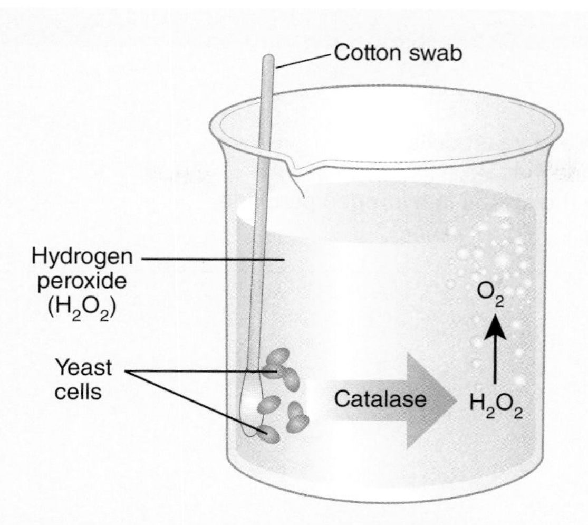

FIGURE 76.1 Yeast cells produce catalase, which reacts with hydrogen peroxide (H_2O_2) to produce oxygen (O_2) gas and water (H_2O).

- To investigate this principle, proceed as follows:
    - One person should dip the thumb and forefinger into the yeast suspension, then pick up a washed, uncontaminated coin.
    - After rubbing the coin, the person should pass the coin to a series of individuals who will become "infected" by handling the coin.
    - The coin should be returned after the transfers have been completed.
- Now the catalase tests can be performed (Section I, Steps 3 and 4).
- Determine which individuals acquired the yeast cells from the fomite and were thus "infected" by the index patient.
- Record your conclusions in the Results section.

III. Other Variations

1. Many other variations to the above procedures may be employed to study the principle of fomite transmission.
    - One variation is to use several different coins (e.g., a penny, a nickel, and a dime) and contaminate only certain ones.
    - An epidemiological study may then be instituted to determine which persons became infected by handling which coin.
2. Another variation of the procedure is to test the effectiveness of handwashing to interrupt the epidemic.
    - To perform this exercise, several of the infected individuals should wash their hands after handling the "infected" coins.
    - Swab samples are then taken, and a study is conducted to determine whether washing reduced or eliminated the microbial population.

Quick Procedure

Fomite Transmission

1. Contaminate a coin with harmless yeast cells.
2. Pass the coin among several individuals.
3. Swab the fingertips and place the swabs in hydrogen peroxide.
4. Observe the tube for bubbling.

Name: ______________________ Date: __________ Section: __________

A Simulated Epidemic: Microbial Transmission via Fomites

EXERCISE RESULTS 76

RESULTS

MICROBIAL TRANSMISSION BY FOMITES

I. Simulation of Disease Spread

Observations and Conclusions

____

____

____

____

____

II. Identification of Index Patient

Observations and Conclusions

____

____

____

____

____

Questions

1. What characteristics of yeasts make this organism suitable for use in an exercise such as this?

2. Name several microbial diseases whose agents are transmitted by fomites, and specify which fomites might be involved.

A Microbial Hunt

EXERCISE 77

Materials and Equipment

- Suspension of yeast cells (commercial yeast such as "active dry" yeast or yeast cake, or from a broth culture of *Saccharomyces cerevisiae*)
- Sterile cotton swabs and forceps
- Sterile saline for moistening swabs
- Tubes of hydrogen peroxide (H_2O_2)

During the course of an epidemiological investigation, public health officials often are called on to sample various surfaces or objects to determine whether they were the source of a disease outbreak. For example, recovering microorganisms from eating utensils may indicate unsanitary conditions exist and persons eating with that batch of utensils are at risk for contracting intestinal disease.

A MICROBIAL HUNT

PURPOSE: to identify the source of a disease outbreak or epidemic.

One method for testing objects and surfaces is to use moistened sterile cotton swabs to recover microorganisms from a surface, then cultivate the microorganisms under laboratory conditions. Another method is to use a distinctive marker enzyme to identify the microorganism. In this exercise, tests will be conducted to locate the source of microorganisms responsible for the "epidemic" that is currently taking place. As in Exercise 76, the microorganism involved will be yeast cells, and the marker enzyme will be catalase.

PROCEDURE

1. Prior to the laboratory session, the instructor will have painted several surfaces and objects in the lab with a suspension of yeast cells, the test organism used in the study.
    - The surfaces painted could include a door handle, a piece of equipment, a telephone, a table top, or any other object.
    - The instructor may wish to set the stage for your investigation by pointing out certain known facts about the epidemic.
    - Alternatively, you may wish to make inquiries or ask certain salient questions.

2. Once you have decided on the surface or object to be sampled, obtain a sterile cotton swab, a tube of sterile saline, and a tube of hydrogen peroxide.
   - Take the swab and moisten it with sterile saline.
   - Then use the swab to rub the surface vigorously.
   - Now drop the swab into a tube of hydrogen peroxide (H_2O_2).
3. Observe the tubes for signs of oxygen gas release, which will be manifested by bubbling as FIGURE 77.1 shows.
   - The bubbling results from catalase released by yeast cells on the swab. (Nothing will happen in the tube if yeast cells were not picked up on the swab.)
   - The bubbling may be vigorous if a large number of yeast cells are recovered, or it may be mild if there are fewer cells.
   - Therefore, you may ascertain whether the surface was heavily or mildly contaminated. This may help you decide which surface was a primary source of microorganisms and which ones were secondary.
   - Prepare a negative control with moistened swabs and a positive control with swabs dipped into the yeast suspension.
   - Note your observations in the Results section, and draw your conclusions.
4. Your findings of contaminated surfaces may be confirmed in several ways. For example, it may be possible to cultivate the yeast cells on Sabouraud dextrose agar or potato dextrose agar.
   - To do this, swab the surfaces with moistened cotton swabs and rub the swabs across plates of these media.
   - Cultivation at room temperature for several days should reveal yeast colonies.

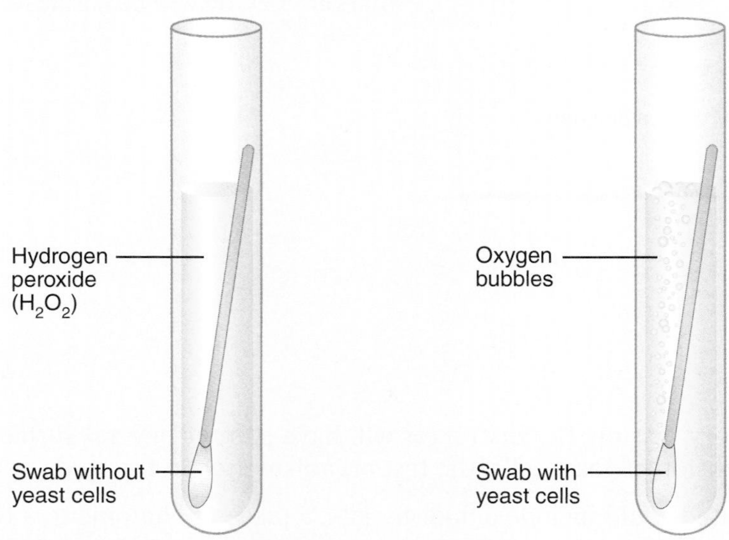

A When the swab contains no yeast, no changes occur in the hydrogen peroxide.

B When the swab contains yeast cells, the latter produce catalase that breaks down the hydrogen peroxide and releases oxygen bubbles.

FIGURE 77.1 The catalase test for detecting yeast cells.

- Microscopic observations can be performed to confirm the presence of yeasts.
- It may also be possible to take scrapings from the suspected surface and prepare slides for microscopic observation even before yeasts are cultivated.
- These options should be discussed with the instructor before you proceed.

Quick Procedure

Hunting For Microorganisms

1. Contaminate a surface with harmless yeast cells.
2. Swab the surface.
3. Place the swab in hydrogen peroxide.
4. Observe the tube for bubbling.

Name: ______________________ Date: __________ Section: __________

A Microbial Hunt

EXERCISE RESULTS **77**

RESULTS

A MICROBIAL HUNT

Observations and Conclusions

Questions

1. Why is it unnecessary to use sterile hydrogen peroxide in the procedures of this exercise?

2. Hypothesize why yeast cells naturally produce catalase.

Transmission of Microorganisms through Toilet Paper

EXERCISE

78

Materials and Equipment

- Roll of toilet paper
- Yeast suspension
- Plate of any type of agar
- Sterile cotton swabs and saline solution
- Tubes of hydrogen peroxide

Infectious diseases such as hepatitis A, viral gastroenteritis, typhoid fever, shigellosis, and numerous others can be transmitted by infected hands when a person fails to wash thoroughly after using the toilet. One of the drawbacks of toilet paper is its porosity, a factor that permits microbes to reach the finger surface even though the skin surface is not touched.

TRANSMISSION OF MICROORGANISMS THROUGH TOILET PAPER

PURPOSE: to simulate the spread of disease-causing organisms through hand contact.

In this exercise, yeast cells will be used as a test organism and the presence of catalase will be used as a marker to detect them. A simulated "wiping" will be performed, followed by a test for "infectious organisms."

PROCEDURE

1. Obtain a plate of agar of any type and pour onto its surface about 10 ml of the yeast suspension.
   - Tilt the plate to cover the surface with the yeast.
   - If there is excess yeast fluid, pour it off into a beaker of disinfectant.
   - The "infected" plate will represent the anal area after using the toilet.

2. Wash your hands thoroughly and completely to remove all traces of yeast cells contacted during preparation of the plate.
   - To ensure that your hands are clean, perform the following procedure.
     - Obtain a sterile cotton swab, a tube of sterile water, and a tube of hydrogen peroxide.
     - Moisten the swab with the sterile water, and vigorously rub your fingers where they were in contact with the yeast suspension or agar plate.
     - Drop the swab into a tube of hydrogen peroxide and watch for the presence or absence of bubbling.
   - No bubbling should occur, an indication that there are no yeast cells on the finger surface.
3. Now obtain several sheets of toilet paper and hold them as you might do after using the toilet.
   - Wipe the surface of the agar (try not to gouge the agar) to simulate wiping the anal area.
   - Discard the toilet paper in the disinfectant designated by the instructor.
4. A fellow student may assist you in the following.
   - Obtain another sterile cotton swab, tube of sterile water, and tube of hydrogen peroxide.
   - Moisten the swab with the sterile water, then vigorously rub the fingers where they held the toilet paper.
   - Drop the swab into a tube of hydrogen peroxide and watch for the presence or absence of bubbling.
5. If the tube begins to bubble, there is evidence for the presence of yeasts because these organisms produce the catalase responsible for the bubbles.
   - Assuming there were no yeasts present before you began the "wiping," the data points to the transmission of yeasts through the toilet paper.
   - On the other hand, if there is no bubbling, then you may determine that yeasts were not transmitted through the toilet paper.
   - Write your conclusions in the Results section.
6. As an alternative to toilet paper, a sheet of sanitary toilet bowel rim paper may be used.

Name: ______________________ Date: __________ Section: __________

Transmission of Microorganisms through Toilet Paper

EXERCISE RESULTS **78**

RESULTS

TRANSMISSION OF MICROORGANISMS THROUGH TOILET PAPER

Observations and Conclusions

____

____

____

____

____

Questions

1. How can one increase or decrease the ability of toilet paper to transmit organisms?

____

____

____

____

2. What steps might be taken to interrupt the spread of infectious microorganisms in an epidemic even before the microorganisms have been identified?

____

____

____

____

____

Disinfection of Drinking Water

EXERCISE 79

Materials and Equipment

- Inoculating loop
- 1 quart of clear water
- Fresh soil sample
- Commercial tincture of iodine solution
- 1-ml pipettes and mechanical pipetters
- Sterile culture dishes and melted nutrient agar

Globally, clean drinking water is an important public health concern. Contaminated or unclean water can lead to a variety of intestinal infections. The United States has even guaranteed federal protection of our water supplies through the Safe Drinking Water Act. Clear water suspected of containing bacteria can be disinfected for drinking purposes by using a commercially available tincture of iodine **tincture of iodine**; that is, 2–7% elemental iodine, along with potassium iodide or sodium iodide, dissolved in a mixture of ethanol and water. The Centers for Disease Control and Prevention (CDC) recommends 5 drops of tincture of iodine per quart of water, with 30 minutes contact time before consumption.

DISINFECTION OF DRINKING WATER

PURPOSE: to test the effectiveness of a tincture of iodine to disinfect drinking water.

In this exercise, the efficiency of water disinfection with iodine will be tested.

PROCEDURE

1. Obtain three sterile culture dishes, and label each on the bottom side with your name and the date.
   - Designate one dish "pretreatment," another "post-treatment," and the third "control."
2. In a clean vessel, obtain 1 quart of clear water such as might be used for drinking.
   - Using a sterile pipette and a mechanical pipetter, transfer 1 ml of the water to the culture dish marked "control."
   - Add sufficient melted nutrient agar to cover the bottom of the plate.

- Mix the water and agar by gently rotating the plate in a wide arc, and then allow the agar to harden.
- Discard the pipette in the available disinfectant.

3. Contaminate the water with a small pinch of fresh soil.
   - Mix thoroughly by stirring and shaking so that the water remains relatively clear and the contamination is not evident to the casual observer.
4. Using another sterile pipette and mechanical pipetter, transfer 1 ml of the water to the culture dish designated "pretreatment."
   - Again, add sufficient melted nutrient agar to cover the bottom of the plate.
   - Mix the water and agar by gently rotating the plate in a wide arc, and allow the agar to harden.
   - Discard the pipette in the available disinfectant.
5. Add 5 drops of commercial tincture of iodine to the remaining contaminated water, and allow the water to stand at room temperature for 30 minutes.
6. Using a fresh sterile pipette, transfer 1 ml of the water to the culture dish designated "post-treatment."
   - Add sufficient agar to cover the bottom of the plate, and gently mix it with the water. Allow the agar to harden.
7. Incubate the three plates inverted at 37°C for 24 to 48 hours, then refrigerate until observed.
8. Examine the plates for bacterial colonies, and prepare representations of the plates in the appropriate spaces of the Results section.
   - Observe the number of colonies in each plate, and determine whether the iodine reduced the bacterial content of the water by expressing the number of bacterial colonies per milliliter of water.
   - Express your opinion on the value of iodine as a disinfectant in water.

Name: ______________________ Date: __________ Section: __________

Disinfection of Drinking Water

EXERCISE RESULTS 79

RESULTS

DISINFECTION OF DRINKING WATER

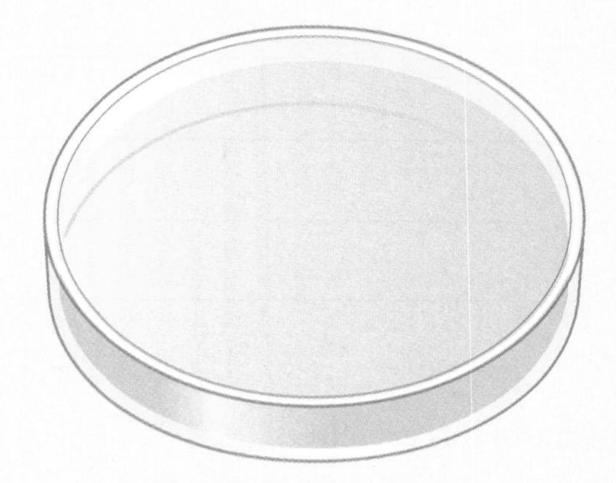

Nutrient Agar Plate Control

Nutrient Agar Plate Pretreatment

Colonies/ml: ______________________ ______________________

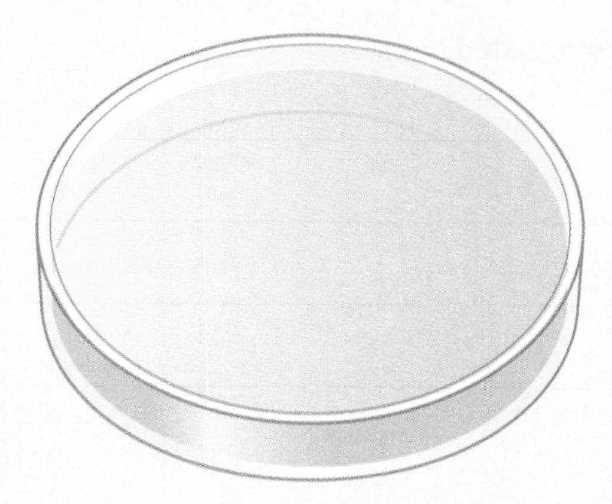

Nutrient Agar Plate Post-treatment

Colonies/ml: ______________________

Observations and Conclusions

Notes:

Questions

1. Explain several variables that influence the outcome of this exercise.

2. How does water become contaminated?

Microbiology of Foods: Food Preservation with Salt and Garlic

EXERCISE 80

Materials and Equipment

- Raw hamburger
- Tubes of normal nutrient broth
- Tubes of nutrient broth containing 1%, 5%, and 10% salt
- Raw garlic and garlic press

Most foods provide an excellent growth medium for microorganisms. The supply of organic matter is plentiful, the water content is usually sufficient, and the pH is generally neutral or only slightly acidic. The result is **food spoilage**, which leads to an economic loss to the manufacturer and a waste of money to the consumer, as well as a threat to health. In this exercise, tests will be conducted to determine how spoilage may be prevented with salt and garlic.

High-salt environments exert an inhibitory effect on bacterial growth by stimulating the flow of water out of the organisms by the process of osmosis. This causes the microorganisms to shrink and disintegrate. Foods such as salted beef and cod, bacon, and ham are preserved in this way. In this exercise, a piece of food will be suspended in broths containing various concentrations of salt, and the growth of bacteria will be determined. In addition, the antimicrobial effect of garlic will be tested.

FOOD PRESERVATION WITH SALT AND GARLIC

PURPOSE: to examine the effects of salt and garlic as preservatives.

PROCEDURE

I. Effectiveness of Salt

Treat incubated tubes carefully because bacterial pathogens may have increased in numbers to dangerous levels.

1. Select one tube of normal nutrient broth and tubes of nutrient broth containing 1%, 5%, and 10% salt.
   - Label each of the tubes with your name, the date, and the salt concentration of the broth.

2. Aseptically inoculate each tube with approximately 1 gram of raw hamburger meat, being careful to minimize airborne contamination. The meat need not be weighed, but the sample should be about the size of a pea.
   - A clean and lightly flamed spatula or knife should be used.
   - Incubate the four tubes at 37°C for 24 to 48 hours.
3. Observe the tubes for the presence of bacteria (seen as turbidity in the tubes), and note your results in TABLE 80.1 of the Results section, using (+++) for heavy growth (very turbid), (++) for moderate growth (less turbid), (+) for trace of growth (lightly turbid), and (–) for absence of growth (not turbid).
   - Be careful to distinguish bacterial turbidity from meat particles.
   - Note whether the amount of turbidity decreases as the salt concentration increases, and write your observations on the effect of salt as a food preservative.
   - Prepare Gram stains (Exercise 15) from loopfuls of the various broths, and observe the types of bacteria present in the hamburger meat.
   - Note whether the type of bacteria changes as the salt concentration increases.
   - Place representations in the appropriate spaces in the Results section.

II. Antimicrobial Effectiveness of Garlic

1. Much has been written in recent years about the inhibitory effects of garlic and how garlic can be used therapeutically to kill bacteria.
   - To test this principle, you will need two tubes of nutrient broth, a sample of hamburger meat, and a sample of freshly-squeezed garlic.
2. As in Steps I.1 and I.2 above, inoculate two tubes of nutrient broth with 1-gram samples of raw hamburger meat.
   - One tube is the control and will receive no further treatment.
   - The second tube, the experimental, will receive approximately 0.25 grams of fresh garlic, including the juice and pulp.
   - Incubate the tubes at 37°C for approximately 48 hours.
3. Examine the tubes for the presence or absence of bacterial growth (turbidity) and record your observations in the Results section.
   - The absence of bacteria in the experimental tube (no turbidity) provides evidence for the inhibitory effect of the garlic.
   - The presence of equal amounts of turbidity in the control and experimental tubes indicates noninhibition.
   - The reduction of turbidity in the experimental tube indicates some inhibition.

Note that the experimental conditions may be varied to provide different results than you obtained.

Name: ______________________ Date: __________ Section: __________

Microbiology of Foods: Food Preservation with Salt and Garlic

EXERCISE RESULTS 80

RESULTS

FOOD PRESERVATION WITH SALT AND GARLIC

I. Antimicrobial Effectiveness of Salt

TABLE 80.1 Presence of Growth in Nutrient Broth Tubes

Salt Concentration	No Added Salt	1% Salt	5% Salt	10% Salt
Growth (turbidity)				

Observations and Conclusions

____

____

____

____

____

Stained Smears from Nutrient Broth Tubes

Salt conc.: ____________ ____________ ____________

Magnif.: ____________ ____________ ____________

II. Antimicrobial Effectiveness of Garlic

Growth of Bacteria in Nutrient Broth

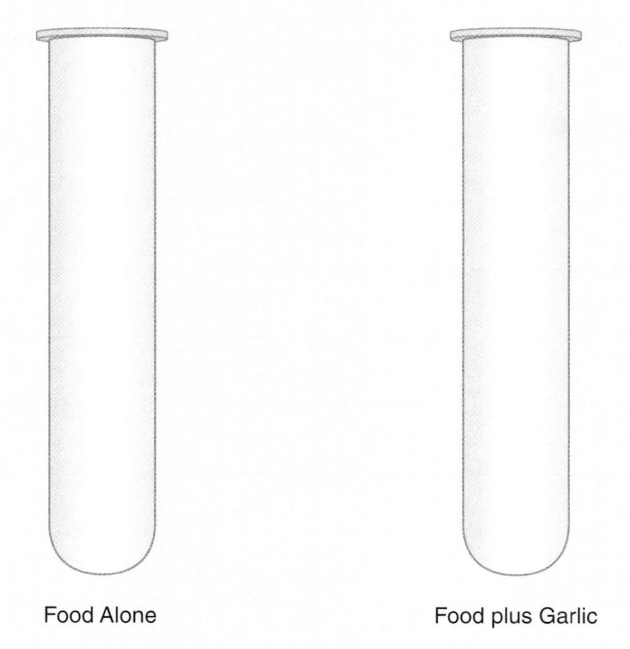

Observations and Conclusions

____

____

____

____

____

Questions

1. Suggest variations in the garlic procedure, and indicate how they will provide further information on garlic's effects on bacteria.

    ____

    ____

    ____

    ____

    ____

Microbiology of Foods: Standard Plate Count

EXERCISE 81

Materials and Equipment

- Raw hamburger
- Other food samples, such as potato salad, fresh vegetables, cold cuts, egg salad, rice pudding, or salad fixings
- Sterile culture dishes and blenders
- Sterile 1.1-ml pipettes and mechanical pipetters
- Sterile 180-ml water samples
- Sterile 99-ml water dilution blanks
- Weighing apparatus
- Melted nutrient agar
- 10 fresh mushrooms and a bottle of salad dressing
- Microscope

The extent of bacterial contamination in foods may be determined by the **standard plate count procedure**. In this process, food is ground with fluid in a blender to suspend the microorganisms. Samples of the fluid then are diluted and placed in culture dishes with growth medium. After incubation, the number of colonies is counted and multiplied by the dilution factor to yield the total number of bacteria per gram of food sample. The underlying principle is that each bacterium will form a visible colony. The test is similar to one that will be performed with milk samples in Exercise 83.

STANDARD PLATE COUNT

PURPOSE: to determine the bacteria present in raw hamburger and other foods.

In this exercise, samples of foods including hamburger meat and mushrooms will be tested for their bacterial content.

PROCEDURE

I. The Bacterial Count in Ground Beef

1. To ensure reliable results, follow aseptic procedures throughout this exercise.
   - Use sterile instruments whenever possible, and take precautions to limit airborne contamination.
   - The procedure may be performed in pairs due to the volume of materials necessary.
   - The instructor will assign the type of food to be tested, and may give special directions on the use of the 1.1-ml pipettes that are commonly used for this type of test.

2. Obtain four sterile culture dishes and label them on the bottom side with your name, the date, and the designations 1:10, 1:100, 1:1,000, and 1:10,000.
    - Select one sterile 180-ml water sample, one sterile 99-ml water dilution blank, one sterilized blender, and a 1.1-ml pipette and mechanical pipetter.
3. Aseptically weigh 20 grams of the food sample to be tested.
    - Combine the food with 180 ml of sterile water in a sterile blender as shown in FIGURE 81.1A.
    - Blend for 3 minutes or as directed by the instructor.
    - The resulting fluid represents a 1:10 dilution of the food sample.
    - Allow the large particles of food to settle before proceeding.
    - If a blender is not available, vigorous shaking will provide adequate suspension of the organisms.
4. Using a mechanical pipetter, pipette 0.1 ml of the blended material to the 1:100 plate and 1.0 ml to the 1:10 plate, as shown in FIGURE 81.1B.
    - Be careful to avoid airborne contamination of the plate by lifting the lid only high enough to permit entry of the pipette.
5. Pipette 1.0 ml of the blended material to the 99-ml water dilution blank (FIGURE 81.1C), and draw up and release some material several times to wash out the pipette.
    - Shake the bottle for 2 minutes to effect even distribution.
    - Aseptically pipette 0.1 ml of this material to the 1:10,000 plate, and 1.0 ml to the 1:1,000 plate (FIGURE 81.1D).
6. Aseptically pour into each plate enough melted nutrient agar to cover the bottom of the plate.
    - Rotate the plates ten times in a wide arc on the laboratory desk to ensure even mixture of the sample and nutrient agar.

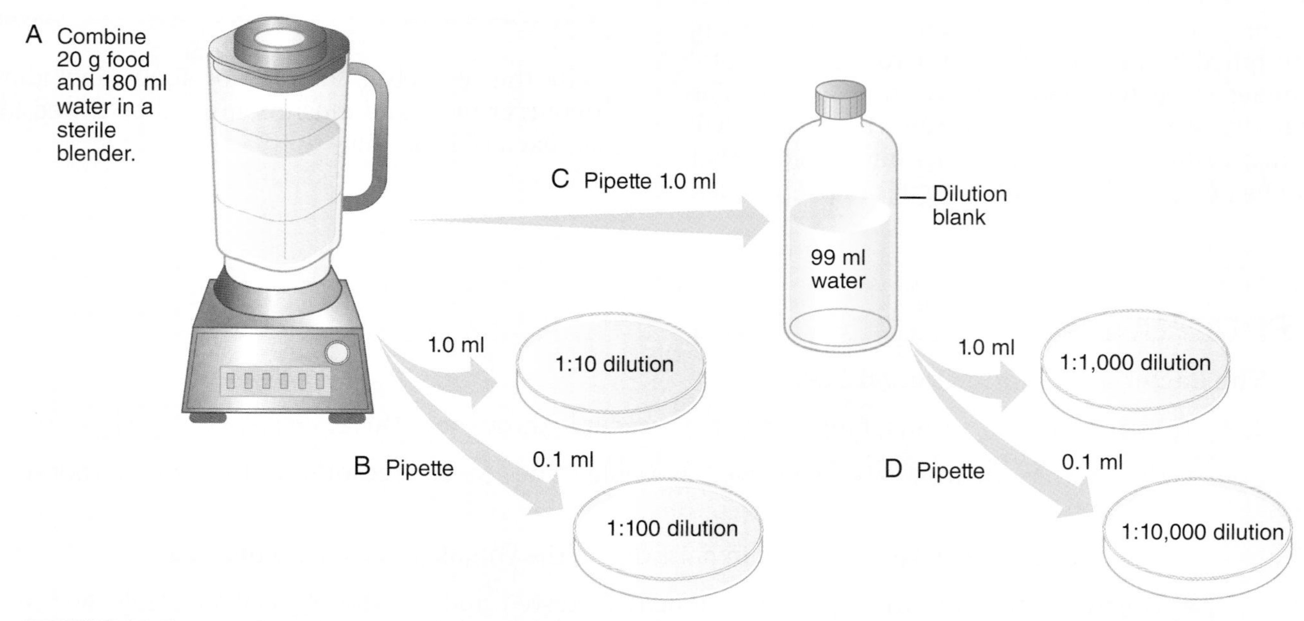

FIGURE 81.1 Standard plate count procedure using a food sample.

- Allow the agar to harden, and incubate the plates in the inverted position at 37°C for 24 to 48 hours.

7. The instructor will demonstrate the use of the Quebec colony counter for performing plate counts (or a dissecting microscope can be used).
    - Place each plate on the colony counter, determine the number of colonies per plate, and enter your results in the chart.
    - Be sure to count surface as well as subsurface colonies.
    - Be careful to avoid counting food particles, which generally appear as irregular spots on the plates.
    - If a plate contains significantly more than 300 colonies, it is unnecessary to obtain an exact count. In such a case, enter the acronym TNTC (too numerous to count) in TABLE 81.1 in the Results section.
8. From your results, select the colony count that falls between 30 and 300.
    - This is the "valid" count. Counts over 300 are considered invalid because overcrowding may cause two or more bacteria to form a single colony; counts under 30 are invalid because the chance of sampling error is significant.
    - Multiply the valid count by the dilution factor of the plate (i.e., 10, 100, 1,000, or 10,000). This is the total plate count per gram of food sample.
    - Consult with the instructor if more than one colony count falls between 30 and 300, if all colony counts are below 30, or if all colony counts are over 300.
    - Enter your observations on the bacterial contamination of the food in Table 81.1.
    - Obtain plate counts for other foods from fellow students, and enter these results in TABLE 81.2.
    - At the direction of the instructor, prepare Gram stains from the colonies on the plates to examine the bacterial content of the food.
    - Record examination results in the Results section.

II. The Bacterial Content of Marinated Mushrooms

1. Mushrooms are a valuable food sample for bacterial testing because they are grown in rich organic soil, where the bacterial content is normally high.
    - A plate count, as described in the previous section, is a useful way to determine the number of bacteria per gram of mushrooms.
    - In addition, the effectiveness of the acid in a marinade can be demonstrated to be a method for lowering the bacterial count.
2. Select ten fresh, unprocessed mushrooms and divide them into two equal groups.
    - Slice the first group of five and place them in a clean plastic bag.
    - Then cover the mushrooms in the bag with any commercial dressing to be used as a marinade. The dressing should contain vinegar (Italian dressing is a good choice).
    - The mushrooms should be left undisturbed for one hour.
    - During that time interval, the other five mushrooms should be sliced and set aside.

3. At the end of the hour, weigh out one gram of the marinated mushrooms and add it to a sterile 99-ml water dilution blank.
    - Perform a plate count as described in Steps I.1 to I.8 in the previous procedure.
    - Then weigh out one gram of the unmarinated mushrooms and add it to another sterile 99-ml water dilution blank.
    - Perform a plate count on this sample as well.
    - Set all the plates aside to incubate in the inverted position for 24 to 48 hours.
4. Perform calculations as cited previously to complete the plate counts.
    - Compare the results and draw your conclusions on the effect of the marinade on the bacterial count.
    - Enter your conclusions in TABLE 81.3 and TABLE 81.4 of the Results section.

Quick Procedure

Standard Plate Count (Food)

1. Blend food sample with sterile water.
2. Pipette 1.0- and 0.1-ml samples to culture dishes.
3. Pipette 1.0 ml to a 99-ml water blank and shake.
4. Pipette 1.0- and 0.1-ml samples of diluted food to culture dishes.
5. Add liquid nutrient agar to all culture dishes and mix.
6. Incubate.
7. Perform colony counts and locate valid count.
8. Multiply valid count by dilution factor.

Name: ______________________ Date: __________ Section: ___________

Microbiology of Foods: Standard Plate Count

EXERCISE RESULTS **81**

RESULTS

STANDARD PLATE COUNT

I. The Bacterial Count in Ground Beef

TABLE 81.1 Plate Counts at Various Dilutions

Dilution	1:10	1:100	1:1,000	1:10,000
Plate Count				

_________ × _________ = _________ bacteria per gram of hamburger
(valid count) (dilution factor)

Stained Smears of Flora from Hamburger Meat

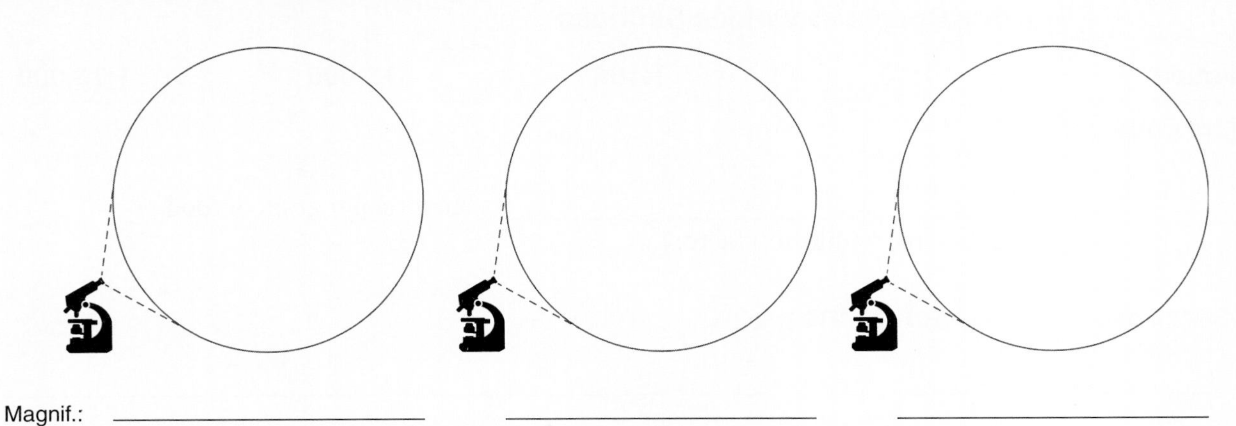

Magnif.: _______________ _______________ _______________

TABLE 81.2	Summary of Standard Plate Counts	
Student	Standard Plate Count	Sample Tested
1.		
2.		
3.		
4.		
5.		

II. The Bacterial Content of Mushrooms

Unmarinated Mushrooms

TABLE 81.3	Plate Counts at Various Dilutions			
Dilution	1:10	1:100	1:1000	1:10,000
Plate Count				

__________ × ____________ = __________ bacteria per gram of food
(valid count) (dilution factor)

Marinated Mushrooms

TABLE 81.4	Plate Counts at Various Dilutions			
Dilution	1:10	1:100	1:1000	1:10,000
Plate Count				

__________ × ____________ = __________ bacteria per gram of food
(valid count) (dilution factor)

Observations and Conclusions

____

____

____

____

____

Questions

1. Why must a culture plate be considered invalid if it contains less than 30 or more than 300 colonies?

2. Does a high standard plate count in food indicate that the food should not be eaten? Why or why not?

3. Explain some common errors that may cast doubt on the accuracy of a plate count.

4. Consider other preservatives that can be used to reduce bacterial counts and suggest variations of this procedure.

Microbiology of Foods: Fermentation of Wine and Beer

EXERCISE 82

Materials and Equipment

- Cotton-plugged tubes of grape juice supplemented with 5% sugar
- Tubes of malt extract broth with cotton plugs
- Cultures of *Saccharomyces cerevisiae* or commercial yeast
- Balloon
- Wire or rubber band
- Inoculating loop

Certain microorganisms may be beneficial to the food industry because they bring about fermentation in the food and lead to consumable products. This aspect will be demonstrated by using organisms to produce fermented beverages.

The production processes for wine and beer are highly specialized and sophisticated. However, the fundamental process may be observed in the laboratory by fermenting grape juice and malt extract broth with yeasts, as shown in this exercise.

FERMENTATION OF WINE AND BEER

PURPOSE: to examine the fundamental process of wine and beer fermentation.

PROCEDURE

I. Basic Fermentation

1. Select tubes of sterile grape juice and malt extract broth containing cotton plugs. The cotton plug will permit the carbon dioxide that is produced to escape.
   - Inoculate each tube with a heavy loopful of *Saccharomyces cerevisiae* or commercial yeast, and incubate the tubes at approximately 30°C for several days to a week as directed by the instructor.
   - Include uninoculated control tubes.
2. Shake the experimental tubes lightly and note the foaming at the surface. The foaming indicates that carbon dioxide has been produced and is now escaping from the liquid.

- Note the winelike aroma in the grape juice and the beerlike odor of the malt extract broth.
- At the instructor's direction, taste the liquid to determine its quality, and prepare stains of the sediment at the bottom of the tubes to observe the yeast cells.

Never eat or drink anything produced in laboratory exercises except at the direction of the instructor.

- Enter your observations in the Results section.

II. Quantitative Estimation of Fermentation

1. The extent of fermentation over a period of time can be studied by measuring the amount of gas produced in the fermentation flask.
   - For this exercise, you will need a balloon and a flask of grape juice inoculated with yeast.
   - Stretch out the balloon to loosen it.
   - You will also need graph paper for the calculations.
2. Place the balloon over the mouth of the flask and secure it tightly with a wire, rubber band, or other device.
   - Set the flask aside to allow the juice to ferment and observe the balloon at regular time periods over a number of hours.
   - The balloon will open and expand with gas.
   - The circumference of the balloon should be measured and noted in TABLE 82.1 of the Results section.
   - The expansion will be rapid at first as the oxygen is used up and carbon dioxide is quickly produced.
   - It will eventually slow as the changeover from respiration to fermentation occurs. (Less carbon dioxide is produced during fermentation than during respiration.)
3. Prepare a graph comparing the circumference of the balloon as it relates to time. This graph will help quantitate the progress of the fermentation.
4. The above exercise can be varied in numerous ways to study various aspects of fermentation.
   - For example, different carbohydrate sources can be used (grape juice, apple juice, orange juice) to see whether there is a difference in the rate of fermentation and whether the yeast shows a preference.
   - The fermentation can be conducted at different temperatures (refrigerator, room, 37°C, 55°C) to determine the optimum temperature for the fermentation.
   - The concentration of the grape juice can be varied to study the effect of substrate concentration on fermentation activity.
   - Inhibitors such as metals or drugs can be incorporated into the grape juice to study their effect on the fermentation.

Name: ______________________ Date: __________ Section: __________

Microbiology of Foods: Fermentation of Wine and Beer

EXERCISE RESULTS 82

RESULTS

FERMENTATION OF WINE AND BEER

I. Basic Fermentation

Observations and Conclusions

____

____

____

____

____

II. Quantitative Estimation of Fermentation

TABLE 82.1 Balloon Size at Various Time Periods

Experimental Conditions	Time	Circumference
(describe)		

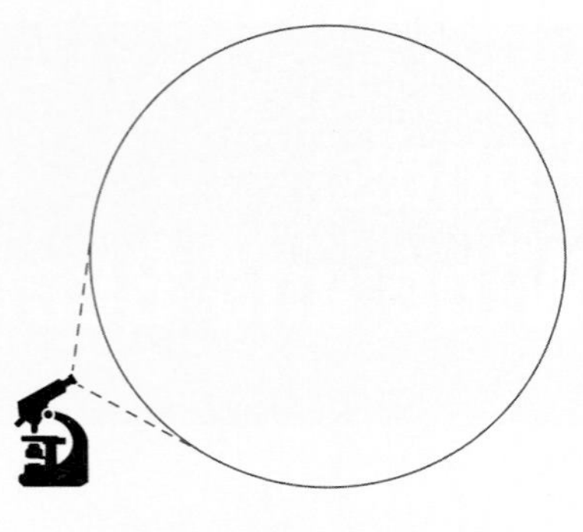

Microbial Flora in Wine/Beer

Magnif.: ______________________

Observations and Conclusions

____

____

____

____

____

Questions

1. Suppose a balloon were tied over the mouth of a tube of grape juice inoculated with yeast. What would take place? Explain.

____

____

____

____

____

2. Suggest other variables to be tested.

____

____

____

Microbiology of Milk and Dairy Products: Standard Plate Count of Milk

EXERCISE 83

Materials and Equipment

- Samples of pasteurized and unpasteurized milk
- Sterile culture dishes
- Sterile 1.1-ml pipettes and mechanical pipetters
- Sterile 99-ml water dilution blanks
- Melted nutrient agar

Milk is an extremely nutritious food. It contains proteins, fats, carbohydrates, vitamins, and minerals, and as such, it is also an ideal growth medium for microorganisms. These microorganisms may induce spoilage in milk and dairy products, and they may cause disease if pathogenic microorganisms enter the milk from the cow or during processing.

This exercise will explore the standard plate count procedure to determine the bacteria content of milk.

The **standard plate count procedure** is used to determine the total number of bacteria in a milliliter of milk. This test is used in the grading of milk. Various dilutions of milk are prepared and placed in culture dishes, and a growth medium such as nutrient agar is added. After incubation, a colony count is performed, and the valid count is multiplied by the dilution factor to give the total plate count. This count represents the original number of bacteria in the milk. The process is very similar to that performed with the food sample in Exercise 81.

STANDARD PLATE COUNT OF MILK

PURPOSE: to become familiar with the standard plate count procedure for bacteria in milk.

PROCEDURE

1. The instructor will assign a milk sample and may recommend that this exercise be done in pairs due to the large volume of materials.
   - Directions on the use of the special 1.1-ml pipettes also may be given.
   - If the milk is to be pasteurized in class, it should be held at exactly 63°C for 30 minutes and then cooled before using it.

2. Obtain a sample of milk, two sterile 99-ml water dilution blanks, a sterile 1.1-ml pipette and mechanical pipetter, and six sterile culture dishes.

To avoid contamination, only lift the cap of the plate slightly when permitting entry of the pipette.

   - Label the dishes on the bottom side with your name, the date, and the following designations: 1, 1:10, 1:100, 1:1,000, 1:10,000, and 1:100,000. Refer to FIGURE 83.1.

3. Aseptically pipette 0.1 ml of milk into the plate labeled 1:10, and 1.0 ml of the milk into the plate marked 1, as shown in Figure 83.1A.
   - Be careful to avoid airborne contamination of the plate by lifting the cap only slightly enough to permit entry of the pipette.

4. Aseptically pipette 1 ml of the milk into a dilution blank containing 99 ml of sterile water (Figure 83.1B).
   - Shake to mix the contents thoroughly.
   - Draw up and release some diluted milk several times to wash out the pipette.
   - From this bottle, pipette 0.1 ml of diluted milk into the 1:1,000 plate, and 1.0 ml into the 1:100 plate as shown in Figure 83.1C.

5. Now pipette 1 ml from the first dilution blank to a second 99-ml dilution blank, and mix and wash out the pipette as before (Figure 83.1D).
   - Shake to mix the contents thoroughly.
   - From the second blank, pipette 0.1 ml into the 1:100,000 plate, and 1.0 ml into the 1:10,000 plate, as shown in Figure 83.1E.
   - The pipette may now be discarded, as directed by the instructor.

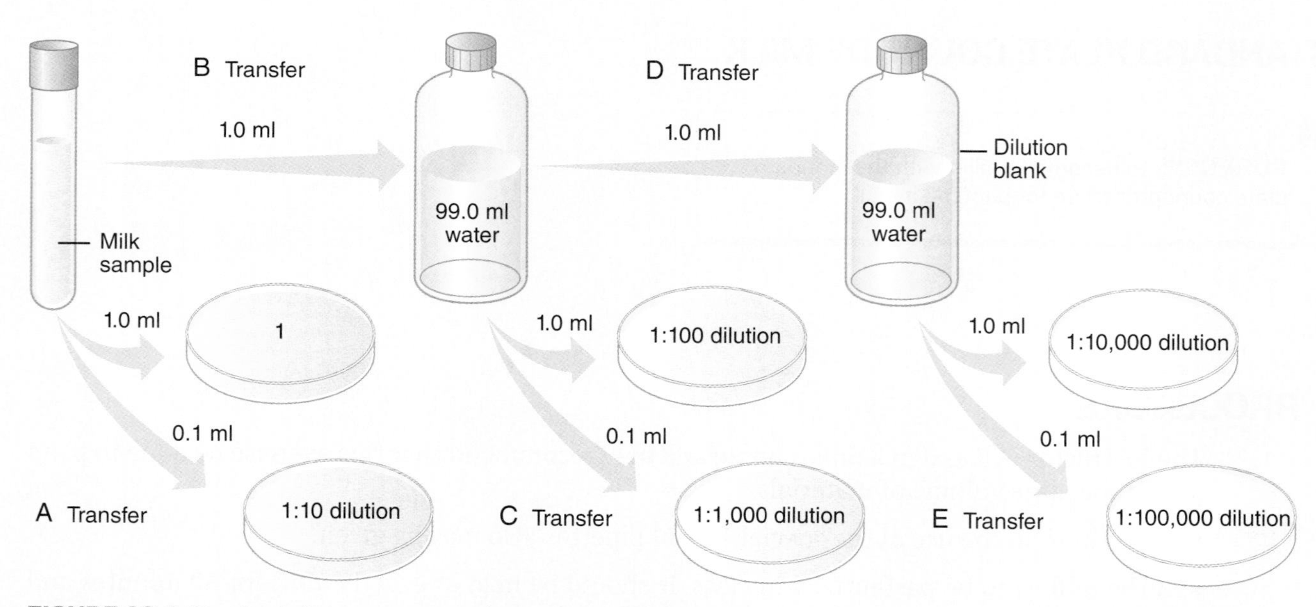

FIGURE 83.1 Standard plate count procedure using a milk sample.

6. Aseptically pour into each plate enough melted nutrient agar to cover the bottom of the plate.
   - Mix the medium with the diluted milk by gently rotating the plates ten times in a wide arc on the laboratory bench.
   - After the medium has hardened, invert the plates and incubate them at 37°C for 24 to 48 hours.
7. Observe the plates, and count all surface and subsurface colonies on the Quebec colony counter as directed by the instructor.
   - If it is apparent that a plate has well over 300 colonies, designate it as "TNTC" (too numerous to count) and go on to the next plate. Enter your results in TABLE 83.1 of the Results section.
   - Select the "valid" count; that is, the one that falls between 30 and 300.
   - Multiply this count by the dilution factor of the plate (i.e., 1, 10, 100, 1,000, 10,000, or 100,000), and enter the total plate count per milliliter of milk in the Results section. If two colony counts fall between 30 and 300, this generally indicates a pipetting or other technique error. Consult with the instructor on how to proceed.
   - Make your observations on the quality of the milk tested, and obtain plate count results from your classmates on other milk samples.

Quick Procedure

Standard Plate Count (Milk)

1. Pipette 1.0- and 0.1-ml milk samples to culture dishes.
2. Pipette 1.0 ml milk to a 99-ml water blank and shake.
3. Pipette 1.0- and 0.1-ml samples of diluted milk to culture dishes.
4. Pipette 1.0 ml diluted milk to a second 99-ml water blank and shake.
5. Pipette 1.0- and 0.1-ml samples of diluted milk to culture dishes.
6. Add liquid nutrient agar to all dishes and mix.
7. Incubate.
8. Perform colony counts and locate valid count.
9. Multiply valid count by dilution factor.

Name: ______________________ Date: __________ Section: __________

Microbiology of Milk and Dairy Products: Standard Plate Count of Milk

EXERCISE RESULTS 83

RESULTS

STANDARD PLATE COUNT OF MILK

Milk Sample Tested: ______________________

TABLE 83.1	Plate Counts at Various Dilutions					
Dilution	**1**	**1:10**	**1:100**	**1:1,000**	**1:10,000**	**1:100,000**
Plate Count						

__________ × __________ = __________ total bacteria per ml of milk
(valid count) (dilution factor)

Observations and Conclusions

____

____

____

____

____

Questions

1. What might happen if agar media were not permitted to cool sufficiently before pouring it into plates during the plate count procedure?

____

____

____

2. What would happen if airborne contamination were to occur during Step 3 of the procedure?

___

___

___

Microbiology of Milk and Dairy Products: Coliform Plate Count of Milk

EXERCISE 84

Materials and Equipment

- Samples of pasteurized and unpasteurized milk
- Sterile culture dishes
- Sterile 1.1-ml pipettes and mechanical pipetters
- Sterile 99-ml water dilution blanks
- Melted violet red bile agar or other growth medium
- Quebec colony counter

Milk contains proteins, fats, carbohydrates, vitamins, and minerals, and as such, it is an ideal growth medium for microorganisms. These microorganisms may induce spoilage in milk and dairy products and may cause disease if pathogenic microorganisms enter the milk from the cow or during processing.

This exercise will explore the coliform plate count procedure to test for fecal contamination of milk.

COLIFORM PLATE COUNT OF MILK

PURPOSE: to use the coliform bacteria count as an indicator of the fecal contamination of milk.

Coliform bacteria are gram-negative rods that ferment lactose to acid and gas under specific incubation conditions. Because these organisms are commonly found in the intestine, their presence may be used as an indicator of the fecal contamination of milk and, therefore, its sanitary quality.

A plate count of coliform bacteria is performed in essentially the same way as the standard plate count (Exercise 83) except that a selective and differential medium is used. Examples of these media are MacConkey agar, deoxycholate agar, and violet red bile agar. In this exercise, we will use **violet red bile agar**, which contains bile salts to inhibit most noncoliform bacteria, and neutral red, which is taken up by the lactose-fermenting coliforms to yield violet-purple colonies. A thin layer of medium is added after the initial layer has solidified to simulate the partially anaerobic environment favored by coliform bacteria.

PROCEDURE

1. This procedure may be performed in pairs at the direction of the instructor. The milk sample tested in Exercise 83 should be used for this test.
2. Obtain a sterile 99-ml water dilution blank, a sterile 1.1-ml pipette and mechanical pipetter, and four sterile culture dishes.
   - Label the dishes with the designations 1, 1:10, 1:100, and 1:1,000.

3. Using a sterile pipette, carefully transfer 0.1 ml of the milk sample into the 1:10 plate, and 1.0 ml of the milk into the 1 plate.
4. Aseptically pipette 1.0 ml of milk into the 99-ml water dilution blank.
   - Wash the pipette by drawing the solution up and down several times, and shake the bottle to mix its contents.
   - Transfer 0.1 ml of the contents of the 1:1,000 plate, and 1.0 ml to the 1:100 plate.
   - Discard the pipette as directed by the instructor.
5. Aseptically pour **violet red bile agar** or other selective medium into the plates, and mix the plates by rotating them in a wide arc.
   - Allow the medium to solidify.
   - Pour a thin layer of the medium on top of the initial layer.
   - Wait several minutes until the top layer has hardened, and then invert the plates and incubate them at 37°C for 24 to 48 hours.
6. Assay the plates on the Quebec colony counter, and count those surface and subsurface colonies that are violet-purple. Do not count plates that contain obviously more than 300 colonies; instead, designate them "TNTC" (too numerous to count).
   - Enter your results in TABLE 84.1, select the count between 30 and 300, and multiply the "valid" colony count by the dilution factor to obtain the total number of coliform bacteria per ml of milk.
   - Obtain results from other students for different milk samples, and enter your observations from the results in TABLE 84.2.
   - Gram stains may be made from colonies of coliform bacteria, and gram-negative rods should be apparent.

Quick Procedure

Coliform Plate Count (Milk)

1. Prepare various dilutions of the liquid (e.g., milk).
2. Place measured samples of the dilutions in culture dishes.
3. Add selective agar and mix.
4. Incubate plates.
5. Count colonies and multiply by dilution factor.

Name: ______________________ Date: __________ Section: __________

Microbiology of Milk and Dairy Products: Coliform Plate Count of Milk

EXERCISE RESULTS 84

RESULTS

COLIFORM PLATE COUNT OF MILK

Milk Sample Tested: ______________________

TABLE 84.1	Plate Counts at Various Dilutions			
Dilution	**1**	**1:10**	**1:100**	**1:1,000**
Plate Count				

__________ × __________ = __________ coliform bacteria per ml of milk
(valid count) (dilution factor)

TABLE 84.2	Summary of Standard and Coliform Counts of Milk		
Student	**Standard Plate Count**	**Coliform Count**	**Type of Milk**
1.			
2.			
3.			
4.			
5.			

Stained Bacterial Smears from Milk

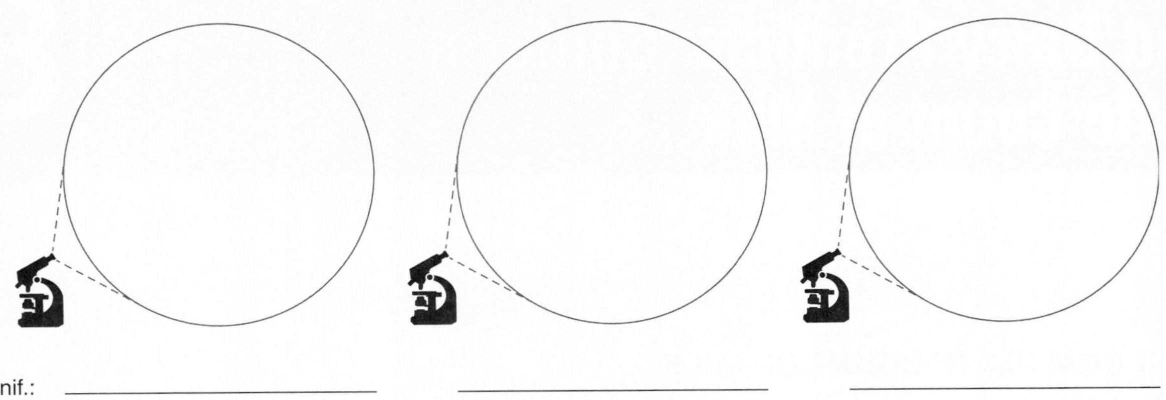

Magnif.: ______________ ______________ ______________

Observations and Conclusions

____

____

____

____

____

Questions

1. Identical quantities of milk are used for standard and coliform plate count procedures, but the final counts are different. Why is this so?

____

____

____

____

Microbiology of Milk and Dairy Products: Methylene Blue Reduction Test

EXERCISE **85**

Materials and Equipment

- Milk samples
- Methylene blue thiocyanate solution
- Sterile 1.0-ml and 10-ml pipettes and mechanical pipetters
- Sterile test tubes with rubber stoppers

This exercise will explore some of the laboratory procedures used in the detection of bacteria in milk, specifically the methylene blue reduction test.

METHYLENE BLUE REDUCTION TEST

PURPOSE: to use the methylene blue reduction test to measure the relative numbers of bacteria in milk.

The **methylene blue reduction test** is a rapid procedure for determining the relative numbers of bacteria in a sample of milk. A solution of methylene blue is added to a milk sample, and the time needed for the blue color to disappear is determined. The removal of the oxygen from milk and the formation of reducing substances during the fermentation of lactose causes the color to disappear. The number of bacteria present is proportional to the amount of time necessary for the color to disappear.

PROCEDURE

1. Using a mechanical pipetter, aseptically pipette 10 ml of the milk sample to a sterile test tube having a rubber stopper, and label the tube with your name.
2. Pipette 1.0 ml of the methylene blue solution to the milk, and tighten the rubber stopper.
   - Carefully invert the tube three times to mix its contents.
   - Place the tube in the rack in the water bath at 37°C, and record the time.
3. Observe the tube at 30-minute intervals for the loss of blue color. As this takes place, the milk will return to its normal white.
   - After each reading, invert the tube three times and return it to the incubator.
   - When four-fifths of the tube has become white, the test is completed.

TABLE 85.1 Interpretation of Methylene Blue Reduction Test

Description of Reaction	Incubation Time with Methylene Blue	Quality of Milk
No reduction of blue color	After 8 hours	Excellent
Reduction of blue color to colorless	6 - 8 hours	Good
Reduction of blue color to colorless	2 - 6 hours	Fair
Reduction of blue color to colorless	Less than 2 hours	Poor

- Refer to TABLE 85.1, and determine the quality of the milk according to the reduction time for methylene blue.
- Enter your data and observations in the Results section.

4. A rough version of this test can be performed with methylene blue stain (such as used for simple staining) and various milk samples, including soil-contaminated milk and milk that has stood at room temperature for a day or two. Fresh milk should be used as a control.
   - Place 10-ml samples of the milk samples in test tubes and add a drop of the methylene blue solution to each tube.
   - Set aside the tubes to incubate at 37°C or at room temperature, refrigerator temperature, or another temperature depending on the experimental conditions desired.
   - The blue color will disappear from the milk in proportion to the bacterial content of the milk. Although the quality of the milk cannot be determined by industry standards, the principle of methylene blue reduction can be demonstrated, and the relative bacterial content of the milk can be assessed.

Name: ______________________________ Date: __________ Section: __________

Microbiology of Milk and Dairy Products: Methylene Blue Reduction Test

EXERCISE RESULTS **85**

RESULTS

METHYLENE BLUE REDUCTION TEST

Milk sample tested:

____

Time for methylene blue reduction:

____

Quality of milk:

____

Observations and Conclusions

____

____

____

____

____

Questions

1. Is it possible that the methylene blue color disappears rapidly in the reduction test but that the milk is still safe to drink? Support your answer.

Microbiology of Milk and Dairy Products: Preparation of Cheese

EXERCISE 86

Materials and Equipment

- Quart of fresh milk and rennin tablets; alternately, a container of large-curd, unflavored cottage cheese
- Colander, cheesecloth, salt, and spoon
- Small piece of blue cheese
- Plastic container

Even in the form of cheese, milk contains proteins, fats, carbohydrates, vitamins, and minerals, and as such, it is also an ideal growth medium for microorganisms. This exercise will show how microorganisms may be used to yield dairy products by focusing on the processes employed for preparing blue cheese.

PREPARATION OF CHEESE

PURPOSE: to recognize the role of microorganisms in cheese making.

Cheese is produced from the **casein** portion of milk. Casein represents a family of phosphoproteins that making up 80% of the proteins in cow's milk. The casein may be precipitated as a curd by the enzyme **rennin** as well as by the acid produced during bacterial growth. The curd is an **unripened cheese**, which may be sold as cottage cheese or ricotta. Microbial growth yields the broad variety of familiar **ripened** (aged) **cheeses**. In this section, the organism *Penicillium roqueforti* will be used to make blue cheese.

PROCEDURE

1. Milk curds may be prepared as follows:
   - Place a quart of fresh milk in a large container and add a rennin tablet as directed by the package instructions.
   - Stir and warm the mixture with gentle heat, but do not allow the milk to become too hot. Curds will begin to form in a few minutes.
   - When curdling is complete, empty the curds into a colander lined with cheesecloth.
   - An alternative to milk curds is to empty a package of large-curd, unflavored cottage cheese into the cheesecloth in the colander.

2. Squeeze as much liquid from the curds as possible, and thoroughly mix a small piece of blue cheese into the curds using a spoon. The blue cheese is a source of *Penicillium roqueforti*, which is used for ripening.
3. An inoculum of *Penicillium* may also be made as follows:
   - Mix a small piece of blue cheese with cubes of fresh bread in a plastic bag.
   - Set the cubes aside at room temperature for several days, and note how they become covered with green mold.
   - Dry the cubes in a warm, dry environment, and crush them to produce a green powder containing mold spores.
   - A sample of the powder may be mixed with the milk curds as an inoculum.
4. A teaspoon of buttermilk may also be used as an inoculum to prepare a different type of cheese.
5. Shape the inoculated curds into a compressed mass, and once again express as much liquid as possible.
   - Salt the outside lightly to add flavor and inhibit bacterial growth.
   - Replace the cheesecloth covering.
6. Place the cheese into a plastic container, cover tightly, and incubate at room temperature for one week or less to stimulate fungal growth.
   - Place the cheese in the refrigerator to continue the ripening.
   - If cottage cheese has been used, the original package may be used for storage.
   - The excess water may be removed as it appears in the container.

The cheese made in this exercise could potentially have harmful organisms growing in it. Taste it only at the direction of the instructor.

7. With the concurrence of the instructor, taste the cheese at weekly intervals, and note the development of blue veins of *Penicillium* mold and the characteristic blue cheese flavor and aroma.
   - Your instructor may make plans for a "Microbe Appreciation Day," at which time the cheese and various other microbial products may be sampled.

Name: ______________________ Date: __________ Section: __________

Microbiology of Milk and Dairy Products: Preparation of Cheese

EXERCISE RESULTS **86**

RESULTS

PREPARATION OF CHEESE

Observations

____

____

____

____

____

Questions

1. In addition to odor, spoiled milk is identified by the formation of small curds. Why would you eat the curds as cheese, but not as spoiled milk?

____

____

____

____

2. Could other types of milk (nonfat, powdered, buttermilk) be used to make cheese? Support your answer.

____

____

____

____

Microbiology of Milk and Dairy Products: Preparation of Yogurt

EXERCISE 87

Materials and Equipment

- Incubator set at 55°C
- Fresh milk
- Box of powdered milk
- Unflavored commercial yogurt
- Thermometers
- Styrofoam cups and lids

Milk, in the form of yogurt, is extremely nutritious. **Yogurt** is a type of sour milk. The two starter cultures necessary for its formation are *Streptococcus thermophilus* and *Lactobacillus bulgaricus*. When these bacteria are incubated in concentrated milk, they ferment the lactose and produce lactic acid, which brings on yogurt's characteristic sour taste. Evaporation thickens the yogurt, and some proteins coagulate under the acidic conditions. In a few short hours, the yogurt is ready to enjoy.

PREPARATION OF YOGURT

PURPOSE: to recognize the role of bacteria in yogurt production.

This exercise will show how microorganisms may be used to yield dairy products by focusing on the processes employed for preparing yogurt.

PROCEDURE

1. In a suitable container, heat one quart of milk to about 77°C (170°F).
    - Stir the milk often, and use a thermometer to check its temperature.
    - Let the milk cool to about 55°C (130°F).
2. Add one cup of powdered milk and mix thoroughly.
    - Add one-third cup of unflavored commercial yogurt. Fresh or frozen fruit may also be added now or (preferably) after the incubation period.
    - Mix thoroughly.
3. Pour the thickened milk into Styrofoam cups, and cover the cups tightly with their lids.

4. Incubate the cups at 55°C for 6 to 8 hours. A Styrofoam picnic cooler filled with several inches of hot water can be used for this purpose. It is also possible to keep the cups warm by wrapping hot towels around them.
    - After the incubation period, add fruit, and refrigerate the yogurt for several hours.
5. At the direction of the instructor, taste the yogurt, and compare its taste and texture to commercial brands.
    - Air-dried, heat-fixed smears of the yogurt may be prepared to observe the streptococci and *Lactobacillus* essential to its production.

Name: ______________________ Date: __________ Section: __________

Microbiology of Milk and Dairy Products: Preparation of Yogurt

EXERCISE RESULTS 87

RESULTS

PREPARATION OF YOGURT

Observations

____

____

____

____

____

Stained Bacterial Smears from Yogurt

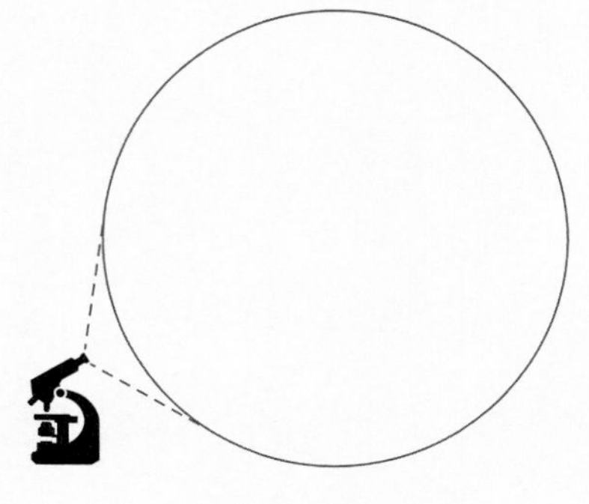

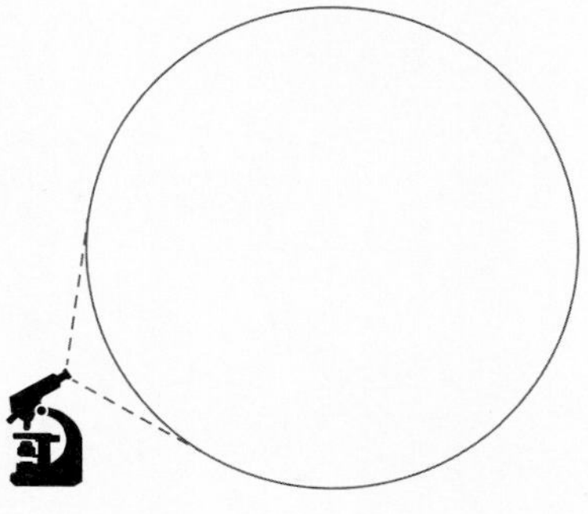

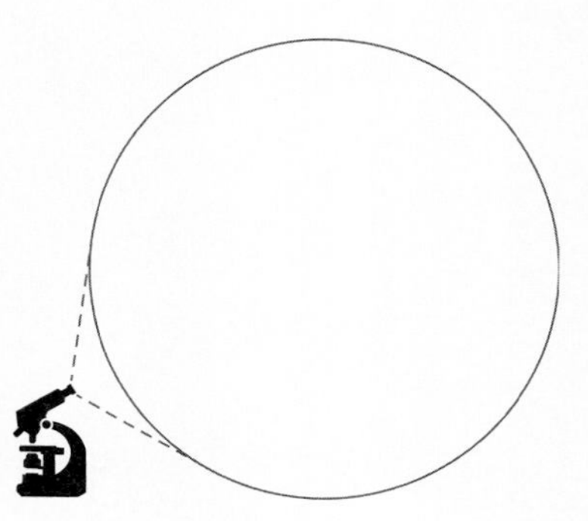

Magnif.: ____________   ____________ ____________

Questions

1. Compare the yogurt made in the lab to the commercially prepared yogurt. How are they alike? How are they different? What do you assume accounts for the difference?

2. How would this experiment change if you used other types of milk (nonfat, powdered, buttermilk)?

Microbiology of Milk and Dairy Products: Natural Bacterial Content of Milk

EXERCISE 88

Materials and Equipment

- Various types of milk (e.g., whole milk, skim milk, buttermilk, and boiled milk)
- Erlenmeyer flasks
- pH paper

This exercise will explore some of the laboratory procedures used in the detection of bacteria in milk.

THE NATURAL BACTERIAL CONTENT OF MILK

PURPOSE: to examine the natural population of bacteria in milk.

The natural population of bacteria in milk is plentiful and diverse. Various types of bacteria emerge in a milk sample when it is placed at different temperatures. For example, certain bacterial species such as *Lactobacillus* and *Streptococcus* produce lactic acid from the lactose and make the milk sour. The acid can build up and cause the protein to curdle. The result is curds and whey.

Other bacteria, notably of the genera *Bacillus*, *Micrococcus*, and *Proteus*, digest the casein in milk. The casein then curdles out of the milk. This curd is a sweet curd because no acid has been produced. In dairy plants, the curd is removed and sold as cottage cheese.

Bacteria also may attack the milk's butterfat. Members of the genera *Pseudomonas* and *Achromobacter* produce the enzymes for fat hydrolysis and the milk becomes sour from fatty acid accumulation. Species of *Micrococcus* may give a yellow tint to the milk because they produce yellow pigments. Anaerobic bacteria such as *Clostridium* grow within the curds and produce gas. The gas may tear apart the curds and cause the curd to explode out of the container.

One of the most direct ways of studying the bacterial population of milk is to set milk samples under various environmental conditions and determine which populations emerge and their sequence of emergence.

PROCEDURE

1. Obtain a number of Erlenmeyer flasks and measure out 100-ml samples of the milk.
   - The instructor will indicate which forms of milk are available for testing in the lab. These may include whole milk, skim milk, buttermilk, and boiled milk.
   - Set the flasks aside at a specified environment (room temperature, body temperature, refrigerator temperature, or other).

2. Observe the flasks daily or at regular intervals as noted by the instructor.
   - At each check, make note in TABLE 88.1 of the physical appearance of the milk, including any color changes that have occurred.
   - Note whether any curds have formed and if any clear fluid has separated from the curds in the milk. If gas has been produced by the bacteria, it will force apart the curds.
3. At each check, use a piece of pH paper to determine the pH of the milk and record it in Table 88.1.
   - Perform microscopic studies using the Gram stain (Exercise 15) to determine which are the predominant forms at that time.
   - Enter your results in the Results section.
4. An interesting variation to the above procedures is to add a pinch of fresh soil to 100 ml of milk in an Erlenmeyer flask.
   - Incubate the milk at room temperature or at 37°C, according to the instructor's directions.
   - As time passes, you will note unique bacterial, physical, and chemical changes occurring rapidly in the flask. The effects are brought about by the considerable numbers of soil bacteria, which change the flora of the milk.
   - Perform cultural studies by isolating bacteria on plates of nutrient agar and making microscopic observations.
   - Microscopic observation may also be made from a sample taken directly from the flask.

Name: ______________________________ Date: __________ Section: __________

Microbiology of Milk and Dairy Products: Natural Bacterial Content of Milk

EXERCISE RESULTS 88

RESULTS

THE NATURAL BACTERIAL CONTENT OF MILK

Type of Milk Cultured:

____

TABLE 88.1	Observations of Developing Bacteria				
Incubation Time	Color	Curd Formation	Gas Production	Acidity	Other Characteristics

Additional Observations:

Stained Bacterial Smears from Milk

Magnif.:

Incuba. Time:

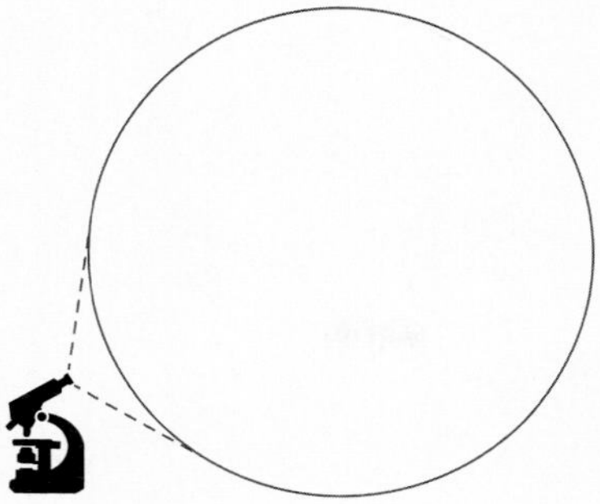

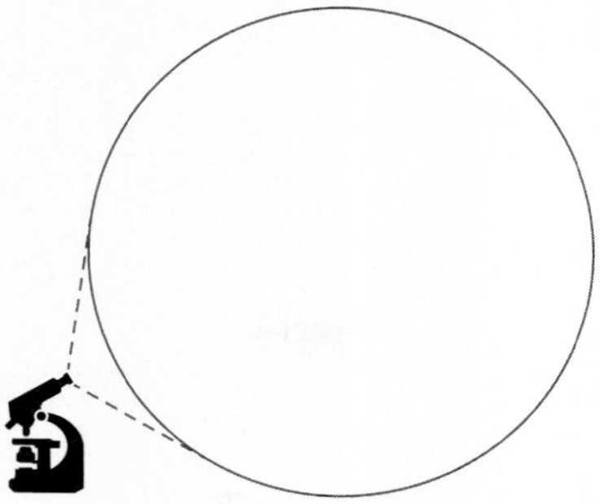

Magnif.: ______________ ______________

Incuba. Time: ______________ ______________

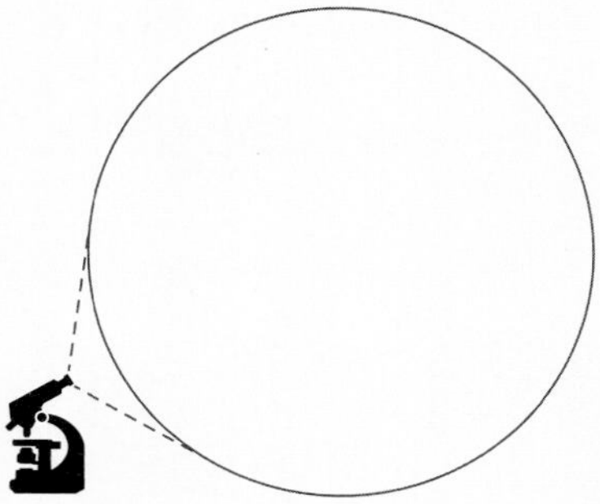

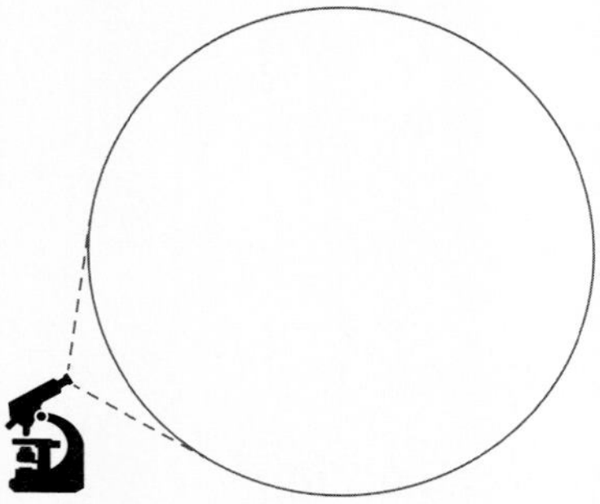

Magnif.: ______________ ______________

Incuba. Time: ______________ ______________

Questions

1. The FDA Pasteurized Milk Ordinance of 2009 states that Grade A pasteurized milk should have no more than 20,000 bacterial cells/ml, with coliforms not to exceed 10/ml. Would the milk tested in this exercise fit those standards?

2. Which other tests are available for testing the bacterial content of milk, and why might they be preferred to those performed in this exercise?

Microbiology of Water: Presumptive Test by the MPN Method

EXERCISE 89

Materials and Equipment

- Water samples for testing
- Double-strength lactose broth in large test tubes containing inverted vials
- Single-strength lactose broth in normal-size test tubes containing inverted vials
- 10-ml pipettes and mechanical pipetters
- 1-ml pipettes

Water may be the vehicle for transferring a broad variety of microbial diseases, including typhoid fever, cholera, amoebiasis, and traveler's diarrhea. Because it is virtually impossible to test for the causative agents of these diseases, certain indicator organisms have been recognized as signals of human waste in waters. An important group of indicator organisms are the **coliform bacteria**, a collection of gram-negative, nonsporeforming rods that ferment lactose to acid and gas. One coliform, *Escherichia coli*, is found almost exclusively in human and animal waste. Its presence in water provides evidence that other intestinal organisms also may be present and that the water may be unfit for human consumption.

PRESUMPTIVE TEST BY THE MPN METHOD

PURPOSE: to test for the presumed presence of coliform bacteria by the most probable number method.

The bacterial examination of water is broken into three tests. The first, or **presumptive test**, is a screening test to sample water for the presence of coliform organisms. In the presumptive test, water samples are inoculated to tubes of lactose broth, and the tubes are incubated and observed for acid and gas production. If acid and gas are produced, the water is presumed to contain coliform organisms, and additional tests are performed (see Exercises 90 and 91). If acid and gas are not formed from lactose, the water is presumed safe for consumption.

The presumptive test also provides a statistical estimate of the number of coliform bacteria present. This estimate is called the **most probable number (MPN)** of coliform bacteria. Though inexact, the MPN gives an idea of the extent of bacterial pollution and is used instead of the standard plate count method (Exercise 83) when precise figures are not required.

PROCEDURE

1. Select five tubes of double-strength lactose broth and 10 tubes of single-strength lactose broth.
   - Label each tube with your name and the date.
   - You will also need a mechanical pipetter, one 10-ml pipette and one 1.0-ml pipette.
   - Collect the water sample in a closed container and shake vigorously to effect even distribution of any organisms present.
   - About 60 ml of water will be needed for testing.
2. Using a mechanical pipetter, aseptically pipette 10 ml of water into each double-strength lactose broth tube. Double-strength lactose is used to provide for dilution by the water sample.
3. Pipette 1 ml of water into five single-strength lactose broth tubes, and 0.1 ml into the remaining five tubes.
   - Label the tubes accordingly.
4. Incubate the 15 tubes at 37°C for 24 hours. The 24-hour time is critical, and tubes should be refrigerated if observations cannot be made at this time.
5. Examine the tubes, and determine the number in each set of inoculations that show at least 10% gas in the inverted vial.
   - Record the number of positive tubes per set in TABLE 89.1 of the Results section.
   - Refer to the table of most probable numbers in **Appendix C** to determine the MPN of coliform bacteria per 100 ml of water.
   - For example, if two tubes in set 1, one tube in set 2, and one tube in set 3 showed gas, then the results would be read 2-1-1 and, according to the table, the MPN would be 9.2 coliform bacteria per 100 ml of water, with 95% confidence that the actual number is between 3.4 and 22.
6. If no tubes show gas production after 24 hours, the presumption is that coliform bacteria are not present in the water.
   - The tubes may be reincubated, and if gas is formed during the second 24-hour incubation period, the presence of coliforms is considered doubtful.

Quick Procedure

MPN Test

1. Pipette 10-ml samples of water to three tubes of double-strength lactose broth.
2. Pipette 1-ml samples to three tubes of single-strength lactose broth.
3. Pipette 0.1-ml samples to three tubes of single-strength lactose broth.
4. Incubate all tubes at 37° C for 24 hours.
5. Evaluate tubes for gas production and compare to table.

Name: ______________________ Date: ____________ Section: ____________

Microbiology of Water: Presumptive Test by the MPN Method

EXERCISE RESULTS **89**

RESULTS

PRESUMPTIVE TEST BY THE MPN METHOD

Source of Water:

____

TABLE 89.1 Number of Lactose Broth Tubes Containing Gas

Inoculum	Set #1 : 10 ml	Set #2 : 1.0 ml	Set #3 : 0.1 ml
Tubes with gas			

Most probable number = ______________ coliform bacteria/100 ml of water

Observations and Conclusions

____

____

____

____

____

Questions

1. List the advantages of the presumptive test by the MPN method over the standard plate count procedure.

2. Why must double-strength lactose broth be used in the MPN test when the inoculum of water consists of 10 ml?

3. Is water showing a negative presumptive test after 24 hours of incubation be presumed safe to drink? Explain.

Microbiology of Water: Confirmed Test

EXERCISE

90

Materials and Equipment

- Positive presumptive test tubes from Exercise 89
- Plates of Levine's EMB agar and Endo agar
- Inoculating loop

If gas-producing organisms are detected in the presumptive test, a **confirmed test** is performed to confirm the presence of coliform bacteria because the gas may have been produced by a gram-positive organism such as *Clostridium perfringens*. Samples of broth from a positive presumptive test are streaked on selective and differential media such as **Levine's eosin methylene blue (EMB) agar** and **Endo agar**. Both media contain lactose for the coliform organisms. EMB agar has methylene blue, which is inhibitory to gram-positive bacteria. Endo agar contains sodium sulfite and basic fuchsin, which are taken up by coliform bacteria to form pink to rose-red colonies.

CONFIRMED TEST

PURPOSE: to confirm the presence of coliform bacteria using selective and differential media.

PROCEDURE

1. Select the tube from Exercise 89 that shows a positive presumptive test and contains the smallest inoculum of water.
   - Obtain or prepare plates of EMB agar and Endo agar, and label each on the bottom side with your name, the date, the medium, and the source of material.
2. Inoculate the EMB agar with a loopful of material from the positive presumptive test using the streak plate method described in Exercise 5.
   - Treat the Endo agar plate in the same way. Incubate the plates in the inverted position at 37°C for 24 to 48 hours.

3. Examine the EMB agar plate for small colonies with dark centers and a greenish metallic sheen in reflected light. These are probably colonies of *E. coli*. Colonies of *Enterobacter aerogenes,* another important coliform, are usually larger and more pink and mucoid.
   - The presence of *E. coli* provides substantial evidence that the water contained human wastes, since this is an intestinal organism.
   - *E. aerogenes* is commonly found in the intestine, but it has also been isolated from grains and soil.
   - At the direction of the instructor, the two organisms may be differentiated by the IMViC series described in Exercise 48.
   - Enter a labeled representation of the plate in the Results section, and draw your conclusions.
4. Observe the Endo agar plate for pink to rose-red colonies, which signify coliform bacteria.
   - Other organisms may be present, but their colonies will be faintly pink or colorless.
   - Enter a labeled representation in the assigned space with appropriate conclusions.

Name: ______________________ Date: __________ Section: __________

Microbiology of Water: Confirmed Test

EXERCISE RESULTS

90

RESULTS

CONFIRMED TEST

EMB Agar

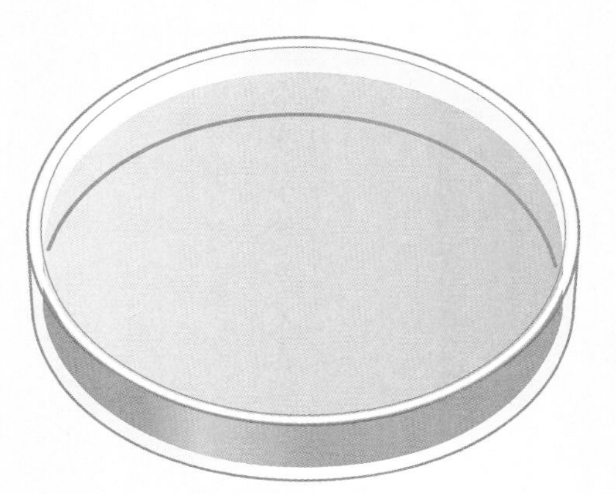

Endo Agar

Observations and Conclusions

____

____

____

____

____

Questions

1. What factors cause EMB agar to be considered a selective as well as a differential medium?

____

____

____

____

Microbiology of Water: Completed Test

EXERCISE

91

Materials and Equipment

- Inoculating loop
- Nutrient agar slants
- Gram stain reagents
- Tubes of lactose broth containing inverted vials
- Incubator at 45°C

The bacteriological analysis of water is continued in a completed test by performing Gram stains on samples from colonies isolated on confirmed test media to determine the morphology and Gram reaction of the isolate and by rechecking for lactose fermentation to gas at the higher incubation temperature of 45°C tolerated by coliforms.

COMPLETED TEST

PURPOSE: to complete the water analysis of coliforms by Gram staining and lactose fermentation.

PROCEDURE

1. Select a tube of lactose broth, and inoculate it with a sample from a coliform colony isolated on EMB agar or Endo agar.
   - Gas should be evident after 24 hours of incubation at 45°C.
   - Enter your results in the Results section, and draw your conclusions.
2. Transfer a sample of organisms from an isolated colony onto a nutrient agar slant, and incubate it at 37°C for 24 to 48 hours.
   - Gram stain a sample of the growth (see Exercise 15 and watch **Microbiology Video: Preparing a Gram Stain** if you need to review the procedure) and examine it for gram-negative nonsporeforming rods.
   - Enter representations in the Results section and draw conclusions.

Name: ______________________________ Date: __________ Section: __________

Microbiology of Water: Completed Test

EXERCISE RESULTS 91

RESULTS

COMPLETED TEST

Evidence of gas after 24-hour incubation period:

____

Gram Stain

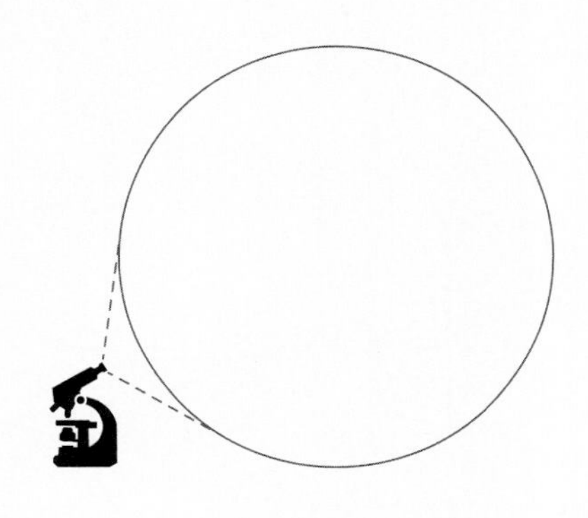

Source: ____________________

Magnif.: ____________________

Observations and Conclusions

____

____

____

____

____

Microbiology of Water: Membrane Filter Technique

EXERCISE 92

Materials and Equipment

- Membrane filter apparatus with 0.45 p.m. filters and absorbent pads
- Sterile 50-mm culture dishes
- Sterile forceps and sterile graduated cylinders
- Flask of Endo broth
- 5-ml pipettes
- Water samples

In the **membrane filter technique**, a large volume of water is tested with equipment that is generally portable. This factor has added to the value of the method. A sample of water is filtered through a special apparatus, and organisms trapped on the filter form colonies during incubation. The colony count gives a reliable estimate of the original number of bacteria present. By using a differential and selective medium (**Endo broth**), the number of pink to rose-color coliform colonies may be determined.

MEMBRANE FILTER TECHNIQUE

PURPOSE: to estimate bacterial prevalence based on filtration and bacterial growth on selective and differential media.

PROCEDURE

1. The instructor will describe the important features of the membrane filter apparatus (FIGURE 92.1) and the special materials to be used.
   - A 100-ml sample of water will be tested for coliform bacteria, and special precautions to avoid contamination will be explained.
2. Place a sterile absorbent pad in a sterile 50-mm culture dish, and add 2 ml of Endo broth to saturate the pad.
3. Obtain a membrane filter apparatus (Figure 92.1) and, using sterile forceps, place a membrane filter in the chamber with the grid side up.
   - The pore size of the filter should be 0.45 p.m.
4. Carefully measure out 100 ml of a water sample in a sterile graduated cylinder, and pour it into the filter funnel.
   - Allow the water to pass through the filter into the collecting flask.

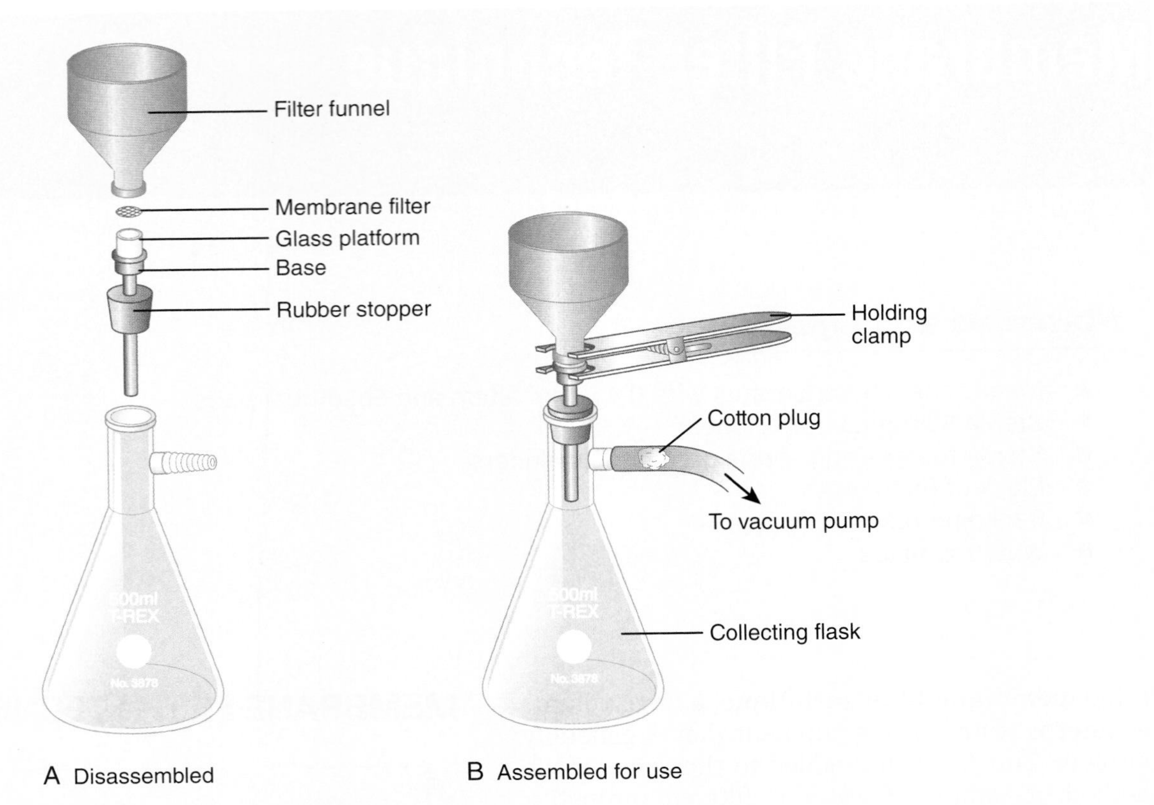

FIGURE 92.1 The membrane filter apparatus.

5. Aseptically remove the filter, and place it with the grid side up on the soaked pad in the culture dish.
   - Incubate the plate at 37°C for 24 hours in the upright position.

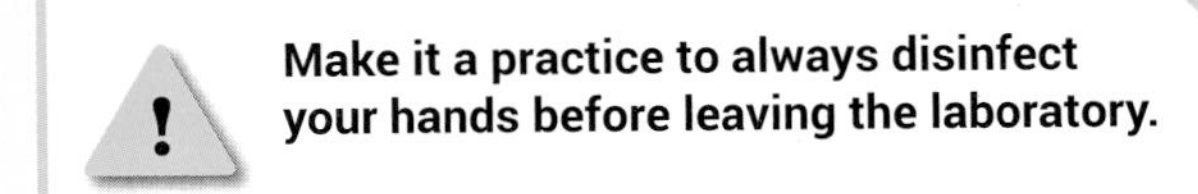

6. Count the coliform colonies that are pink or red with a metallic sheen, and express the colony count per 100 ml of water in the Results section.
   - Record your observations on the presence and number of coliform bacteria in the sample of water tested.
   - Include a diagram of the plate.

Quick Procedure

Membrane Filter Technique

1. Filter 100 ml water through the membrane filter apparatus.
2. Place the filter on pad soaked with Endo broth.
3. Incubate.
4. Evaluate and count colonies on plate.

Name: ______________________ Date: __________ Section: __________

Microbiology of Water: Membrane Filter Technique

EXERCISE RESULTS 92

RESULTS

MEMBRANE FILTER TECHNIQUE

Membrane Filter

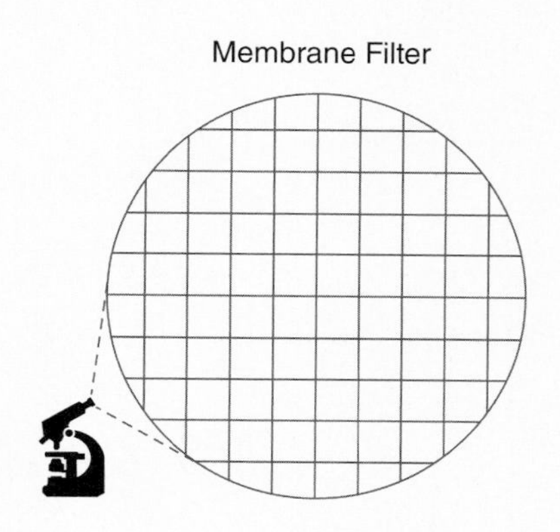

Source of water:

____

Colony count = ______________ coliform bacteria/100 ml of water

Observations and Conclusions

____

____

____

____

____

Questions

1. Why are laboratory tests for bacterial pollution of waters directed to the detection of coliform bacteria and, especially, *Escherichia coli*?

Microbiology of Water: Preparation of a Biofilm

EXERCISE 93

Materials and Equipment

- Inoculating loop
- Beakers of various environmental waters
- Slides, string, waterproof tape, and pencils
- 60-watt lamp
- Microscope

There are many interesting microorganisms of diverse shapes and sizes existing in the aquatic environment. Often, these organisms are difficult to cultivate in the laboratory because the correct methods have not yet been established. However, the aquatic organisms may be captured in a **biofilm**, which is a population of microorganisms adhering to a surface. In this exercise, a biofilm is developed on a slide and along the glass wall of a beaker. The aquatic microorganisms may then be observed for comparison and study. Many photosynthetic bacteria and cyanobacteria will be captured.

PREPARATION OF A BIOFILM

PURPOSE: to prepare and observe microorganisms that compose a biofilm.

PROCEDURE

1. Obtain a sample of river water, pond water, sewage water, or other water as specified by the instructor. Birdbath, fountain, or aquarium water may be used.
   - Place the water in a beaker until it is nearly full.
2. Thoroughly clean a glass slide and attach a piece of string to one end using waterproof tape.
   - Then extend a glass or wooden rod across the mouth of the beaker for use as a scaffold.
   - Tie the end of the string to the rod and lower the slide into the water, thereby suspending it (FIGURE 93.1).
   - Several slides may be prepared in this manner.

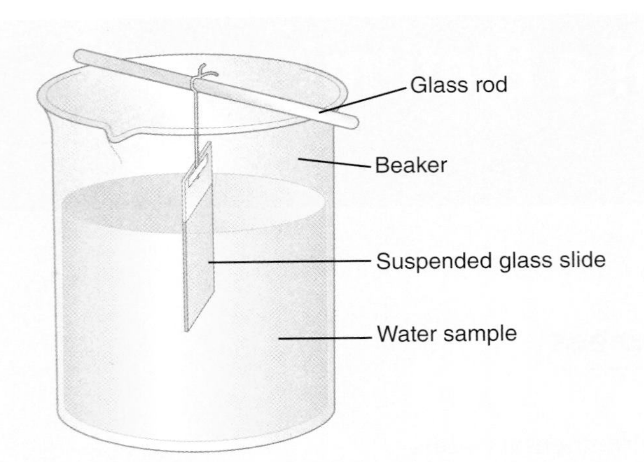

FIGURE 93.1 Apparatus for obtaining a biofilm on a slide.

3. Set the beaker aside at room temperature or other temperature designated by the instructor.
   - Over a period of days, bacteria and other microorganisms will cling to the slide because of their natural tendency to attach to solid objects.
   - Additional water may be added from time to time to replace water lost to evaporation.
4. At regular intervals, remove a slide from the beaker, wipe one side clean, and permit the other side to air-dry.
   - Then heat-fix the slide and stain it by the simple stain technique (Exercise 13) or the Gram stain technique (Exercise 15).
   - Oil immersion observations should reveal many unique and unusual forms of bacteria, algae, and other microorganisms.
   - Record your observations in the Results section.
5. A biofilm also may be prepared along the side of a beaker by encouraging the growth of photosynthetic organisms.
   - To accomplish this, fill a beaker half-full with a type of water as noted in Step 1 above.
   - Then set up a 60-watt lamp and shine it directly on the side of the beaker.
   - After some days, a biofilm of green organisms will begin to develop along the glass.
   - Samples may be obtained and placed on slides for staining and observation.
   - Many algae will be seen, and photosynthetic bacteria will also appear, depending on the source of the water used.
   - Water lost to evaporation should be replaced from time to time.

Name: ______________________ Date: __________ Section: __________

Microbiology of Water: Preparation of a Biofilm

EXERCISE RESULTS 93

RESULTS

PREPARATION OF A BIOFILM

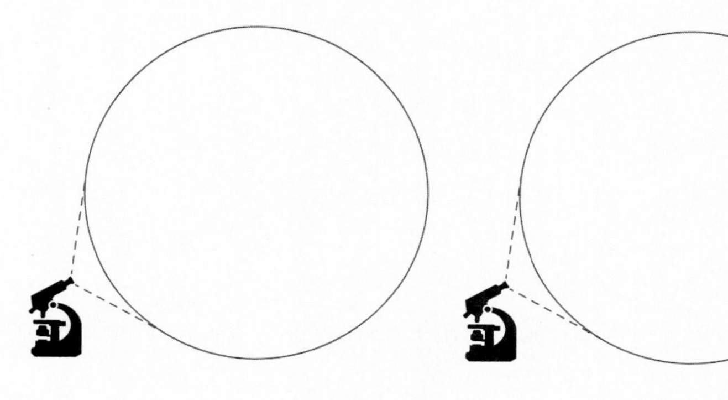

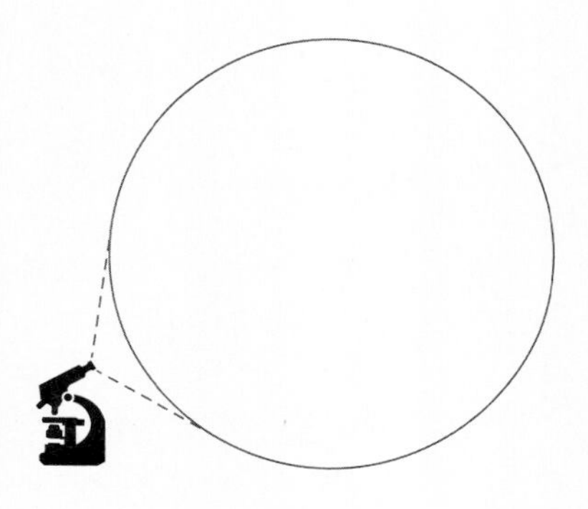

Source: ______________ ______________ ______________

Magnif.: ______________ ______________ ______________

Observations and Conclusions

____

____

____

____

____

Questions

1. Depending on the source of a biofilm, the community of microbes may be considered beneficial or harmful. This decision depends upon how a biofilm affects humans and the environment. How might a soil or aquatic biofilm benefit or harm the environment, and how might a human biofilm in the colon benefit or harm the human?

 ____

 ____

 ____

 ____

Microbiology of Soil: Isolation of *Rhizobium* from Legume Roots

EXERCISE
94

Materials and Equipment

- Leguminous plants, such as peas, beans, or clover
- Crystal violet or methylene blue
- Culture plates of Emerson agar containing cycloheximide

Microorganisms enhance the quality of life through their activities in the soil. The bacteria, fungi, and other microbes forming a soil biofilm play substantial roles in the carbon, sulfur, and nitrogen cycles in the environment. In addition, many soil microorganisms produce the antibiotics used in medicine.

In this exercise, bacteria isolated from the roots of plants will be observed.

ISOLATION OF *RHIZOBIUM* FROM LEGUME ROOTS

PURPOSE: to isolate *Rhizobium* and observe cell morphology.

Members of the genus ***Rhizobium*** are gram-negative bacteria that live in a symbiotic relationship with legumes (plants that bear their seeds in pods). The bacteria live in nodules on the legume roots and trap atmospheric nitrogen and convert it to ammonia. The excess ammonia is released by the bacteria and the compound becomes available to the plant as part of the nitrogen cycle in the soil.

In this exercise, *Rhizobium* species will be observed from root nodules and cultivated on **Emerson agar**. This medium contains yeast extract and glucose, which encourage the growth of rhizobia, as well as cycloheximide, a drug that inhibits eukaryotic protein synthesis and therefore, mold growth.

PROCEDURE

1. Obtain a fresh legume plant, and examine the roots for nodules, which contain *Rhizobium* species.
   - Select one nodule, and clean it thoroughly before proceeding.
2. Place a drop of water on a clean slide, and crush the nodule in the water, using a convenient instrument such as the edge of a second slide. This will produce a milky suspension.

3. Transfer a loopful of the suspension to a slide, air-dry and heat-fix the smear, and perform a simple stain using crystal violet (Exercise 13).
   - Observe the pleomorphic rods under oil immersion, and note the X, Y, star, club, and other shapes characteristic of *Rhizobium* species.
   - Enter a representation in the Results section.
4. Obtain a second loopful of the suspension, and streak it for isolated colonies on **Emerson agar** as described in Exercise 5.
   - Incubate the plate at 25°C or room temperature for 1 week.
   - Examine the plate for colonies with a glistening appearance that are white at the edges.
   - Prepare a Gram stain from a *Rhizobium* colony, and note whether small gram-negative rods are present.
   - Draw representations in the Results section.

Name: ______________________ Date: __________ Section: __________

Microbiology of Soil: Isolation of *Rhizobium* from Legume Roots

EXERCISE RESULTS **94**

RESULTS

ISOLATION OF *RHIZOBIUM* FROM LEGUME ROOTS

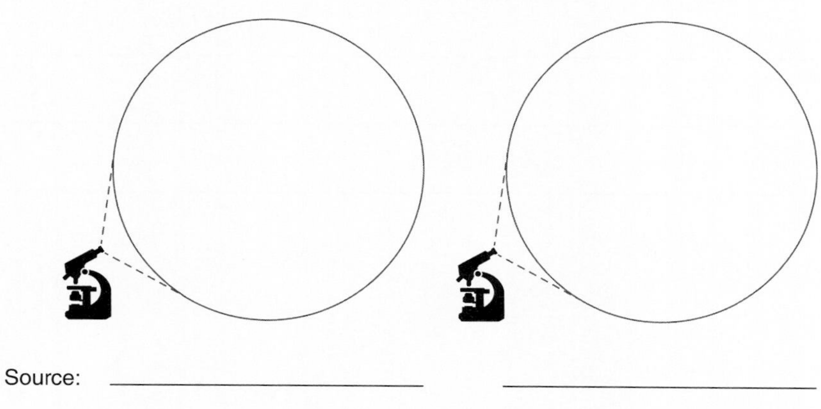

Source: ______________________ ______________________

Magnif.: ______________________ ______________________

Observations and Conclusions

____

____

____

____

____

Questions

1. Why are *Rhizobium* species from root nodules considered pleomorphic bacteria?

2. Explain the relationship that exists between *Rhizobium* species and leguminous plants.

Microbiology of Soil: Ammonification by Soil Microorganisms

EXERCISE **95**

Materials and Equipment

- Tubes of peptone broth
- Samples of rich (garden) soil
- Depression plates
- Nessler's reagent in dropper bottles
- Test tubes

An important aspect of the **nitrogen cycle** takes place when soil bacteria convert protein to amino acids and then digest the amino acids to yield ammonia. This process, known as **ammonification**, is critical to the cycle because the ammonia is next converted to nitrite, or the nitrogen is released to the atmosphere.

In this exercise, ammonification will be demonstrated by incubating soil bacteria in a protein solution and then testing for the presence of ammonia. Ammonia will be detected using **Nessler's reagent**, which is an aqueous solution of potassium iodide, mercuric chloride, and potassium hydroxide. The normally pale solution becomes yellow to brown in color in the presence of ammonia.

AMMONIFICATION BY SOIL MICROORGANISMS

PURPOSE: to detect the process of ammonification.

PROCEDURE

1. Suspend a 1-gram sample of rich soil in several milliliters of water in a test tube.
   - Obtain a tube of peptone broth, and label it in the prescribed manner.
2. Inoculate the peptone broth with two loopfuls of the soil suspension, and incubate the tube at room temperature for 48 hours. An uninoculated control tube should be included.
3. Test the broth for the presence of ammonia by adding a loopful of broth to a drop of Nessler's reagent in one area of a depression plate or on a slide.
   - Mix the reactants well.
   - Combine a loopful of the uninoculated control broth with a second drop of Nessler's reagent in a second area and mix.

4. Observe for a faint yellow color in the experimental mixture, which indicates a "trace" amount of ammonia.
    - A deep yellow color points to a "moderate amount" of ammonia, while a brown color indicates a "large amount" of ammonia.
    - The control broth should show no color change.
    - Note the incubation period tested and the amount of ammonia produced in TABLE 95.1 of the Results section.
    - Repeat this test after additional incubation periods, and indicate the relative amount of ammonia present.

Name: ______________________ Date: __________ Section: __________

Microbiology of Soil: Ammonification by Soil Microorganisms

EXERCISE RESULTS 95

RESULTS

AMMONIFICATION BY SOIL MICROORGANISMS

TABLE 95.1	Amount of Ammonia Present after Various Incubation Periods		
Incubation time	Hr	Hr	Hr
Final solution color			

Observations and Conclusions

____

____

____

____

____

Questions

1. What does peptone contain that is necessary for the demonstration of ammonification by soil microorganisms?

____

____

____

____

Microbiology of Soil: Isolation of *Streptomyces* from Soil

EXERCISE 96

Materials and Equipment

- Plates of Emerson agar with cycloheximide
- Sample of rich (garden) soil
- Sterile cotton swabs and disinfectant

Streptomyces species commonly found in the soil produce many of the antibiotics used in medicine. These organisms grow well on **Emerson agar** with cycloheximide to inhibit the proliferation of molds.

ISOLATION OF *STREPTOMYCES* FROM SOIL

PURPOSE: to isolate *Streptomyces* species and detect antibiotic production.

In this exercise, *Streptomyces* species will be isolated from the soil.

PROCEDURE

1. Suspend a sample of soil in water as described in Exercise 95.
   - Prepare or select a plate of Emerson agar, and label it in the prescribed manner.
2. Obtain a loopful of the soil suspension, and streak it for isolated colonies as outlined in Exercise 5.
   - Incubate the plate at room temperature for 1 week.
3. Examine the plate for hard, white colonies shaped like tiny volcanoes. These are possible *Streptomyces* species.
   - The material in the colony will be dry and difficult to remove, but a section should be pried from the edge of the colony and examined in a wet mount preparation under the high-power (40×) lens.
   - Make a comparison to fungal hyphae and place a representation in the Results section. Be careful not to confuse *Streptomyces* spores with bacterial cocci.
   - When your observations are complete, place the wet mount in a beaker of disinfectant to kill the cells.

Name: ______________________ Date: __________ Section: __________

Microbiology of Soil: Isolation of *Streptomyces* from Soil

EXERCISE RESULTS **96**

RESULTS

STREPTOMYCES FROM THE SOIL

Wet Mount of *Streptomyces*

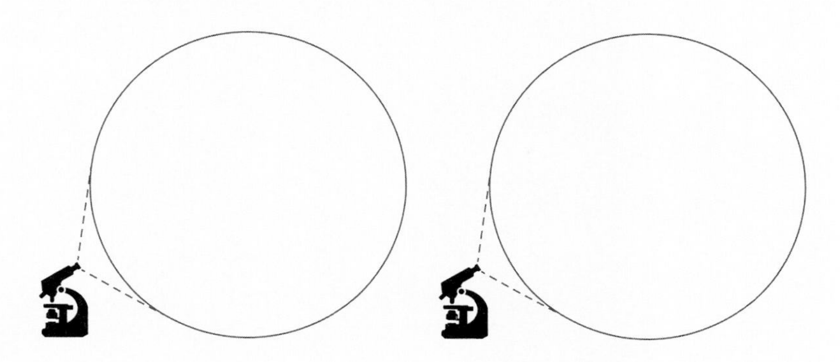

Observations and Conclusions

____

____

____

____

____

Questions

1. Compare the microscopic appearance of *Streptomyces* species with that of the fungi.

____

____

____

____

Microbiology of Soil: Antibiotic Production by *Streptomyces*

EXERCISE 97

Materials and Equipment

- Plates of Emerson agar with cycloheximide
- Broth tubes of selected bacterial species
- Sterile cotton swabs and disinfectants
- Inoculating loop
- Rich (garden) soil samples
- Plates from Exercise 96

The ability to produce antibiotics may be examined in isolated species of **Streptomyces** by incubating the organism with certain test organisms and determining whether growth inhibition of the test organism takes place. The method is outlined in this exercise.

ANTIBIOTIC PRODUCTION BY *STREPTOMYCES*

PURPOSE: to perform an inhibition test to screen for antibiotic production.

PROCEDURE

1. Obtain or prepare a plate of Emerson agar, and label it in the prescribed manner.
2. Using the bacteriological loop, take a section of the *Streptomyces* colony identified in Exercise 96, and make a single streak across the center of the plate, as shown in FIGURE 97.1.

Avoid carrying culture dishes containing bacterial colonies plus additives over to your instructor to give him or her a look. Ask the instructor to stop by your desk to see your results.

- Because the inoculum is dry, slight pressure may be needed to ensure that the organisms have been deposited on the plate.
- Incubate the plate at room temperature for 3 to 4 days until the growth of the *Streptomyces* is apparent.

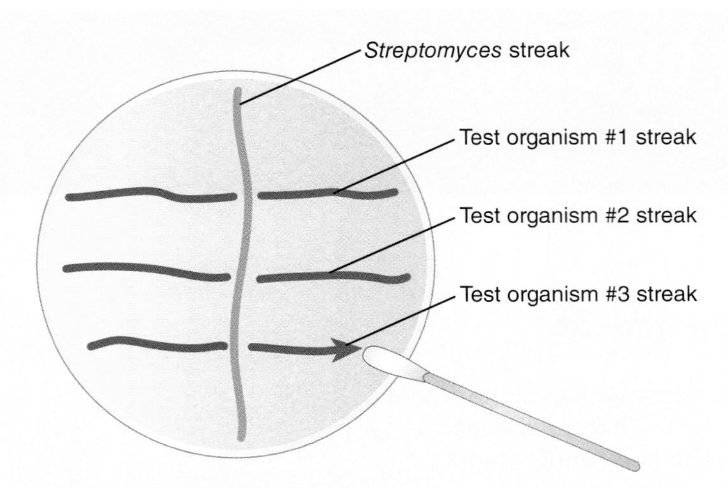

FIGURE 97.1 Procedure for testing antibiotic activity of *Streptomyces* on various test organisms.

3. After the initial incubation period, make single streaks of available test organisms at right angles to the *Streptomyces* streak, using different sterile cotton swabs for each organism (**FIGURE 97.1**).
   - Be careful not to touch the *Streptomyces* growth with the swab.
   - Reincubate the plate for an additional week at room temperature.
4. Examine the growth of the test organisms where the streak crosses that of *Streptomyces*.
   - Note a reduction in the amount of growth of the test organism, indicative of antibiotic activity.
   - Illustrate your results in the appropriate space.
5. It may also be possible to test antibiotic activity of various soil bacteria by swabbing a plate of Emerson agar with a test organism and then lightly sprinkling a tiny amount of soil over the growth. As soil organisms proliferate, certain ones will produce antibiotic substances that inhibit the growth of the test organism.
   - Colonies of soil bacteria will be surrounded by clear **zones of inhibition** where antibiotic activity has taken place.
   - Place a representation of the plate in the Results section.

Name: ______________________ Date: __________ Section: ___________

Microbiology of Soil: Antibiotic Production by *Streptomyces*

EXERCISE RESULTS **97**

RESULTS

ANTIBIOTIC PRODUCTION BY *STREPTOMYCES*

Antibiotic Production on Emerson Agar

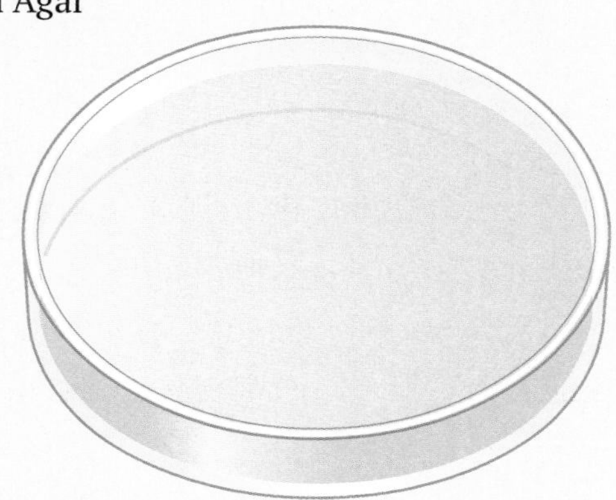

Observations and Conclusions

____

____

____

____

____

Questions

1. Why would species of *Streptomyces* normally produce antibiotics in soil?

____

____

____

____

Microbiology of Soil: Plate Count of Soil Bacteria

EXERCISE **98**

Materials and Equipment

- Samples of rich (garden) soil
- Sterile culture dishes
- Sterile 1.1-ml pipettes and mechanical pipetters
- Sterile 99-ml water dilution blanks
- Melted nutrient agar

Some samples of rich soil may contain over a million bacterial cells per gram. The bacteria add to the fertility of the soil and, with the fungi, serve as the **primary decomposers** of organic material. In this section, a determination will be made of the numbers of bacterial cells in a soil sample.

PLATE COUNT OF SOIL BACTERIA

PURPOSE: to quantify soil bacteria.

PROCEDURE

1. Obtain two sterile 99-ml water dilution blanks, a sterile 1.1-ml pipette, and four sterile culture dishes. Label the dishes with your name, the date, and the designations 1:100, 1:1,000, 1:10,000, and 1:100,000.
2. Weigh 1 gram of rich soil and add it to the first dilution blank.
   - Shake the bottle vigorously for 2 minutes.
   - Allow the large particles to settle.
   - Using a mechanical pipetter, pipette 0.1 ml of the fluid to the plate labeled 1:1,000, and 1.0 ml to the plate labeled 1:100.
   - Be careful to avoid airborne contamination.
3. Using a mechanical pipetter, pipette 1 ml from the first dilution blank to the second 99-ml dilution blank, then draw up and release the fluid several times to wash out the pipette.
   - Shake the bottle vigorously for 2 minutes.
   - Pipette 0.1 ml to the 1:100,000 plate, and 1.0 ml to the 1:10,000 plate.
   - The pipette may now be discarded.

4. Aseptically add to each plate enough melted nutrient agar to cover the bottom.
   - Mix the medium and fluid by gently rotating the plates in a wide arc.
   - After the medium has solidified, invert the plates and incubate them at 37°C.
   - Soil bacteria tend to grow rapidly, so a 24-hour incubation time or less should be sufficient.
   - Refrigerate the plates until observed.
5. Observe the plates, and perform a plate count as described in Step 7 of Exercise 83.
   - Enter the plate count per gram of soil in TABLE 98.1 of the Results section.
   - Consult with the instructor if you wish to examine the various types of bacteria that have appeared in the plates.

Quick Procedure

Standard Plate Count (Soil)

1. Blend soil sample with sterile water in first dilution blank and shake.
2. Pipette 1.0- and 0.1-ml samples of diluted soil to culture dishes.
3. Pipette 1.0 ml diluted soil to a second 99-ml water blank and shake.
4. Pipette 1.0- and 0.1-ml samples of diluted soil from the second water blank to culture dishes.
5. Add liquid nutrient agar to all culture dishes and mix.
6. Incubate.
7. Perform colony counts and locate valid count.
8. Multiply valid count by dilution factor.

Name: ______________________ Date: __________ Section: __________

Microbiology of Soil: Plate Count of Soil Bacteria

EXERCISE RESULTS 98

RESULTS

PLATE COUNT OF SOIL BACTERIA

Source of soil:

____

TABLE 98.1	Plate Counts at Various Dilutions			
Dilution	1:100	1:1,000	1:10,000	1:100,000
Plate count				

__________ × __________ = __________ total bacteria per gram of soil
(valid count) (dilution factor)

Observations and Conclusions

____

____

____

____

____

Questions

1. Why is there such a diversity of bacteria in soil?

Microbiology of Soil: Microbial Ecology in the Soil—The Winogradsky Column

EXERCISE **99**

Materials and Equipment

- 100-ml graduated cylinder
- Plug-in light with 60-watt bulbs
- Calcium sulfate
- Calcium carbonate
- Glass rods (or other device) for tamping
- Pasteur pipettes
- Aluminum foil
- Glass slide
- Microscope

Soil microorganisms live in a complex ecological system where changing environments stimulate the emergence of various microbial populations including archaea, bacteria, cyanobacteria, protists, fungi, and other microscopic forms. The microbial populations fulfill essential roles in the ecology of the environment. For example, photosynthetic bacteria and algae in soil trap energy from the sun and use it to synthesize carbohydrates, which are used as energy sources by all other forms of life. Anaerobic microorganisms help digest sewage and garbage in landfills, converting the organic material into useful compost. Autotrophic microorganisms of the soil synthesize their foods from simple materials such as carbon dioxide, ammonia, methane, and water, thus providing a way of converting simple chemicals to complex organic matter used by animals and plants.

MICROBIAL ECOLOGY IN THE SOIL—THE WINOGRADSKY COLUMN

PURPOSE: to study the emergence of microbial populations in a constructed ecosystem.

This exercise studies the emergence of microbial populations in an ecosystem known as the **Winogradsky column** (named for Sergei Winogradsky, a noted soil microbiologist of the early 1900s). The column is actually a small vertical pond formed by packing a cylinder with mud, shredded paper, salts, and water. The cylinder will be sealed and exposed to light, and different microbial populations will be encouraged to emerge.

PROCEDURE

1. Collect a sample of rich mud and water from the bottom of a pond or lake.
   - Remove the large particles and stones from the sample.
2. Obtain a 100-ml graduated cylinder and fill it with about one inch of shredded paper or paper towels mixed with distilled water.
3. Mix approximately 100 grams of mud with about 10 grams of calcium sulfate and 10 grams of calcium carbonate.
   - Pour enough of the mixture into the column to fill it about two-thirds full. These substances provide sulfur and carbon to autotrophic organisms.

4. With a thick glass rod or other device, tamp the surface of the mud to pack the column tightly and eliminate any air bubbles.
   - Pond water may be used to displace the air trapped in the mud.
   - Then add enough water to fill the column to about one inch from the top of the cylinder.
5. Place a piece of plastic wrap over the top of the column to prevent evaporation of the water.
   - Cover the entire column with aluminum foil to prevent light from penetrating the column. This covering will eliminate algae in the ecosystem and encourage other populations to emerge.
6. Place the column at room temperature and allow it to remain for four to seven days.
7. Remove the aluminum foil and place the column near a light source having a 60-watt bulb.
   - The light should be approximately 12 to 20 inches from the column.
   - Permit the column to remain here for several days or weeks.
8. During the extended incubation period, various colors will develop in the column as various pigmented organisms emerge (FIGURE 99.1).
   - Watch the column for the development of brown, green, or red patches and make a note of where they occur in the column. These patches contain various photosynthetic bacteria. The bacteria are using the light as an energy source to synthesize their carbohydrates.

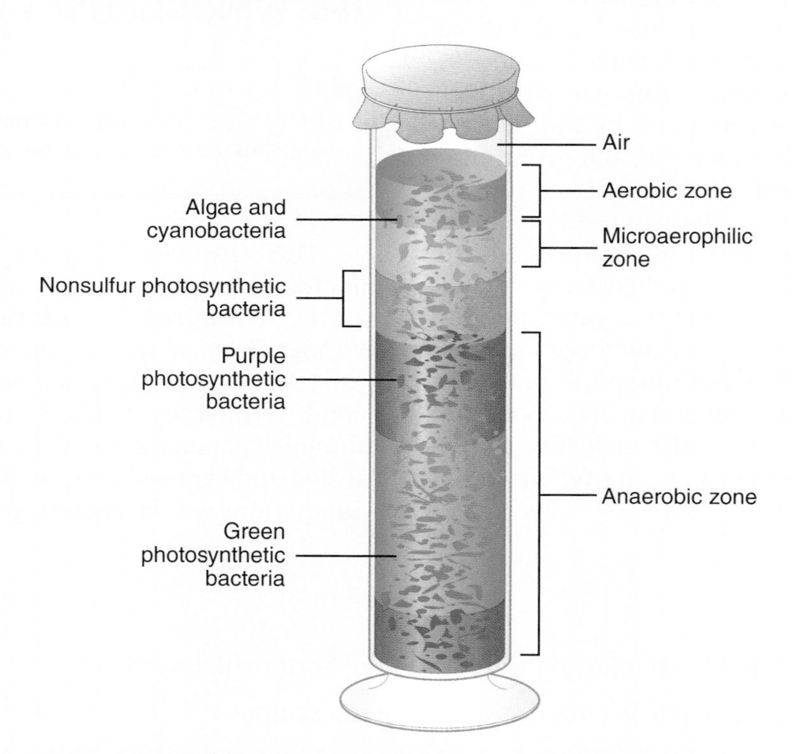

FIGURE 99.1 A Winogradsky column showing the different gaseous zones and types of pigmented organisms that develop in these zones.

9. Using a Pasteur pipette, fish some of the colored growth from the Winogradsky column, and place it on a clean glass slide.

- Place a cover glass over the drop and observe the organisms under the low power (10×) and high power (40×) lenses of the microscope.
- Stained slides may be made for oil immersion microscopy at the direction of the instructor.
- Enter your results in the Results section.

10. Note: It is a good idea to let the column develop for an unusually long time—a month or a year is not unheard of. You will see an unusual array of colors and an equally interesting number of microbial populations as they flourish and die off.

Name: ______________________ Date: __________ Section: __________

Microbiology of Soil: Microbial Ecology in the Soil—The Winogradsky Column

EXERCISE RESULTS **99**

RESULTS

MICROBIAL ECOLOGY IN THE SOIL—THE WINOGRADSKY COLUMN

Stained Smears of Microbial Samples from the Column

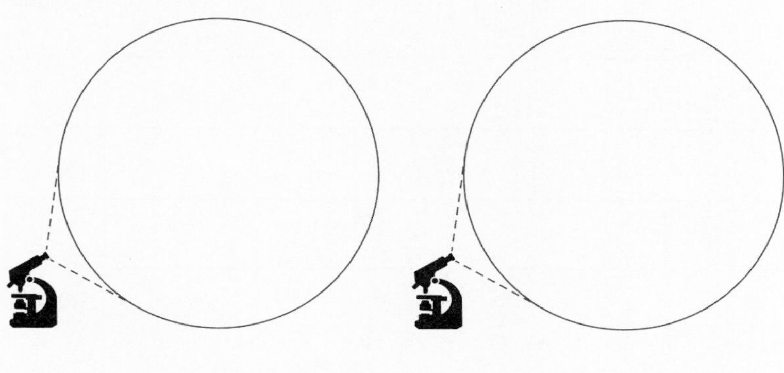

Magnif.: ______________ ______________

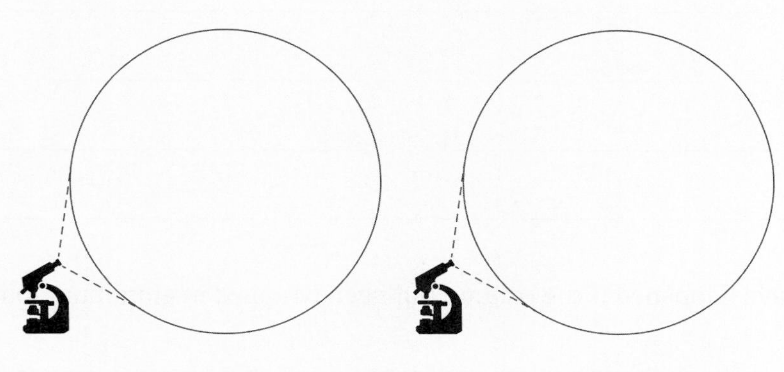

Magnif.: ______________ ______________

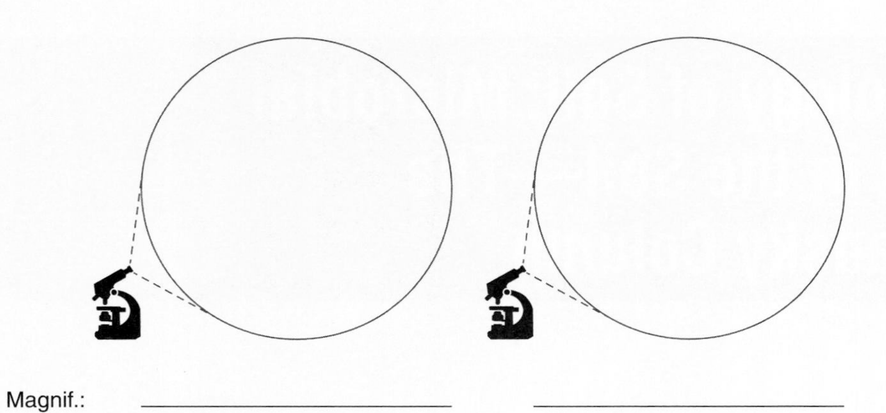

Magnif.: ____________________ ____________________

Observations and Conclusions

____

____

____

____

____

Questions

1. What is the function of material (mud, newspaper, etc.) added in the Winogradsky column?

____

____

____

____

2. What would have happened if the column not been wrapped in aluminum foil for the first incubation?

____

____

____

____

Appendix A

Preparation of Stains and Diagnostic Reagents

STAINING REAGENTS

Acid-Fast Stain

Carbolfuchsin (*Ziehl-Neelsen acid-fast technique*)

Solution A:

Basic fuchsin (90% dye content)	0.3 g
95% ethyl alcohol	10.0 ml

Solution B:

Phenol	5.0 g
Distilled water	95.0 ml

Mix solutions A and B.

Carbolfuchsin (*Kinyoun acid-fast technique*)

Prepare Ziehl-Neelsen carbolfuchsin.

Add 1 drop of tergitol 7 to 35 ml of stain.

Also available from commercial suppliers.

Acid alcohol

Concentrated hydrochloric acid	3.0 ml
95% ethyl alcohol	97.0 ml

Methylene blue (*Ziehl-Neelsen acid-fast technique*)

Methylene blue (90% dye content)	0.1 g
Distilled water	100.0 ml

Brilliant green

Brilliant green	1.0 g
Distilled water	100.0 ml

Add 0.01 g sodium azide and mix.

Capsule Stain

Crystal violet (1%)

Crystal violet (85% dye content)	1.0 g
Distilled water	100.0 ml

Gram Stain

Crystal violet

Solution A:

Crystal violet (90% dye content)	2.0 g
95% ethyl alcohol	20.0 ml

Solution B:

Ammonium oxalate	0.8 g
Distilled water	80.0 ml

Mix solutions A and B, and filter.

Gram's iodine

Potassium iodide	2.0 g
Distilled water	300.0 ml
Iodine crystals	1.0 g

Store in dark brown bottle.

Ethyl alcohol (95%)

100% ethyl alcohol	95.0 ml
Distilled water	5.0 ml

Safranin

Safranin O (90% dye content)	0.25 g
95% ethyl alcohol	10.0 ml
Distilled water	100.0 ml

Filter final solution.

Negative Stain

Congo red

Congo red	1.0 g
Distilled water	100.0 ml

Nigrosin

Nigrosin (water soluble)	10.0 g
Distilled water	100.0 ml

Immerse in a boiling waterbath for 30 minutes.

Add 0.5 ml of formalin (40%) and filter two times through double filter paper.

Simple Stain

Crystal violet

Same as Gram stain

Methylene blue

Solution A:

Methylene blue (90% dye content)	0.3 g
95% ethyl alcohol	30.0 ml

Solution B:

Potassium hydroxide	0.01 g
Distilled water	100.0 ml

Mix solutions A and B.

Spore Stain

Malachite green

Malachite green oxalate	5.0 g
Distilled water	100.0 ml

Safranin

Same as Gram stain

Blood Smear Stain

Wright's stain

Wright's stain powder	0.3 g
Glycerin	3.0 ml

Grind together in a mortar. Then, gradually add 100 ml of acetone-free methyl alcohol.

Store in a tightly stoppered bottle, and filter before use.

Also available from commercial suppliers.

DIAGNOSTIC REAGENTS – BIOCHEMICAL TESTS

Catalase Test

Hydrogen peroxide (3%)

Available from commercial suppliers.

Indole Test

Kovac's reagent

Para-dimethylaminobenzaldehyde	5.0 g
Amyl alcohol	75.0 ml

Heat the mixture at 55°C until crystals dissolve. Allow the mixture to cool.

Concentrated hydrochloric acid	25.0 ml

Also available from commercial suppliers.

Methyl Red Test

Methyl red solution

Methyl red	0.1 g
95% ethyl alcohol	300.0 ml

Dilute with distilled water to 500 ml.

Oxidase Test

Oxidase reagent

Dimethyl-*p*-phenylenediamine hydrochloride	1.0 g
Distilled water	100.0 ml

Prepare shortly before use, refrigerate, and discard darkened solutions.

Voges-Proskauer Test

Barritt's A solution (VP reagent #1)

Alpha-naphthol	6.0 g
95% ethyl alcohol	100.0 ml

Dissolve with constant stirring. Caution should be exercised since alpha-naphthol in considered carcinogenic. Avoid contact with human tissues.

Also available from commercial suppliers.

Barritt's B solution (VP reagent #2)

Potassium hydroxide	16.0 g
Distilled water	100.0 ml

Also available from commercial suppliers.

Starch Hydrolysis Test

Gram's iodine

Same as Gram stain

DIAGNOSTIC REAGENTS – OTHER SOLUTIONS

Ammonia Detection

Nessler's reagent

Potassium iodide	50.0 g
Distilled water	35.0 ml

Add saturated aqueous mercuric chloride until a slight precipitate is seen.

Potassium hydroxide (50% aqueous)	400.0 ml

Dilute to 1,000 ml with distilled water. Let stand for one week and then decant supernatant into dark brown bottle.

Also available from commercial suppliers.

Methyl Cellulose

Methyl cellulose	10.0 g
Distilled water	90.0 ml

Reduction Test (for milk)

Methylene blue thiocyanate solution

Methylene blue thiocyanate (90% dye content)	0.05 g
Distilled water	100.0 ml

Also available as tablets from commercial suppliers.

Saline Solution (0.85%)

Sodium chloride	8.5 g
Distilled water	1.0 L

Appendix B

Using Enterotube® II*

REACTIONS			REMARKS
	Negative	**Positive**	
			Compartment 1
Glucose	Red	Yellow	**Glucose (GLU):** The end products of bacterial fermentation of glucose are either acid, or acid and gas. The shift in pH due to the production of acid is indicated by change in the color of the indicator in the medium from red (alkaline) to yellow (acidic). Any degree of yellow should be interpreted as a positive reaction; orange should be considered negative.
Gas Production	Wax not lifted	Wax lifted	**Gas production (GAS):** This is evidenced by definite and complete separation of the wax overlay from the surface of the glucose medium, but not by bubbles in the medium. Since the amount of gas produced by different bacteria varies, the amount of separation between medium and overlay will also vary with the strain being tested.
			Compartment 2
Lysine	Yellow	Purple	**Lysine decarboxylase (LYS):** Bacterial decarboxylation of lysine, which results in the formation of the alkaline end product cadaverine, is indicated by a change in the color of the indicator in the medium from pale yellow (acidic) to purple (alkaline). Any degree of purple should be interpreted as a positive reaction. The medium remains yellow if decarboxylation of lysine does not occur.
			Compartment 3
Ornithine	Yellow	Purple	**Ornithine decarboxylase (ORN):** Bacterial decarboxylation of ornithine, which results in the formation of the alkaline end product putrescine, is indicated by a change in the color of the indicator in the medium from pale yellow (acidic) to purple (alkaline). Any degree of purple should be interpreted as a positive reaction. The medium remains yellow if decarboxylation of ornithine does not occur.

(*continued*)

REACTIONS			REMARKS
	Negative	Positive	
			Compartment 4
H_2S	Beige	Black	**H_2S production (H_2S):** Hydrogen sulfide is produced by bacteria capable of reducing sulfur-containing compounds, such as peptones and sodium thiosulfate present in the medium. The hydrogen sulfide reacts with the iron salts also present in the medium to form a black precipitate of ferric sulfide usually along the line of inoculation. Some *Proteus* and *Providencia* strains may produce a diffuse brown coloration in this medium, however, this should not be confused with true H_2S production, i.e., presence of black color. Note: The black precipitate may fade or revert back to negative if Enterotube® II is read after 24 hr incubation.
Indole formation	Colorless	Pink–red	**Indole formation (IND):** The production of indole from the metabolism of tryptophan by the bacterial enzyme tryptophanase is detected by the development of a pink to red color after the addition of Kovacs' indole reagent, which is injected into the compartment after 18 to 24 hr incubation of the tube.
			Compartment 5
Adonitol	Red	Yellow	**Adonitol (ADON):** Bacterial fermentation of adonitol, which results in the formation of acidic end products, is indicated by a change in color of the indicator present in the medium from red (alkaline) to yellow (acidic). Any sign of yellow should be interpreted as a positive reaction; orange should be considered negative.
			Compartment 6
Lactose	Red	Yellow	**Lactose (LAC):** Bacterial fermentation of lactose, which results in the formation of acidic end products, is indicated by a change in color of the indicator present in the medium from red (alkaline) to yellow (acidic). Any sign of yellow should be interpreted as a positive reaction; orange should be considered negative. This test is useful to confirm the lactose reaction of colonies taken from various enteric differential media, e.g., *Salmonella-Shigella* (SS), Mac-Conkey (MAC), Eosin Methylene Blue (EMB).
			Compartment 7
Arabinose	Red	Yellow	**Arabinose (ARAB):** Bacterial fermentation of arabinose, which results in the formation of acidic end products, is indicated by a change in color of the indicator present in the medium from red (alkaline) to yellow (acidic). Any sign of yellow should be interpreted as a positive reaction; orange should be considered negative.

REACTIONS			REMARKS
	Negative	**Positive**	
			Compartment 8
Sorbitol	Red	Yellow	**Sorbitol (SORB):** Bacterial fermentation of sorbitol, which results in the formation of acidic end products, is indicated by a change in color of indicator present in the medium from red (alkaline) to yellow (acidic). Any sign of yellow should be interpreted as a positive reaction; orange should be considered negative.
			Compartment 9
Voges-Proskauer	Colorless	Red	**Voges-Proskauer (VP):** Acetylmethylcarbinol (acetoin) is an intermediate in the production of butylene glycol from glucose fermentation. The production of acetoin is detected by the addition of two drops of a 20% w/v aqueous solution of potassium hydroxide containing 0.3% w/v of creatine and three drops of a 5% w/v solution of alpha naphthol in absolute ethyl alcohol. The presence of acetoin is indicated by the development of a red color within 20 min. However, most positive reactions are evident within 10 min.
			Compartment 10
Dulcitol	Green	Yellow	**Dulcitol (DUL):** Bacterial fermentation of dulcitol, which results in the formation of acidic end products, is indicated by a change in color of the indicator present in the medium from green (alkaline) to yellow or pale yellow (acidic).
PA	Green	Black–smoky gray	**Phenylalanine deaminase (PA):** This test detects the formation of pyruvic acid from the deamination of phenylalanine. The pyruvic acid formed reacts with a ferric salt present in the medium to produce a characteristic black to smoky gray color. Ferric chloride need not be added since the medium already contains an iron salt.
			Compartment 11
Urea	Beige	Red-purple	**Urea (UREA):** Urease, an enzyme possessed by various microorganisms, hydrolyzes urea to ammonia causing the color of the indicator in the medium to shift from yellow (acidic) to red-purple (alkaline). The urease test is strongly positive for *Proteus* species and may be evident as early as 4 to 6 hr after incubation; it is weakly positive (light pink color) after 18 to 24 hr incubation for *Klebsiella* and *Enterobacter* species.

(*continued*)

REACTIONS			REMARKS
	Negative	Positive	
			Compartment 12
Citrate	Green	Blue	**Citrate (CIT):** This test detects those organisms which are capable of utilizing citrate, in the form of its sodium salt, as the sole source of carbon. Organisms capable of utilizing citrate produce alkaline metabolites which change the color of the indicator from green (acidic) to deep blue (alkaline). Any degree of blue should be considered positive.

Courtesy BD Bioscience.

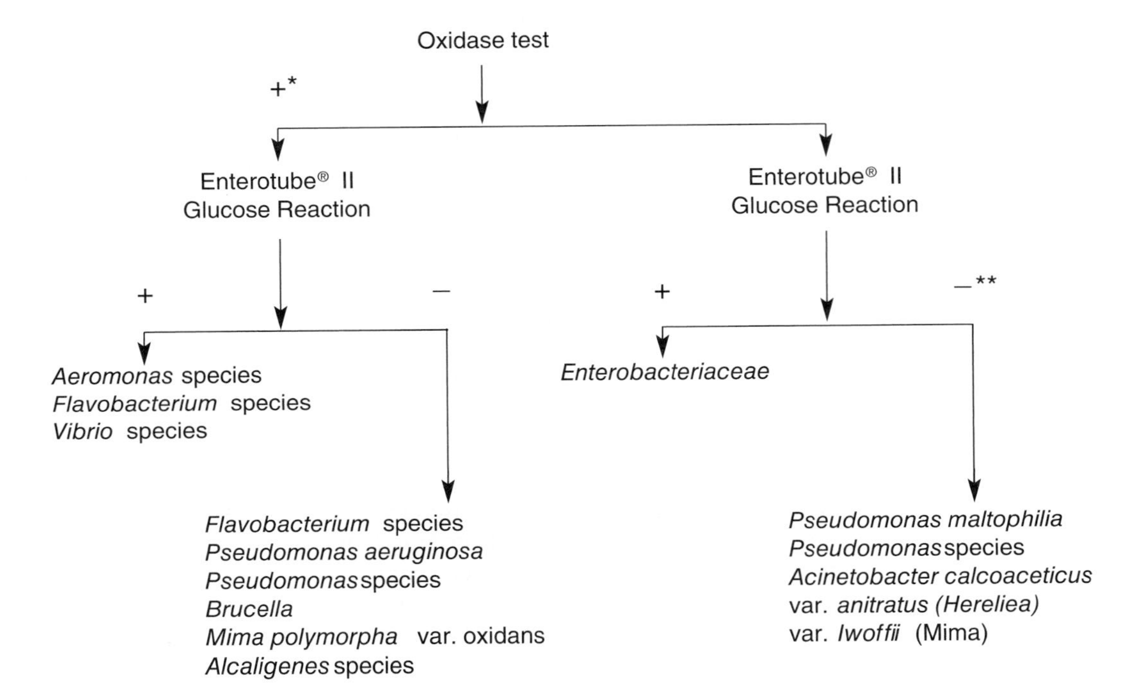

* This side of the flow diagram may be found useful if an oxidase-positive is inadvertently inoculated into the Enterotube® II. It is suggested that Oxi/Ferm® Tube be used for their identification.

** Common biotypes of this group of organisms may be identified with Enterotube® II.

Courtesy BD Bioscience.

Appendix C

The MPN Index per 100 ml for Combinations of Positive and Negative Presumptive Test Results When Five 10-ml, Five 1-ml, and Five 0.1-ml Portions of Sample are Used.

Number of Tubes with Positive Results					
Five of 10 ml each	Five of 1 ml each	Five of 0.1 ml each	MPN Index per 100 ml	Confidence Limit	
				Low	High
0	0	0	<1.8	–	6.8
0	0	1	1.8	0.09	6.8
0	1	0	1.8	0.09	6.9
0	2	0	3.7	0.7	10
1	0	0	2	0.1	10
1	0	1	4	0.7	10
1	1	0	4	0.7	12
1	1	1	6.1	1.8	15
1	2	0	6.1	1.8	15
2	0	0	4.5	0.79	15
2	0	1	6.8	1.8	15
2	1	0	6.8	1.8	17
2	1	1	9.2	3.4	22
2	2	0	9.3	3.4	22
2	3	0	12	4.1	26
3	0	0	7.8	2.1	22
3	0	1	11	3.5	23
3	1	0	11	3.5	26
3	1	1	14	5.6	36
3	2	0	14	5.7	36
3	2	1	17	6.8	40
3	3	0	17	6.8	40
4	0	0	13	4.1	35
4	0	1	17	5.9	36
4	1	0	17	6	40

(continued)

Number of Tubes with Positive Results					
Five of 10 ml each	Five of 1 ml each	Five of 0.1 ml each	MPN Index per 100 ml	Confidence Limit	
				Low	High
4	1	1	21	6.8	42
4	1	2	26	9.8	70
4	2	0	22	6.8	50
4	2	1	26	9.8	70
4	3	0	27	9.9	70
4	3	1	33	10	70
4	4	0	34	14	100
5	0	0	23	6.8	70
5	0	1	31	10	70
5	0	2	43	14	100
5	1	0	33	10	100
5	1	1	46	14	120
5	1	2	63	22	150
5	2	0	49	15	150
5	2	1	70	22	170
5	2	2	94	34	230
5	3	0	79	22	220
5	3	1	110	34	250
5	3	2	140	52	400
5	3	3	180	70	400
5	4	0	130	36	400
5	4	1	170	58	400
5	4	2	220	70	440
5	4	3	280	100	710
5	4	5	430	150	1100
5	5	0	240	70	710
5	5	1	350	100	1100
5	5	2	540	150	1700
5	5	3	920	220	2600
5	5	4	1600	400	4600
5	5	5	≥1600	700	–

Glossary

1N hydrochloric acid (HCl): an aqueous solution of hydrogen chloride gas (HCl) with a normality of one (giving off a single hydrogen atom).

Absorbance: a measure of the capacity of a substance to absorb light of a specified wavelength.

Acid-alcohol: a mixture of an acid and alcohol, used to bleach or decolorize samples that contain a mixture of acid-bleaching resistant cells.

Acid-fast technique: a staining process in which mycobacteria resist decolorization with acid alcohol.

Acid-resistant: materials stable towards the destructive action of acids.

Acidic stains: stains that carry a negative charge and therefore are repelled by the negatively charged bacteria (e.g., nigrosin, congo red, and India ink).

Adaptive immunity: a response to a specific immune stimulus that involves immune defensive cells and frequently leads to the establishment of host immunity.

Agglutination: the clumping or adhering of particles that occurs if an antigen is mixed with its corresponding isoagglutinin.

Ames test: a diagnostic procedure used to detect potential cancer-causing agents in humans by the ability of the agent to cause mutations in bacterial cells.

Ammonification: the process in which microorganisms break down organic matter into ammonia.

Antibiotic-resistant organisms: bacteria that cannot be controlled or killed by antibiotics.

Antibiotics: chemical inhibitors produced by microorganisms.

Antigen: a chemical substance that stimulates the production of antibodies by the body's immune system.

Antiseptic: a chemical used to reduce or kill pathogenic microorganisms on a living object, such as the surface of the human body.

***Bacillus*:** a genus of rod-shaped, gram-positive bacteria.

Bacterial colony characteristics: the characteristics that reflect the genetic composition of a bacterium and serve as markers for different bacterial species.

Bacterial genetics: the subfield of genetics devoted to the study of bacteria.

Bacterial growth: the increase in numbers of viable vegetative cells in a population.

Bacterial smear: a thin layer of bacteria placed on a slide for staining.

Bacteriophages: viruses that infect bacteria.

Basic stain: a positively charged colored substance that is used to stain cells (such as methylene blue or crystal violet).

Basophils: a type of white blood cell with granules that functions in allergic reactions.

Binary fission: an asexual process in bacterial and archaeal cells by which a cell divides to form two new cells while maintaining genetic constancy.

Biofilm: a complex community of microorganisms that form a protective and adhesive matrix that attaches to a surface, such as a catheter or industrial pipeline.

Biogeochemical cycle: a pathway or cycle through which a chemical substance passes through the atmosphere, hydrosphere, lithosphere and biosphere (e.g., the water cycle).

Blood agar: a nutrient culture medium that is enriched with whole blood and used for the growth of certain strains of bacteria.

Blood smear: a blood test used to identify abnormalities in blood cells.

Blood type: a classification of blood based on the presence or absence of the A and B antigens.

Brightfield microscopy: an optical configuration of the light microscope that magnifies an object by passing visible light directly through the lenses and object; the object appears dark against a light background.

Brownian motion: a phenomenon caused by molecules striking the organisms and displacing them briefly.

Capsule: a layer of polysaccharides and proteins secreted by certain bacteria, including many pathogens.

Carbolfuchsin: a mixture of an aqueous solution of phenol and an alcoholic solution of fuchsin.

Carcinogens: cancer-causing agents.

Casein: the major protein in milk.

Catalase: an enzyme that catalyzes the decomposition of hydrogen peroxide.

Chemotherapeutic agents: a chemical compound used to treat diseases and infections in the body.

Chocolate agar: an enriched growth medium used to isolate and cultivate fastidious organisms.

Citrate test: a test that determines the ability of organisms to utilize citrate as a carbon source.

***Clostridium*:** a genus of anaerobic, endospore-forming, rod-shaped bacteria that frequently inhabit the intestinal tract.

Coagulase: an enzyme produced by some staphylococci that catalyzes the formation of a fibrin clot.

Coliform bacteria: a gram-negative, nonspore-forming, rod-shaped cell that ferments lactose to acid and gas and usually is found in the human and animal intestine; high numbers in water is an indicator of contamination.

Colonies: individual organisms that have formed separate masses of bacteria.

Compound microscope: a basic tool of the microbiology laboratory. This precision instrument contains a series of lenses allowing a specimen to be magnified up to a thousand times (1,000×).

Confirmed test: a test carried out to confirm findings.

Conjugation: the transfer of genetic material between bacterial cells by direct cell-to-cell contact or by a bridge-like connection between two cells.

Counterstain: a stain with color contrasting to the principal stain, making the stained structure more easily visible.

Crystal violet: a synthetic violet dye, related to rosaniline, used as a stain in microscopy and as an antiseptic in the treatment of skin infections.

Cysteine: a sulfur-containing amino acid that occurs in keratins and other proteins, often in the form of cystine, and is a constituent of many enzymes.

Cysteine desulfurase: a protein that delivers sulfur to a variety of metabolic pathways, involved in the biosynthesis of iron-sulfur clusters, thionucleosides in tRNA, thiamine, biotin, lipoate and pyranopterin (molybdopterin).

Denaturation: a process caused by heat or pH in which proteins lose their function due to changes in their 3-D structure.

Deoxyribonucleic acid (DNA): the genetic material of all cells and many viruses.

Depression slide: a glass slide that has a concavity in one surface over which a cover glass can be placed and that is used in biology for hanging-drop cultures and for the microscopic study of small specimens.

Differential media: a growth medium in which different species of microorganisms can be distinguished visually.

Differential white blood cell count: a test that determines the percentage of each type of white blood cell present in a blood sample, which can also detect immature cells and abnormalities.

Disinfection: the process of cleaning something, especially with a chemical, in order to destroy bacteria.

Disinfectant: a chemical used to kill or inhibit pathogenic microorganisms on a lifeless object such as a tabletop.

DNA repair enzymes: enzymes that recognize and correct physical damage in DNA, caused by exposure to radiation, UV light or reactive oxygen species.

DNase: an enzyme that catalyzes the hydrolysis of DNA into oligonucleotides and smaller molecules.

Durham tube: smaller test tubes inserted upside down in another test tube, used to detect production of gas by microorganisms.

ELISA: *See* **enzyme-linked immunosorbent assay**.

EMB agar: (Eosin-methylene blue) an agar used for the isolation and differentiation of lactose fermenters from nonlactose fermenters.

Emerging infectious disease: a new disease or changing disease that is seen within a population for the first time; *see also* **reemerging infectious disease**.

Emerson agar: a medium used to isolate and cultivate Actinomycetaceae, Streptomycetaceae, fungi, and molds.

End products: that which is produced as the final result of an activity or process.

Endo agar: a slightly selective and differential agar for the isolation, cultivation, and differentiation of gram-negative microorganisms.

Endospore: an extremely resistant dormant cell produced by some gram-positive bacterial species.

Enteric bacteria: bacteria of the intestines.

Enriched media: growth media in which special nutrients must be added to get an species to grow.

Enzyme: a reusable protein molecule that brings about a chemical change while itself remaining unchanged.

Enzyme-linked immunosorbent assay (ELISA): A serological test in which an enzyme is used to detect an individual's exposure to a pathogen.

Enzyme systems: a group of protein compounds that are produced by living cells and that act together as catalysts in specific biochemical reactions.

Eosinophils: a type of white blood cell with granules that stains with the dye eosin and plays a role in allergic reactions and the body's response to parasitic infections.

Epidemic: Referring to a disease that spreads more quickly and more extensively than normally expected.

***Escherichia coli*:** bacteria found in the environment, foods, and intestines of people and animals.

Eukaryotic organisms: an organism containing a cell nucleus with multiple chromosomes, a nuclear envelope, and membrane-bound compartments.

Fermentation: the chemical breakdown of a substance by bacteria, yeasts, or other microorganisms, typically involving effervescence and the giving off of heat.

Flagella: protein appendages that facilitate motion (motility) of bacteria.

Flatworms: any of various parasitic and nonparasitic worms of the phylum Platyhelminthes.

Fomite: an inanimate object, such as clothing or a utensil, that carries disease organisms.

Fruiting body: the general name for a reproductive structure of a fungus from which spores are produced.

Fungi: one of the four kingdoms in the domain Eukarya; composed of the molds and yeasts.

Generation time: the time interval for a cell population to double in number.

Genetic recombination: the process of bring together different segments of DNA.

Genetic variability: a measure of the tendency of individual genotypes in a population to vary from one another.

Glycocalyx: a viscous polysaccharide material covering many prokaryotic cells to assist in attachment to a surface and impart resistance to desiccation.

Gradient plate technique: a method for isolating antibiotic-resistant bacteria mutants by exposing an agar plate containing concentration gradient of antibiotic to an inoculation of bacteria to be tested.

Gram-negative: term used to indicate the loss (negative) of crystal violet-iodine complex.

Gram-positive: term used to indicate the retention (positive) of crystal violet-iodine complex.

Gram stain technique: a commonly used method of staining used to differentiate bacterial species into two large groups (gram-positive and gram-negative).

Gram's iodine: iodine most commonly used as the second step in a Gram stain. Gram's iodine is used as mordant to "seal" the crystal violet stain into the cell walls.

Hemagglutination: a type of agglutination that involves blood cells.

Horizontal gene transfer: the movement of genetic material between unicellular and/or multicellular organisms.

Hyphae: branching filaments that make up the mycelium of a fungus.

Immersion oil: a transparent oil with high refractive index used to increase the resolving power of a microscope by immersing both the objective lens and the specimen.

Immunology: the scientific study of the structure and function of the immune system.

IMViC series: a series of four tests (indole, methyl red, Voges-Proskauer, and citrate) used to identify bacterial species, in particular coliforms.

Index patient: (also called index case, or primary case) the first identified case in a group of related cases of a communicable or heritable disease.

Indole test: a biochemical test performed on bacterial species to determine the ability of the organism to convert tryptophan into the indole.

Induced mutations: a mutation caused by exposure to a mutagen.

Innate immunity: an inborn set of the preexisting defenses against infectious agents; includes the skin, mucous membranes, and secretions.

Iodine solution: an aqueous mixture of iodine and potassium iodide.

Kirby-Bauer method: a procedure for determining bacterial susceptibility to an antibiotic by determining if bacterial growth occurs around an antibiotic disk; *see also* **standardized disk test**.

***Lactobacillus*:** a genus of gram-positive, facultative anaerobic or microaerophilic, rod-shaped, non-spore-forming bacteria.

Lactose broth: a medium for the detection of coliform organisms.

Lactose intolerance: the inability to fully digest sugar (lactose) in dairy products.

Lawn of bacteria: a continuous cover of bacteria on the surface of a growth medium.

Lens paper: special fabric used to clean lenses and other glass objects without scratching the surface.

Levine eosin methylene blue (EMB) agar: a selective/differential agar medium that, by inhibiting the growth of gram-positive bacteria, can be used to identify gram-negative, fecal coliform bacteria.

Light (compound) microscope: a core instrument of many microbiology research labs that uses visible light to magnify and resolve objects.

Lowenstein-Jensen medium: a growth medium specially used for culture of *Mycobacterium* species, notably *Mycobacterium tuberculosis*.

Lymphocytes: a type of white blood cell that functions in the immune system.

Lysozyme: an enzyme found in tears and saliva that digests the peptidoglycan of gram-positive bacterial cell walls.

Lytic cycle: the cycle in which the host bacterium disintegrates when new viral particles are released.

MacConkey agar: agar used for the isolation and differentiation of lactose fermenters from nonlactose fermenters.

Magnification: refers to increasing the apparent size of the specimen being observed.

Malachite green: an organic compound that is used as a dyestuff and controversially as an antimicrobial in aquaculture.

Membrane filter technique: when examining water, a technique in which a sample passes through a membrane using a filter funnel and vacuum system, concentrating any organisms on the surface of the membrane.

Methylene blue: a dark-green compound used as a bacteriologic stain and indicator.

Methyl red test: used to identify bacteria producing stable acids by mechanisms of mixed acid fermentation of glucose.

Mitis salivarius agar: a solid medium recommended for use in qualitative procedures for selective isolation of *Streptococcus mitis, Streptococcus salivarius*, and *enterococci*.

Minimal agar plates: recommended for the isolation and characterization of nutritional mutants of *Escherichia coli*.

Mixed populations: a particular species mixed with other species.

Molds: fungi that grow in the form of multicellular filaments (hyphae).

Monocytes: a large, phagocytic white blood cell with an oval nucleus and clear, grayish cytoplasm.

Mordant: a substance used to set dyes on fabrics or tissue sections by forming a coordination complex with the dye which then attaches to the fabric or tissue.

Most probable number (MPN): a method of getting quantitative data on concentrations of discrete items from positive/negative data.

Motility test agar: a semisolid growth medium containing a reduced amount of agar.

Mueller-Hinton agar: a growth medium commonly used for antibiotic susceptibility testing, or to isolate and maintain *Neisseria* and *Moraxella* species.

Multicellular parasites: an organism that lives on or in a host organism and gets its food from or at the expense of its host.

Mutagen(s): permanent and inheritable changes in the genome of an organism.

Mutation: a permanent, heritable change in the nucleotide sequence in a gene or a chromosome; the process in which such a change occurs in a gene or in a chromosome.

Mycelium: the vegetative part of a fungus, consisting of a network of fine white filaments (hyphae).

***Mycobacterium*:** a genus of Actinobacteria, including the causative agents of leprosy and tuberculosis.

Negative control: a group in which no response is expected. It is the opposite of the positive control, in which a known response is expected.

Negative stain: an established method for contrasting a thin specimen with an optically opaque fluid that requires acidic dye.

Negative stain technique: a staining technique that permits one to see unstained bacteria on a stained background and determine their morphology.

Nessler's reagent: an aqueous solution of potassium iodide, mercuric chloride, and potassium hydroxide, used as a test for the presence of ammonia.

Neutrophils: the most common type of white blood cell; functions chiefly to engulf and destroy foreign material, including bacterial cells and viruses that have entered the body.

Nitrogen cycle: the series of processes by which nitrogen and its compounds are interconverted in the environment and in living organisms, including nitrogen fixation and decomposition.

Novobiocin: an aminocoumarin antibiotic produced by the actinomycete *Streptomyces niveus*, used to treat infections by gram-positive bacteria in certain animals.

Objective lenses: the lens or system of lenses in a telescope or microscope that is nearest the object being viewed.

Ocular lens: the eyepiece of a microscope.

Ocular micrometer: a glass disk that fits in a microscope eyepiece that has a ruled scale, which is used to measure the size of magnified objects.

Oil immersion lens: an objective lens designed to work with a drop of liquid (as oil or water) between the lens and cover glass.

Osmotic pressure: the force that must be applied to a solution to inhibit the inward movement of water across a membrane.

Outbreak: a small, localized epidemic.

Oxidase test: used to identify bacteria that produce cytochrome c oxidase, an enzyme of the bacterial electron transport chain.

Paraffin: a flammable, whitish, translucent, waxy material consisting of a mixture of hydrocarbons.

Parasites: an organism that lives and feeds on or in an organism of a different species and causes harm to its host.

Pathogen: microorganism or virus that causes infection or disease.

Phenol red indicator: a water-soluble pH indicator that changes from yellow to red over pH 6.6 to 8.2, turning fuchsia above pH 8.2.

Phenol red lactose broth: a complete medium with lactose and a phenol red pH indicator. Used to determine whether the microbe can use the sugar lactose for carbon and energy.

Photolyase: DNA repair enzymes that repair damage caused by exposure to ultraviolet light.

Photoreactivation: a process that repairs DNA damaged by ultraviolet light using an enzyme that requires visible light.

Pigmentation: organic compounds produced by bacteria that give color to liquid cultures and colonies.

Plaque: a moth-eaten, clear area seen amid the lawn of bacterial growth when bacteriophages lyse their bacterial hosts on an agar surface.

Plaque-forming unit: a measurement of the number of particles capable of forming plaques, per unit volume.

Population growth curve: an empirical model of the growth of a quantity over time.

Positive control: a group that receives a treatment with a known response, so that the positive response can be compared to the unknown response of the treatment.

Pour plate method: a mixed population of bacteria is progressively diluted in three "deep" tubes of liquid nutrient agar.

Prepared slide: a set of slides that have been previously mounted, usually by an individual who is a professional in the field.

Presumptive test: an analysis of a sample which establishes if the sample is or is not a certain substance.

Primary decomposers: a microorganism (primarily bacteria and fungi) that feeds on (decompose) dead plants and animals and the waste from these organisms.

Prokaryotic organisms: organism that has no nuclear membrane, no organelles in the cytoplasm except ribosomes, and has its genetic material in the form of single continuous strands forming coils or loops.

Proteinases: (also endopeptidase) any peptidase that catalyzes the separation of internal bonds in a polypeptide or protein.

Protists: a diverse group of unicellular eukaryotic organisms.

Pure culture: a culture that contains only a single species of organism.

Quadrant (four-way) streak plate method: a relatively inexpensive and rapid method for separating bacteria in a mixed population of high cell density.

Quebec colony counter: a device used to estimate a liquid culture's density of microorganisms by counting individual colonies on an agar plate, slide, mini gel, or Petri dish.

Reemerging infectious disease: A disease showing a resurgence in incidence or a spread in its geographical area; *see also* **emerging infectious disease**.

Refractive index: the light-bending ability of glass, oil, and air—media through which light must pass during image formation.

Rennin: an enzyme secreted into the stomach of unweaned mammals, and in some lower animals and plants, causing the curdling of milk.

Resolving power: the part of the microscope that determines the size of the smallest object that can be seen clearly under specified conditions.

Rh factor: an antigen occurring on the red blood cells of many humans (around 85 percent) and some other primates. It is particularly important as a cause of hemolytic disease of the newborn.

***Rhizobium*:** a genus of gram-negative, nitrogen-fixing bacteria common in soil, particularly in the root nodules of leguminous plants.

Rogosa SL agar: a low pH and high acetate medium used for isolation, enumeration and identification of lactobacilli in oral bacteriology, feces, vaginal specimens and foodstuffs.

Roundworms: a nematode, especially a parasitic one found in the intestines of mammals.

Safranin: a red, water-soluble dye capable of staining all cell nuclei red.

Saline solution: a mixture of sodium chloride in water.

Saprobes: heterotrophic organisms that feed on dead organic matter.

Scientific notation: a way to express inconveniently lengthy decimals using the form $m \times 10^n$.

Selective media: a growth medium that contains ingredients to inhibit certain microorganisms while encouraging the growth of others.

Serial dilutions: a process that, through dilution, sequentially reduces the concentration of a substance or the concentration of microorganisms in a sample.

Serology: a branch of immunology dealing with techniques to identify and measure antigens, and to detect serum antibodies.

***Serratia marcescens*:** a motile, short rod-shaped, gram-negative, facultative anaerobic bacterium in the family Enterobacteriaceae, classified as an opportunistic pathogen.

SIM (sulfide indole motility) agar: A medium containing cysteine and iron ions.

Simple stain technique: a technique for staining a bacterial smear in which only a single stain is used.

Skim milk agar: used for the cultivation and differentiation of microorganisms based on proteolytic activity (hydrolysis of casein).

Slide agglutination technique: (also called VDRL test) a screening procedure used in the detection of syphilis antibodies.

Slide label: a label used to keep track of samples on slides, usually chemical resistant.

Slime layer: the bacterial capsule layer when thin and flowing.

Snyder test: used to determine a person's susceptibility to dental cavities; susceptibility is correlated to fermentation by cariogenic *Lactobacillus* species in the mouth.

Spectrophotometer: an apparatus for measuring the intensity of light in a part of the spectrum, particularly when emitted by a particular substance.

Spirit blue agar: a medium that contains a supply of lipids, allowing the agar to determine the presence of lipase.

Spontaneous mutations: those mutations that take place in nature without human intervention or an identifiable cause.

Standardized disk susceptibility test: a procedure for determining bacterial susceptibility to an antibiotic by determining if bacterial growth occurs around an antibiotic disk. *See also* **Kirby-Bauer method**.

Standard plate count procedure: a direct method to estimate the number of cells in a sample dilution spread on an agar plate.

***Staphylococcus*:** a genus of spherical, gram-positive, parasitic bacteria.

Starch: a polysaccharide consisting of thousands of glucose molecules chemically bonded to one another.

Stormy fermentation: refers to the appearance of the turbid reaction of Clostridium spp in litmus milk with coagulation and gas production, in which milk is converted into a coagulum with entrapped bubbles of gas.

Streak-plate method: a process by which a mixed culture can be streaked onto an agar plate and pure colonies isolated.

***Streptococcus*:** a genus of gram-positive, anaerobic, often pathogenic bacteria that includes the agents of souring of milk and dental decay.

***Streptomyces*:** any of several aerobic bacteria of the genus *Streptomyces*, certain species of which produce antibiotics.

Substrates: the substances upon which an enzyme acts.

Thrombocytes (platelets): a small, colorless, disk-shaped cell fragment without a nucleus, found in blood and involved in clotting.

Titer: a measurement of the amount of antibody in a sample of serum that is determined by the most dilute concentration of antibody that will yield a positive reaction with a specific antigen.

Todd-Hewitt broth: an enrichment medium containing glucose, , as well as sodium carbonate, and disodium phosphate buffers that favor the growth of hemolytic streptococci.

Total magnification: refers to the microscope lens system and is calculated by multiplying the ocular magnification by the objective magnification.

Triple sugar iron (TSI) agar: a differential medium that contains lactose, sucrose, glucose, ferrous sulfate, and the pH indicator phenol red, used to differentiate enterics based on the ability to reduce sulfur and ferment carbohydrates.

Type A: a blood type in which individuals possess A antigens on the red blood cells.

Type B: a blood type in which individuals possess B antigens on the red blood cells.

Type AB: blood type in which individuals possess both A and B antigens on the red blood cells.

Type O: blood type in which individuals possess neither A or B antigens on the red blood cells.

Tube agglutination technique: the process that occurs if an antigen is mixed with its corresponding antibody called isoagglutinin.

Ultraviolet (UV) light: a form of electromagnetic radiation which is invisible to the human eye.

Unripened cheese: a type of cheese which is consumed fresh.

Urea: an end product of protein metabolism in bacteria.

Urease: an enzyme that breaks down urea into ammonia and carbon dioxide.

Vector: (1) an arthropod that transmits the agents of disease from an infected host to a susceptible host. (2) a plasmid used in genetic engineering to carry a DNA segment into a bacterium or other cell.

Virulent bacteriophages: those bacteriophages that destroy their bacterial hosts at the conclusion of the replication cycle.

Virus: an infective agent that typically consists of a nucleic acid molecule in a protein coat, is too small to be seen by light microscopy, and is able to multiply only within the living cells of a host.

Voges-Proskauer test: a test used to detect acetoin in a bacterial broth culture, performed by adding alpha-naphthol and potassium hydroxide to the Voges-Proskauer broth, which has been inoculated with bacteria.

Water quality test: a process to analyze the biological and chemical characteristics of water relative to human consumption needs.

Wet mount: a glass slide holding a specimen suspended in a drop of liquid (as water) for microscopic examination.

White blood cells (leukocytes): colorless blood cells with a nucleus and cytoplasm that help the body fight infection.

Winogradsky column: a transparent cylinder filled with mud, paper, salts, and water to mimic a small vertical pond.

Working distance: the proximity of a slide to the bottom of the objective lens on a microscope.

Wright's stain: a histologic stain that facilitates the differentiation of blood cell types.

Yeasts: fungi that can adopt a single-celled growth habit.

Yogurt: a type of sour milk. The two starter cultures necessary for its formation are *Streptococcus thermophilus* and *Lactobacillus bulgaricus*. When these bacteria are incubated in concentrated milk, they ferment the lactose and produce lactic acid, which brings on yogurt's characteristic sour taste.

Zone of inhibition: the clear region around the paper disk saturated with an antimicrobial agent on an agar surface.

Index

A

Acid-fast stain technique, 227–232
Acid-fast technique, 227
Acidic stains, 85
Adaptive immunity, 412
Agar, 14
Agar deeps, 14
Agar plates, 14
Agar to agar transfer, 19–20
Air-dry, bacterial smear, 75
Alcoholic beverages, 174
Alpha-hemolytic streptococci, 237
Ammonification, soil microorganisms, 531–533
Amylase, 335
Antibiotic resistant bacteria, 373–375
Antibiotic-resistant cells, 373
Antibiotics, 142
Antiseptics and disinfectants, 163–168
Aseptic conditions, 15
Autoclaving, 160

B

Bacillus, 221, 285–289
Bacteria
- antibiotics on, 195–197, 199–200
- biochemical characteristics, 312, 333–334
  - carbohydrate fermentation, 329–331
  - casein hydrolysis, 367–370
  - catalase production, 341–343, 345–346
  - DNA hydrolysis, 347–350
  - fat (triglyceride) hydrolysis, 363–366
  - hydrogen sulfide production, 351–354
  - IMViC series, 355, 357–358
  - lipid digestion, 363–364
  - starch hydrolysis, 335–337, 339–340
  - urea hydrolysis, 359–362
- carbohydrate fermentation test, 333–334
- characteristics of, 309–310
- chemical agents on, 171–172, 174–176
  - alcoholic beverages, 174
  - antiseptics and disinfectants, 163–168
  - inhibit bacterial growth, 173–174
  - lysozyme, 174, 177–180
  - metals, 168–170
  - zones of inhibition, 174
- culture characteristics, 321–323, 325–328
- disinfectants and antiseptics, 193–194
  - and hand washing on skin surface, 185–188
  - on inanimate objects, 181–184
  - mouthwashes and oral bacteria, 189–191
- examination procedures, 308
- identification keys, 308, 311
- physical agents on, 161–162
  - autoclaving, 160
  - cold temperature, 160
  - drying, 155–158
  - heat tolerance, 143–145, 147–148
  - incineration, 159
  - osmotic pressure, 159
  - ultraviolet light, 149–151, 153–154
- structural characteristics, 315–319

Bacterial colony characteristics, streak plate method, 29
Bacterial conjugation, 372, 395–397, 399–400
Bacterial genetics, 372
- antibiotic resistant bacteria, 373–375
- bacterial conjugation, 372, 395–397, 399–400
- bacterial transformation, 401–405, 407–409
- carcinogens and Ames test, 389–391, 393–394
- gradient plate technique, 373–375, 377–379
- mutations
  - and DNA repair, 385–388
  - ultraviolet light, 379–381, 383–384

Bacterial staining techniques
- bacterial smear, preparation of, 73–78
- bacterial structures
  - capsule stain technique, 105–110
  - spore stain technique, 99–104
- Gram stain technique, 91–95, 97–98
- negative stain technique, 85–87, 89–90
- simple stain technique, 79–81, 83–84
- types and uses for, 72

Bacterial transformation, 401–405, 407–409
Bacteriophages, 112, 113
- identification, from sewage, 121–127
- number, estimate, 209–210
  - Petroff-Hausser chamber, 206–207
  - standard plate count, 203–206

Basic growth media, 14–15
Beta-hemolytic streptococci, 237
Binary fission, 211
Biofilm, 439
- preparation, water microbiology, 523–526

Biogeochemical cycles, 439
Biosafety, 4
Biosafety level 1 (BSL-1), 4–5
Biosafety level 2 (BSL-2), 5
Blood agar, 247, 292
Blood type, determination of, 421–423. *See also* Serology
Brightfield microscopy, 50
Broth to agar slant, 19
Broth to broth transfer, 19
Brownian motion, 65
Burner, 15

C

Capsule, 105
Carbohydrate fermentation, 329–331
Carbohydrate fermentation test, 333–334
Carcinogens, 372
Carcinogens and Ames test, 389–391, 393–394
Caseinase, 367
Casein hydrolysis, 367–370
Catalase production, 341–343, 345–346
Catalase test, 286
Cheese, preparation of, 493–495

Chemical agents, bacteria, 171–172, 174–176
alcoholic beverages, 174
antiseptics and disinfectants, 163–168
inhibit bacterial growth, 173–174
lysozyme, 174, 177–180
metals, 168–170
zones of inhibition, 174
Chemical hazard labels, 3–4
biosafety level 1 (BSL-1), 4–5
biosafety level 2 (BSL-2), 5
laboratory biosafety, 4
Chemotherapeutic agents, 142
Chocolate agar, 247
Clostridium, 221, 291–293, 295–298
Cold acid-fast stain technique, 233, 235–236
Cold temperature, 160
Coliform bacteria, 439, 485, 507
Coliform plate count, 485–488
Colonies, 15, 25
Colony and cellular morphology, 223–226
Competent recipient cell, 401
Conjugation, 372
Contaminate, 17
Crossing over, 395
Culture, 14
Culture (Petri) dishes, 15
Culture equipment, 15
Culture medium, 14
Culture plates, 15
Culture spill, 9, 10
Culture transfer techniques, 17–20, 21–23, 37–39
Culture tubes, 15
Cysteine desulfurase, 351

D

Denaturation, 148
Deoxyribonucleic acid (DNA), 347. *See also* DNA hydrolysis
Differential media, 221
Differential white blood cell count, 430
Disk susceptibility test, 195
DNA hydrolysis, 347–350
DNA repair enzymes, 385
DNase, 347
Drinking water, disinfection of, 457–460
Durham tube, 330

E

Emerging infectious diseases, 438
Endospores, 99
Endospores, presence of, 286
End products, 329
Enriched media, 221
Enteric bacteria, 220
IMViC series, 275–277, 279–280
isolation of, 267–270
rapid identification of, 281–284
triple sugar iron (TSI) agar, 271–275
Enzyme-linked immunosorbent assay (ELISA), 433–435
Enzyme systems, 329
Eosinophils, 429
Epidemic diseases, 441
Epidemics, 441
Eukaryotic organisms, 112

F

Fat (triglyceride) hydrolysis, 363–366
Fermentation, 329
Flagella, 65
Flammability, 4
Flatworms, 112
Fomites, 181
Foods microbiology. *See also* Milk and dairy products, microbiology
fermentation of wine and beer, 475–478
preservation with salt and garlic, 461–465
standard plate count, 467–473
Fungi, 112
molds, 125–130
yeasts, 131–133, 135–136

G

Generation time, 211
Genetic recombination, 372
Genetic variability, 372, 395
Glycocalyx, 105
Gradient plate technique, 372, 373–375, 377–379
Gram negative, 91
Gram positive, 91
Gram stain, 286
Green fluorescent protein (GFP), 402
Growth and incubation appliances, 16

H

Hanging drop preparation, 65–69
Hazard labels, 3
Health hazard, 3
Heat-fix, bacterial smear, 76
Heat tolerance, 143–145, 147–148
Hemagglutination, 421
Hemolysis, Brown's classification of, 238
High-frequency recombinants, 395
Horizontal gene transfer, 372
Horizontal (lateral) gene transfer (HGT), 395
Hydrogen sulfide production, 351–354
Hypha, 125

I

Immersion lens, 57
Immersion oil, 57
Immunology, 412. *See also* Enzyme-linked immunosorbent assay (ELISA); Serology
IMViC (indole, methyl red, Voges-Proskauer, and citrate) series, 275–277, 279–280, 355, 357–358
Incineration, 159
Incubator, 16
Indole test, 275, 276
Induced mutations, 379
Inhibit bacterial growth, 173–174
Innate immunity, 412

K

Kirby-Bauer method, 195

L

Laboratory rules and regulations, student safety contract, 11–12
Laboratory safety
best practices in, 9–10
chemical hazard labels, 3–4
biosafety level 1 (BSL-1), 4–5
biosafety level 2 (BSL-2), 5
laboratory biosafety, 4
culture spill, 10
overview, 7
student safety contract, 11–12
Laboratory techniques and skills
agar to agar transfer, 19–20
basic growth media, 14–15
broth to agar slant, 19
broth to broth transfer, 19
culture equipment, 15
culture transfer techniques, 17–20, 21–23, 37–39

growth and incubation appliances, 16
loop sterilization process, 17–18
pure culture techniques, 25–29, 31–32, 33–35
serial dilutions, 45–48
solution transfer, 41–43
subculture transfer techniques, 17–20, 21–23
transfer instruments, 15–16
transfer process, 18
Lactobacillus, 221
caries susceptibility test, 303–306
lactobacilli, isolation of, 299–302
Lawn of bacteria, 117
Legume roots, *Rhizobium* isolation, 527–530
Lens paper, 52
Light (compound) microscope, 50
Lipid digestion, 363–364
Loop sterilization process, 17–18
Lymphocytes, 428
Lysozyme, 174, 177–180
Lytic pathway, 113

M

Magnification, 51
Mannitol salt agar, 255
Material Safety Data Sheet (MSDS), 3
Medical microbiology
Meiosis, 395
Membrane filter technique, 519–521
Metabolism, 329
Metals, 168–170
Methylene blue reduction test, 489–492
Methyl red test, 275, 276
Microbial ecology, soil, 547–549, 551–552
Microbial growth, bacterial growth curve, 215–218
bacterial growth dynamics, 212–213
generation time, growth curve and determining, 213–214
Microbial hunt, 447–449, 451
Microbial transmission
through toilet paper, 453–455
via fomites, 441–446
Microbiology
of foods
fermentation of wine and beer, 475–478
preservation with salt and garlic, 461–465
standard plate count, 467–473
of milk and dairy products
cheese, preparation of, 493–495
coliform plate count, 485–488
methylene blue reduction test, 489–492
natural bacterial content, 501–505
standard plate count, 479–481, 483–484
yogurt, preparation of, 497–500
of soil
ammonification, soil microorganisms, 531–533
legume roots, *Rhizobium* isolation, 527–530
microbial ecology, 547–549, 551–552
soil bacteria, plate count of, 543–546
Streptomyces, antibiotic production by, 539–541
Streptomyces, isolation of, 535, 537–538
Winogradsky column, 547–549, 551–552
of water
biofilm, preparation of, 523–526
completed test, 515, 517
confirmed test, 511–513
membrane filter technique, 519–521
presumptive test by MPN method, 507–510
Microorganisms
basic characteristics of, 112
control of, 142
Microscopy
compound microscope
parts of, 51–56
use, 51–56
motile bacteria, observations of, 65–70
prepared slides, observations of, 57–63
troubleshooting problems, 59–60
Milk and dairy products, microbiology
cheese, preparation of, 493–495
coliform plate count, 485–488
methylene blue reduction test, 489–492
natural bacterial content, 501–505
standard plate count, 479–481, 483–484
yogurt, preparation of, 497–500
Mixed culture, 14
Mixed populations, 25
Moist heat method, 148
Molds, 125–130, 126–128
Monocytes, 429
Most probable number (MPN), 507–510
Motility test agar preparation, 68–70
MSDS. *See* Material Safety Data Sheet (MSDS)
Mueller-Hinton agar plates, 263
Multicellular parasites, 112
Mutagens, 379
Mutations, 372
and DNA repair, 385–388
ultraviolet light, 379–381, 383–384
Mycelium, 125
Mycobacterium, 220
acid-fast stain technique, 227–232
cold acid-fast stain technique, 233, 235–236
colony and cellular morphology, 223–226

N

Natural bacterial content of milk, 501–505
Needles, 15
Neisseria, 220, 247–253
Neutrophils, 428
4-Nitro-o-phenylenediamine (4-NOPD), 389
Nutrient broth, 14

O

Objective lenses, 52
Occupational Safety and Health Administration (OSHA), 3
Ocular lens, 52
Ocular micrometer, 58
Osmotic pressure, 159

P

Pathogens, 131
Petroff-Hausser Chamber, 206–207
Photolyase, 385
Photoreactivation, 385
Physical agents, bacteria, 161–162
autoclaving, 160
cold temperature, 160
drying, 155–158
heat tolerance, 143–145, 147–148
incineration, 159
osmotic pressure, 159
ultraviolet light, 149–151, 153–154
Pipettes, 16, 41
Plaque assay, 203
Plaque formation, 117–120

Plaque-forming unit (PFU), 203
Plaques, 112
Plasmids, 401
Population growth curve, 211
Pour plate method, 33–35
Pressurized steam, 143
Prokaryotic organisms, 112
Protists, 112
Protists and multicellular parasites, 137–140
Pure culture, 14, 25
Pure culture techniques, 25–29, 31–32, 33–35
  pour plate method, 33–35
  streak plate method, 25–29

Q

Quadrant (four-way) streak plate method, 25

R

Reactivity, 4
Red blood cells (erythrocytes), 428
Reemerging infectious diseases, 438
Rennin, 493
Resolving power, 51
Rh factor, 421
Rhizobium, 439
Roundworms, 112

S

Salt and garlic, preservation with, 461–465
Saprobes, 131
Schaeffer-Fulton method, 100–102
Selection marker, 402
Selective media, 221
Serial dilutions, 45–48
Serology, 412
  agglutination techniques, 415–416, 425–426, 431–432
  blood cell identification, 427–430
  blood smear, 427–430
  blood type, determination of, 421–423
  slide agglutination, 413–414
  tube agglutination, 417–420
Skim milk agar, 367
Slide label, bacterial smear, 74
Snyder test, 303
Soil bacteria, plate count of, 543–546
Soil microbiology
  ammonification, soil microorganisms, 531–533
  legume roots, *Rhizobium* isolation, 527–530
  microbial ecology, 547–549, 551–552
  soil bacteria, plate count of, 543–546
  Streptomyces, antibiotic production by, 539–541
  Streptomyces, isolation of, 535, 537–538
  Winogradsky column, 547–549, 551–552
Solution transfer, 41–43
Spectrophotometer, 211, 212
Spirit blue agar, 363
Spontaneous mutations, 373
Standard plate counts, 439, 467–473, 479–481, 483–484
Staphylococcus, 220
  staphylococcal species, differentiation between, 261–263, 265
  Staphylococci
    isolation of, 255–257, 259–260
Starch hydrolysis, 335–337, 339–340
Starch test, 286
Sterilization, 14
Streak plate method, 25–29
Streptococcus, 220
  Streptococci
    from oral cavity, 243–246
    from the upper respiratory tract, 237–242
Streptomyces, 439
  soil, antibiotic production by, 539–541
  soil, isolation of, 535, 537–538
Student safety contract, 11–12
Subculture transfer techniques, 17–20, 21–23
Subculturing, 15, 17
Sulfite polymyxin sulfadiazine (SPS) agar, 292

T

Thioglycolate medium, 292
Thrombocytes, 429
Total magnification, 52
Transfer process, 18
Transformation, 372. *See also* Bacterial transformation
Triple sugar iron (TSI) agar, 271–275
Tuberculosis (TB), 220

U

Ultraviolet (UV) light, 149–151, 153–154, 372, 379–381, 383–384
Unripened cheese, 493
Urea hydrolysis, 359–362

V

Vectors, 438
Violet red bile agar, 485
Virulent bacteriophages, 113
Viruses, bacteriophages on bacteria, 113–116
Voges-Proskauer test, 275, 277

W

Waterbath, 16
Water microbiology
  biofilm, preparation of, 523–526
  completed test, 515, 517
  confirmed test, 511–513
  membrane filter technique, 519–521
  presumptive test by MPN method, 507–510
Water quality, 439
Water quality tests, 439
White blood cells (leukocytes), 428
Wine and beer, fermentation of, 475–478
Winogradsky column, 439, 547–549, 551–552
Wire inoculating loops, 15
Wright's stain, 428

Y

Yeasts, 131–133, 135–136
Yogurt, preparation of, 497–500

Z

Zones of inhibition, 163, 174